THE POLITICAL ECONOMY OF HAZARDS AND DISASTERS

SOCIETY FOR ECONOMIC ANTHROPOLOGY (SEA) MONOGRAPHS

Dolores Koenig, American University
General Editor, Society for Economic Anthropology

Monographs for the Society for Economic Anthropology contain original essays that explore the connections between economics and social life. Each year's volume focuses on a different theme in economic anthropology. Earlier volumes were published with the University Press of America, Inc. (#1–15, 17), Rowman & Littlefield Publishers, Inc. (#16). The monographs are now published jointly by AltaMira Press and the Society for Economic Anthropology (http://sea.org.ohio-state.edu).

No. 18 Jean Ensminger, ed., *Theory in Economic Anthropology*
No. 19 Jeffrey H. Cohen and Norbert Dannhaeuser, eds., *Economic Development: An Anthropological Approach*
No. 20 Gracia Clark, ed., *Gender at Work in Economic Life*
No. 21 Cynthia Werner and Duran Bell, eds., *Values and Valuables: From the Sacred to the Symbolic*
No. 22 Lillian Trager, ed., *Migration and Economy: Global and Local Dynamics*
No. 23 E. Paul Durrenberger and Judith Martí, eds., *Labor in Cross-Cultural Perspective*
No. 24 Richard Wilk, ed., *Fast Food/Slow Food*
No. 25 Lisa Cliggett and Christopher A. Pool, eds., *Economies and the Transformation of Landscape*
No. 26 Katherine E. Browne and B. Lynne Milgram, eds., *Economics and Morality: Anthropological Approaches*
No. 27 Eric C. Jones and Arthur D. Murphy, eds., *The Political Economy of Hazards and Disasters*

To find more books in this series, go to www.altamirapress.com/series.

THE POLITICAL ECONOMY OF HAZARDS AND DISASTERS

EDITED BY
ERIC C. JONES AND
ARTHUR D. MURPHY

ALTAMIRA PRESS
A Division of Rowman & Littlefield Publishers, Inc.
Lanham • New York • Toronto • Plymouth, UK

AltaMira Press
A division of Rowman & Littlefield Publishers, Inc.
A wholly owned subsidiary of The Rowman & Littlefield Publishing Group, Inc.
4501 Forbes Boulevard, Suite 200
Lanham, MD 20706
www.altamirapress.com

Estover Road
Plymouth PL6 7PY
United Kingdom

British Library Cataloguing in Publication Information Available

Library of Congress Cataloging-in-Publication Data

The political economy of hazards and disasters / edited by Eric C. Jones and Arthur D. Murphy.
p. cm. — (Society for economic anthropology (SEA) monographs ; No. 27)
Includes bibliographical references and index.
ISBN-13: 978-0-7591-1309-1 (cloth : alk. paper)
ISBN-10: 0-7591-1309-2 (cloth : alk. paper)
ISBN-13: 978-0-7591-1311-4 (electronic)
ISBN-10: 0-7591-1311-4 (electronic)
1. Disasters—Case studies. 2. Disasters—Economic aspects. 3. Emergency management. I. Jones, Eric C., 1970– II. Murphy, Arthur D.

HV553.P63 2009
363.34—dc22 2008045392

Printed in the United States of America

∞™ The paper used in this publication meets the minimum requirements of American National Standard for Information Sciences—Permanence of Paper for Printed Library Materials, ANSI/NISO Z39.48-1992.

CONTENTS

Acknowledgments

As hosts of the 2007 Society for Economic Anthropology conference "The Political Economy of Hazards and Disasters" and as editors of this volume, we would like to acknowledge several people who made the conference and this volume a success. First, we would like to thank all those who managed to brave the weather "events" across North America that made travel a challenge. That so many managed to attend the meeting and arrive with intellect and humor intact is an indication of the interest in how it is that societies and cultures deal with the political and economic aspects of extreme events. The planning of the event was greatly aided by Sofia Ivanova and Pamela Canady, who ensured that the details were addressed. We also want to recognize Courtney Hill, Phet Keomanyvanh, Amy Morse, and Tess Holder, who greeted attendees with warm smiles and ensured we all knew where we needed to be at our appointed times. A special thanks to the general series editor Dolores Koenig, who has provided valuable, gentle, and insistent guidance in the drafting of this volume. Three anonymous reviewers provided insightful critique and suggestions for our editing. Kari MacLauchlin helped us out with excellent copyediting to prepare the volume for submission. Finally, we wish to acknowledge the support of the University of North Carolina at Greensboro, College of Arts and Sciences and the Department of Anthropology, which provided us with the time and resources needed to plan and execute the conference as well as review and edit this volume. And thank you to Brent K. Marshall (1966–2008) for inspiration as a leader of a new generation of disaster researchers.

List of Tables

List of Figures

ECONOMIC PARAMETERS OF DISASTERS

I

Linking Broad-Scale Political Economic Contexts to Fine-Scale Economic Consequences in Disaster Research 1

ERIC C. JONES AND ARTHUR D. MURPHY

NOW THAT IT IS A CONFIRMED GENERALIZATION that vulnerability to disaster impact is mediated by larger social processes, we find it compelling to link the political economy of disaster with the daily lives of individuals, households, and communities who have experienced extreme events. Because of the ubiquity of hazards and a per capita increase in disasters in some regions of the world (International Federation of Red Cross and Red Crescent Societies 2003), we are called to systematically study how societies incorporate extreme events, whether chronic or one-time/short-lived, into their social, political, and ideological structure. As ethnographers, we are interested in the daily lives of people who are impacted by disasters. As students of economic dynamics, we are specifically interested in how strategies for capital accumulation construct and distribute vulnerability to hazards and cause human disasters.

Our general interest is in how societies distribute vulnerability, and specifically how societies distribute economic vulnerability. This volume builds upon efforts within and without anthropology to address cultural aspects of disaster response. Torry's (1978, 1979:517) anthropological reviews—paralleling similar reviews in geography and sociology at the time—gave examples of the many ways in which cultural groups vary in their responses to disasters, presaging current research on resilience with his assertion that there has always been a ". . .deep and abiding anthropological interest in social system performance under chronic and acute stress." Citing several authors from the mid-1970s, Torry (1979) also noted how economic and political changes at national and international levels were increasing the vulnerability of people in less developed countries.

Researchers have documented differential vulnerability to extreme events and many of the mechanisms that construct this phenomenon. Many researchers today, disillusioned with the lack of progress by national governments around the world in disaster mitigation, want to find ways that individuals, communities, and larger political jurisdictions—below the state level—can lessen the impact of disasters or at least recover more quickly and effectively. Hoffman (1999) theorized that—in addition to the quality of an extreme event and the researcher's time frame of interest—variation in populations' responses to disasters partially depended upon long-held beliefs and behaviors, which provide unique capacities but also specific difficulties (see also Bankoff 2003). Dyer and McGoodwin (1999) referred to this variation in group recovery efforts as "cultures of response." Oliver-Smith (2002) added that, in addition to long-term cultural adaptations, political decision-making helps construct actual vulnerability to hazards. Wisner et al. (2004) suggest that vulnerability depends more upon political and economic factors than upon cultural and social ones—although the authors do consider local presence or lack of various characteristics at the local level as contributing to resilience.

Suitable for our purposes of political economic analysis, Wisner et al. (2004) provide a model and a detailed suite of variables that can help researchers define a range of possible responses for a given population. What should be added to the model is to more fully account for ideology in the everyday economic life of people. Rather than the reduction of ideology to the justification of a social order, ideology to us is very material in its consequences. It is the way that we organize our economic behavior, such as through rituals that celebrate productive activities, or through altruism, competition, and the reciprocity of everyday life (cf. Godelier 1999).

Taking cultural response to disaster as a point of departure, our concern in this volume is to present hazard-related and disaster-related linkages between fine-scale and broad-scale economic behavior, specifically: 1) how vulnerability is economically constructed; 2) how primary producers adapt their production regimes; 3) how traders and merchants adapt their practices; and 4) how the state's political economic objectives play out in recovery efforts. Thus, while the chapters in this volume cover many different economic aspects of disaster, the sections of this book examine class formation, production, distribution, and government intervention. Our objective with this introduction is to provide some basic definitions for political economy and disaster, outline the foci of the volume, and highlight the 13 authors' main contributions.

Linking Elite Strategies to Daily Life

Political economy is: *strategies of the governing elite for the maintenance of power by engaging commoner populations through coherent suites of production/consumption, political focus, and ideological responsibility* (Blanton et al. 1996). Extending Blanton et al.'s (1996) analysis of Mesoamerican civilization cycles, control over production and consumption might be through a focus on elite goods, or perhaps on public goods, while a political focus can be more internal or external, and ideology may emphasize keeping the cosmos in order or insist on loyalty to kin and countrymen. Elites, in concert with the demands of commoners, fashion these foci and adhere to them until their strategies are no longer sufficiently effective at extracting surplus, organizing decision-making, or providing a framework out of which people find meaning. More specifically for the case of disasters, we want to know how various aspects of the political economy provide constraints and opportunities for individuals, households, and communities as they face both the economic impact of an event and then the politics of post-event assistance and recovery.

In addition to defining political economy, we must define disaster. Oliver-Smith, in chapter two, relies on Wisner et al. (2004) for his definition, while others in this volume generally rely on Oliver-Smith (1996, 2002) for their definitions. Combining the insights of these two authors, and adding the bias of our own interest in extreme events and collective action, we see a disaster as a tragedy experienced by a human group at the hands of an identifiable event, mitigated by local capacity and broader intervention, and measured in terms of economic, spiritual, psychological, biological, political, or social impact. The identifiable event or events may be short-lived, long-lived, or intermittent/chronic. It need not be extraordinary or extreme, but usually is. We do not present thresholds to use in deciding whether something is a disaster or not. Perhaps most useful to us is to ask whether or not there appears to be a need for recovery/adaptation. We understand that recovery may be defined in a number of ways, but we minimally understand it as a need of many people to return to a semblance of normalcy. Most of the chapters in this volume deal with recovery in terms of what kinds of economic needs, initiatives, and trajectories exist following a disaster. In more chronic situations, it is appropriate to think of adaptation rather than recovery. Resilience, even less amenable to definition, might capture the intended sense of both adaptation and recovery.

Contributions

In summary, we are looking at the economy of daily life as impacted by elite-led collective responses to disaster. In the second chapter in part I, Oliver-Smith briefly traces the history of anthropological study of disasters and addresses the political economy of disasters by analyzing two case studies. He uses the Pressure and Release model (Wisner et al. 2004) to examine how political economic behaviors and policies of governing elites are linked with the common person's experience of disaster. To our knowledge, this is the first time the model has actually been applied, rather than just used as a reminder to researchers of the types of variables that should be considered. In some sense, his chapter is about the tension between ideology and action, because ideologically and materially people face conditions that override their ability to do what they know could reduce vulnerability at the level of root causes. Leaving aside other factors, the important economic variables in the Pressure and Release model are: 1) access to resources; 2) prevalence of economic ideology; 3) level of local investment; 4) degree of local market development; 5) government budget obligations; 6) resource productivity/degradation; 7) dependence of local jobs on extreme events; 8) income levels; and 9) assistance from local social networks or institutions (see model in Oliver-Smith). Of course, it does not make sense to analyze the economic variables without relation to the other social factors, but we list the variables here so that the reader may look out for them in the chapters that follow.

In part II, the chapters consider how economic class—a product of the larger political economy—influences vulnerability within a society. First, class can help determine who is exposed to a disaster (Williams). It can then influence who is most affected out of those exposed (Underhill). Finally, class can structure access to recovery for those who were affected (Brown-Jeffy and Kroll-Smith). Williams' paleodemographic class analysis of who was exposed to and died from the Johnstown flood includes an examination of the kind of labor force that is left to participate in recovery following a disaster. In considering who was most affected by the impacts of disaster relocation, Underhill focuses on the difference between poorer socially dependent Hurricane Katrina victims interacting with the government for financial recovery and more financially stable socially independent disaster victims interacting with the private sector (e.g., their employers and the insurance industry). Finally, Brown-Jeffy and Kroll-Smith show how, in the San Francisco fire of 1906, class determined how the recovery process unfolded, as well as how various types of recovery-related resources were accessed.

Part III covers risk-related strategies of primary agricultural production in hazardous settings, and these authors look at the economics of how hazards become disasters. All three authors consider agriculturalists' responses to hazards and how larger political economic processes are related to local production strategies. We decided to cover only one type of production so the variables and dynamics at play would be limited and covered more thoroughly. Nelson and Finan show how rural producers in Brazil proceed through a predictable set of risk-minimizing strategies as the hazard grows more extreme, eventually reaching a point where the state must intervene if tragedy is to be avoided. Jaime-Riverón and Pool investigate the diversification of stone tool manufacture activities—especially recycling and reuse—that occurs in response to major resources being cut off by volcanic activity in Mexico one to two millennia ago. Whiteford and Tobin explore how adaptation to volcanic eruptions in Ecuador is molded by the cascading impacts of one aspect of life into another in a way that may produce a disaster, depending upon both government interventions and local traditions/capacity.

Part IV investigates the challenges and opportunities faced in the marketing and distribution of goods in disaster settings. More specifically, the authors consider how traders adapt to hazardous conditions and/or take advantage of new markets in disaster-impacted settings. Matejowsky details the patterns of micro-level behaviors of Philippine street vendors whose business activities are frequently disrupted by hazards and disasters, finding adaptations to hazards include increased vigilance against thieves plus more aggressive deal-making to maintain a flow of clientele. Oka, Kusimba, and Gogte look at the political activities of international Indian Ocean traders that, in the context of hazards presented by warfare and climate change, caused cities in various locales to perish or decline in the early modern period. Finally, Brown discusses how global competition and national marketing strategies interact with risk faced by vanilla farmers in Madagascar.

Part V of this volume more directly addresses how the state intervenes in economic recovery from disasters, including: when governments learn from past disasters (Smart and Smart), the community debate over how to support a major local industry affected by disaster (Trivedi), and what an effective mix of external and internal resources for economic recovery looks like (Dyer).

With some exceptions, particularly in the field of community psychology, very few cross-cultural studies on disasters have been undertaken (see review by Jones and Murphy 2008). Thus, we are pleased that four of the chapters in this volume (Brown-Jeffy and Kroll-Smith; Dyer; Oliver-Smith; Smart and Smart) are cross-national, allowing the authors to use the time-tested method of cross-cultural comparison to illuminate the aspects of hazard and disaster response that depend on long-standing traditions and beliefs within a society and that society's place in a larger regional or world economic system; two other chapters address cultural differences intranationally (Jaime-Riverón and Pool; Underhill).

In terms of the anthropological range of the work presented here, beyond cultural foci that are the norm for this field, two of the chapters are archaeological (Jaime-Riverón and Pool; Oka, Kusimba, and Gogte), one has a linguistic focus (Trivedi), and three others have biological foci (Smart and Smart; Whiteford and Tobin; Williams, which is also ethnohistorical).

The disaster-related events covered in this volume are quite diverse and include: hurricanes/cyclones (Brown-Jeffy and Kroll-Smith; Dyer; Matejowsky; Oliver-Smith; Trivedi; Underhill), floods (Williams), fires (Brown-Jeffy and Kroll-Smith; Smart and Smart), volcanic eruptions (Jaime-Riverón and Pool; Matejowsky; Whiteford and Tobin), drought (Nelson and Finan), warfare (Oka, Kusimba, and Gogte), infectious disease (Smart and Smart), and oil spills (Dyer). The frequency and intensity of extreme events—whether mass violence/warfare, natural hazards, mass migration, or societal crises—and the interrelation of the earth's populations importune us for greater understanding of collective human behavior when pressed to the extreme.

While disaster research has become a relatively mature field of inquiry in the last 30 years, we think greater theoretical organization of both the existing material and of future research questions would be a help. The political economy of hazards and disasters provides one such opportunity to link broader collective processes with the daily life of individuals in their households and communities in order to better understand mechanisms for the creation of vulnerability before, during, and after extreme events. Such a perspective builds upon Klein's (2007) *The Shock Doctrine: The Rise of Disaster Capitalism*, and Gunewardena and Schuller's (2007) *Capitalizing on Catastrophe: Neoliberal Strategies in Disaster Reconstruction*, which discuss the ways in which capitalist-promoting industry and governments have helped shape economies of a number of nation-states following extreme events.

This current volume builds upon those works by including economic formations of various times and places, not only free-market policies and not only global capitalism. The chapters that follow provide examples of a range of disaster-related political economic activities, such that the practices of disaster capitalism clearly find their place among the variety of political economic techniques employed by elites when faced with extreme events.

References

Bankoff, Greg

2003 Cultures of Disaster: Society and Natural Hazard in the Philippines. New York: Routledge.

Blanton, Richard E., Gary M. Feinman, Stephen A. Kowalewski, and Peter N. Peregrine

1996 A Dual Processual Theory for the Evolution of Mesoamerican Civilization. Current Anthropology 31(1):1–14.

Dyer, Christopher L., and J. R. McGoodwin

1999 The Culture of Response: The Political Ecology of Disaster Assistance and its Impact on the Fishing Communities of Florida and Louisiana after Hurricane Andrew. *In* The Angry Earth: Disaster in Anthropological Perspective. Anthony Oliver-Smith and Susanna Hoffman, eds. Pp. 211–231. New York: Routledge.

Godelier, Maurice

1999 The Enigma of the Gift. Chicago: University of Chicago Press.

Gunewardena, Nandini, and Mark Schuller, editors

2008 Capitalizing on Catastrophe: Neoliberal Strategies in Disaster Reconstruction. Lanham, MD: AltaMira Press.

Hoffman, Susanna M.

1999 After Atlas Shrugs: Cultural Change or Persistence after a Disaster. *In* The Angry Earth: Disaster in Anthropological Perspective. Anthony Oliver-Smith and Susanna M. Hoffman, eds. Pp. 302–325. London: Routledge.

Jones, E. C., and A. D. Murphy.

2008 Review of Cross-Cultural Disaster Studies Concerning Latin American Populations. *In* Libro Homenaje a Fernando Pozos. F. Leal Carretero, ed. Guadalajara: University of Guadalajara Press.

Klein, Naomi

2007 The Shock Doctrine: The Rise of Disaster Capitalism. New York: Henry Holt and Company, Inc.

Oliver-Smith, Anthony

1996 Anthropological Research on Hazards and Disasters. Annual Review of Anthropology 25:303–328.

2002 Theorizing Disasters: Nature, Culture, Power. *In* Culture and Catastrophe: The Anthropology of Disaster. Susanna M. Hoffman and Anthony Oliver-Smith, eds. Pp. 23–47. Santa Fe, NM: School of American Research Press.

Torry, W. I.

1978 Natural Disasters, Social Structure and Change in Traditional Societies. Journal of Asian and African Studies 13(3/4):167–83.

1979 Anthropological Studies in Hazardous Environments: Past Trends and New Horizons. Current Anthropology 20(3):517–540.

Wisner, Ben, P. Blaikie, T. Cannon, and I. Davis

2004 At Risk: Natural Hazards, People's Vulnerability and Disasters, 2nd edition. London: Routledge.

International Federation of Red Cross and Red Crescent Societies

2003 World Disasters Report 2003: Focus on Ethics and Aid. Bloomfield, CT: Kumarian Press.

Anthropology and the Political Economy of Disasters

2

ANTHONY OLIVER-SMITH

WRITING THIS CHAPTER ALLOWS ME TO LINK my original interests in economic anthropology with the research focus that has garnered the majority of my efforts over the last 35 years. Although I was originally trained as an economic anthropologist, the occurrence of a major natural disaster in my dissertation research site turned my attention toward human adaptation and adjustment to natural hazards and disasters. However, I have taught Economic Anthropology on a yearly basis for most of my career and over the years have even managed to do some economic anthropological research. I thank the editors of this volume, as well as the Society for Economic Anthropology, for giving me the opportunity in a sense to come full circle, linking those original interests in economic anthropology with my long-term research on hazards and disasters. My objectives with this chapter are twofold: 1) to provide a short description of how *anthropological* political economic approaches to disasters have developed, and 2) to use what I see as the most complete political economic model today for disaster research and apply it comparatively to Hurricanes Mitch and Katrina.

In 1970, when I first entered the field, apart from a scattering of books and a few more articles in the 1930s and 1940s, disaster research was only about 20 years old and almost entirely focused on the developed world. After all, disaster studies grew out of the research on strategic bombing of populations in World War II. There was almost nothing, apart from a few anthropological articles on typhoons and a volcanic eruption in the Pacific

Islands, on the nonindustrialized world and virtually nothing on political economy, which I understand to be the study of the interplay of economic wealth and social and political power in the production and distribution of resources, whether in disasters or reconstruction (the latter about which the field was totally silent).

The Anthropological Study of Disasters

Disasters are densely interwoven with anthropology's research agenda, particularly in regard to adaptation. Since the field's nineteenth century emergence first focused on human biological and cultural evolution, adaptation has been one of the central concepts in anthropology. In order to survive and ensure maintenance, demographic replacement, and social reproduction, human beings interact with nature through a set of material practices that are socially constituted and culturally meaningful (Patterson 1994:223). These practices and beliefs are accomplished through social arrangements and modify the natural and social world in ways that enable, to some degree, the persistence of the society through time—none, I would submit, more than economics, which serves as a major nexus between the natural and the social worlds.

That said, the original focus in disaster research was relatively narrow. Apart from a dissertation by Samuel Prince (1920) on a munitions ship explosion in Halifax harbor in Canada, and a few articles in sociology journals, not much serious social science research was done on disasters until the 1950s, during which two approaches developed.

The first was what has been characterized as the hazards approach, emerging largely from geography, with a focus on the hazards in specific environments and their impacts on people and communities. The temporal and spatial scales employed in their analyses were relatively reduced to the area affected by the hazard and the time frame in which the event unfolded.

The second approach, emerging largely in sociology, with some contributions from political science and psychology, focused on the behavior of individuals, organizations, and institutions in the disaster moments, mostly the threat, impact, and immediate aftermath stages. The spatiotemporal scales were also rather narrow, being focused around the actual event of the disaster impact. The reduced focus and spatiotemporal scales of these traditions notwithstanding, much good research, mostly in North America, Europe, and Australia, was carried out on geographical and sociological topics.

In the 1970s and 80s, anthropologists and cultural geographers both from and working in the developing world found that mainstream disaster studies brought little of value to the task of analyzing disasters in third world contexts. These researchers criticized the essentially passive role prior investigators had assigned to society in risk etiology and the limited attention paid to local, national, and international factors in creating or exacerbating both risk and impact. Researchers from and in the third world called for a rethinking of disasters from a political economic perspective, based on the high correlation between disaster proneness, chronic malnutrition, low income, and famine potential, leading to the conclusion that the root causes of disasters lay more in society than in nature (Hewitt 1983). The concept of vulnerability focused attention on those aspects of society that reduce or exacerbate the impact of a hazard. The concept of vulnerability is, I believe, inherently, of a political-economic nature, although not all vulnerability theorists use it as such. However, the vulnerability/risk approach has changed the way we think about disasters and human-environment relations in fundamental ways.

Vulnerability and the Framing of Disasters

Although there is a popular view of disasters as "acts of God" or "fate" in which nobody is really responsible, researchers now see most natural disasters as more understandable in terms of the "normal" order of things, that is, the conditions of inequality and subordination in the society rather than the accidental geophysical features of a place. This perspective shifted the focus away from the disaster event and toward the "on-going societal and man-environment relations that prefigure" a disaster (Hewitt 1983:24–27).

A disaster is now generally understood as an event/process that overwhelms the capacity of a vulnerable social group, economic activity, or infrastructure to resist and recover (Wisner et al. 2004). Disasters occur at the interface of society, technology, and environment and are fundamentally the outcomes of these features. To adequately analyze disasters, the barrier between human activity and ecosystem activity must be collapsed, thus transforming a relationship of difference into a relationship of mutuality (Oliver-Smith 2002; Wisner et al. 2004). Disasters are therefore no longer defined solely in terms of their natural, technological, or social agents or a triggering event. Vulnerability ties general political economic conditions to very particular environmental forces to understand how basic conditions such as poverty or racism produce susceptibilities to very specific environmental hazards. The definition provided by Wisner et al. (2004:11) is currently among the most utilized:

> By vulnerability we mean the characteristics of a person or group in terms of their capacity to anticipate, cope with, resist, and recover from the impact of a natural hazard (an extreme natural event or process). It involves a combination of factors that determine the degree to which someone's life and livelihood is put at risk by a discrete and identifiable event (or series or "cascade" of such events) in nature or in society.

There are a number of models of vulnerability. The one that I will use here is from Wisner et al. (2004:51; see figure 2.1). I believe it to be the most explicit model from a political economic standpoint. Vulnerability is generated through a causal chain of root causes embedded in ideological, social, and economic systems; the dynamic pressures of a demographic, socioeconomic, or ecological nature; and specific sets of unsafe conditions which, when combined with a natural hazard, produce a disaster.

As can be seen, this approach enlarges the spatial and temporal scales of analysis considerably. Each stage in the progression of vulnerability moves from larger spatial contexts and longer time frames to the more proximate conditions of local vulnerability. This more complex understanding of vulnerability enables researchers to conceptualize how social systems generate the conditions that place different kinds of people, often differentiated along axes of class, race, ethnicity, gender, or age, at different levels of risk from the same hazard and suffering from the same event.

In any environment with existing hazards, a disaster becomes inevitable in the context of a historically produced pattern of vulnerability, evidenced in the location, infrastructure, sociopolitical organization, production and distribution systems, and ideology of a society. A society's pattern of vulnerability is a fundamental feature of a disaster. It conditions the behavior of individuals and organizations throughout the full unfolding of a disaster far more profoundly than will the physical force of the destructive agent. It is only through the pattern of events and processes that emerge in this conditioned conjuncture of a community and hazard that what we call a disaster takes place. In that sense, the disaster agent itself functions as a metonym that encompasses the forces that produce the vulnerability to the systemic agent, the event, and its aftermath, including the process of reconstruction. Thus, when we speak of Hurricane Mitch or Hurricane Katrina, we encompass with those labels a set of social, environmental, and political economic processes that extend deep into the past and far into the future

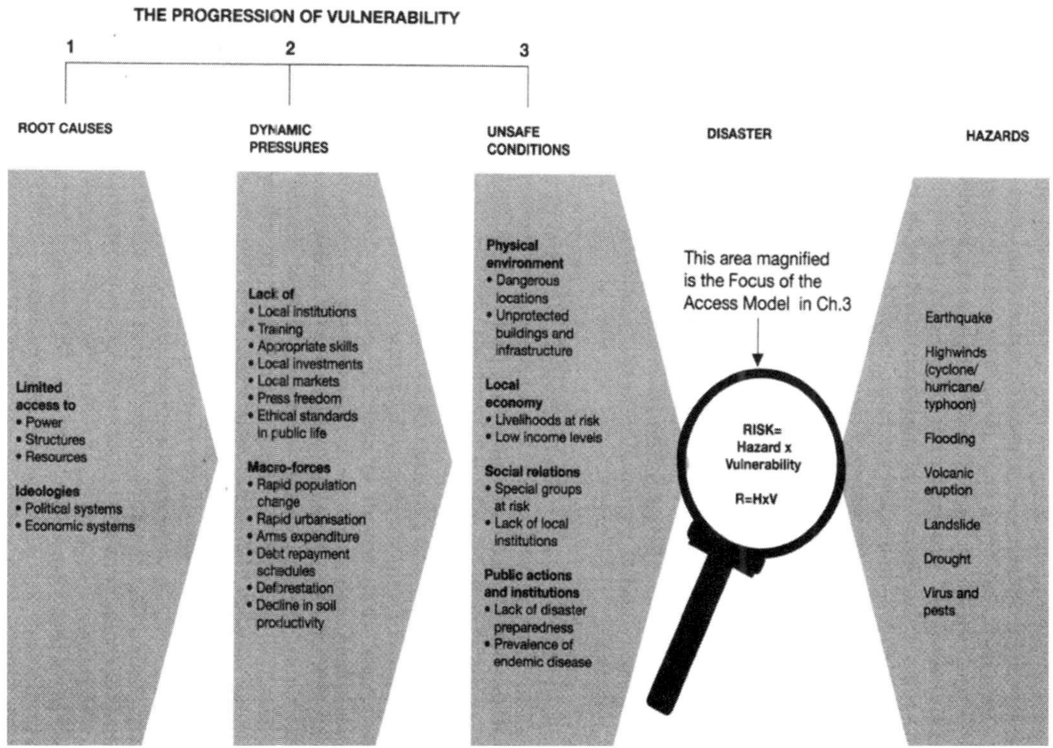

Figure 2.1. Pressure and Release Model for Understanding Disaster-Related Vulnerability
Source: Page 51 of Wisner, Ben, Piers Blaikie, Terry Cannon and Ian Davis 2004 At Risk: Natural Hazards, People's Vulnerability and Disasters, 2nd Edition. London: Routledge (Taylor and Francis). Reproduced by permission of publisher.

as well as spanning beyond local economies and power sectors to regional and global political economic structures.

When social scientists in the 1970s and 1980s began establishing through their research the empirical findings that led to the concept of vulnerability, they created the basis for a powerful model of causation for disaster research and practice. Vulnerability has become an orienting principle for the entire field of disaster studies and management. The concept of vulnerability challenges us to identify those social features and specify those linkages to processes that when coinciding with the onset of a hazard lead to damages and deaths. To the degree that vulnerability is socially constructed, the most salient features of social construction often represent entrenched arrangements of political and economic behavior and ideology. They are not "accidental." They may or may not be intentional. In that sense, we need to recognize that the beginnings of disasters are deeply embedded in the history, ideology, and political economy of a society, or, more specifically, in the human-environment relations and human-human relations and the ideologies that inform them and are reinforced by them. However, not only the processes associated with creating vulnerability but also the process of post-disaster reconstruction reveal how deeply disasters are embedded in the political economy of a region.

My goal in this chapter is to show how an explicit political economy of disaster approach might be applied. I have chosen to do this analysis comparatively, using a disaster from the first world and a disaster from the third world to provide insight into what kinds of variation might be expected from undertaking this approach. Hurricane Mitch in 1998 and Hurricane Katrina in 2005 provide two extraordinary examples to illustrate the assertion that disasters are embedded in the political economy of a region. I rely, though not point by point, on the Wisner et al. (2004) model to guide my analysis of both of these disasters.

Hurricanes Mitch and Katrina in Political Economic Perspective

Certainly, one of the most frustrating, maddening, and tragic things about Katrina was that not only did we know it would happen, but we knew what would happen and why it would happen, at virtually every stage (Laska et al. 2005). And as for Mitch, Susan Stonich's (1992) work in the early 1990s all but predicted what would eventually happen in Honduras in 1998. If we knew all those things, if they were part of our knowledge base, then why did Mitch and Katrina even happen? Our knowledge was good, but it was not sufficient to enable us to deal with the disaster process effectively. The question is why? And further, how does that knowledge base coupled with what happened actually frame the reconstruction process to deal with those threats and vulnerabilities that we know are systemic? These are questions that, at their core, address the basic relationship between ideology and action, or, rephrasing, between knowledge and policy.

There are clear similarities between Hurricane Mitch in 1998 and Hurricane Katrina in 2005, although one occurred in a small, poor, and underdeveloped nation and the other in a large, wealthy, and developed nation. In both circumstances, an intense natural force encountered a society whose environment, infrastructure, and population had been rendered severely vulnerable by social and economic processes deeply embedded in particular approaches to development and the environment. Thus, it was not just a situation of people living in harm's way, but of the active creation of conditions of vulnerability largely through human action on the environment.

Hurricane Mitch

In the case of Mitch, the research of Susan Stonich (1992), Kees Jansen (1998), Manuel Winograd (in press), and others tells us that the vulnerability of Honduras was and is composed of factors rooted in human use of the environment and the socioeconomically derived conditions in which much of the population lives. The local topography and climate were tied to a development model based on agricultural diversification and export-led growth that promulgated practices that degraded the environment, making it more vulnerable to the impact of climatological hazards. More than 50 percent of the country is at risk of landslides or droughts, constituting roughly 60 percent of used lands. More than 25 percent of the country is at risk of flooding (50 percent of used lands), especially in the more populated agricultural areas of San Pedro Sula, Tegucigalpa, and Choluteca (Winograd, in press). The agricultural development model directed toward production of nontraditional agricultural exports (NTAE) brought into use floodplains as well as displacing small farmers to easily eroded hillsides in processes of land concentration to achieve economies of scale for export production. These practices in turn led to a series of dynamic pressures expressed as environmental degradation, land concentration, population displacement, migration, and intense urbanization, ultimately producing the unsafe conditions of vulnerable housing, unstable terrain, contaminated water supplies, disease, malnutrition, and poverty.

The deteriorating environmental situation compounded the social conditions that located Honduras among the poorest of the poor of Latin America. In the 1990s, Honduras had the fourth lowest GDP per capita in Latin America (higher only than Bolivia, Haiti, and Nicaragua). Seventy percent of the total population and 80 percent of the rural population were living below the absolute poverty level. Nationally, undernutrition affected 40 percent of school-age children and, in some areas, 70 percent of preschool children suffer from malnutrition, of which 30 percent reached second or third degree levels. At the beginning of the 1990s, the population was estimated at 5.1 million. By mid-1996, it was at 6.1, more than double the population of 1970. Although fertility rates dropped from 7.4 in 1970 to 5.3 in 1989 and the annual growth rate from 3.7 in 1980 to 2.96 in 1990, 46 percent of the population was less than 20 years old at the time of Hurricane Mitch. Honduras ranked last among Central American countries according to the UN Human Development Index—a composite measure calculated from multiple indicators of life expectancy, educational attainment, and income (Stonich 1992:2–3).

Rapid population growth coupled with scant economic opportunities forced increasing numbers of rural peoples to intensify their efforts in agriculture, in many cases adopting agricultural practices that destroyed forests and degraded soils and other natural resources. In growing numbers, rural peoples were migrating elsewhere, seeking survival either in the rapidly expanding urban areas or in the remaining areas of tropical humid forests of La Mosquitia of Honduras' east coast. The average annual rate of urbanization was roughly 5.5 percent, the highest in Central America. The rate of deforestation was estimated to have ranged from 1.6 to 3.2 percent annually or about 80,000–90,000 hectares per year (Stonich 1992:2–3).

In rural areas, expansion of banana and melon cultivation, shrimp farming, and cattle ranching was associated with an increasing concentration of land ownership and provoked the large-scale dislocation of small farmers from valley bottomlands to steep hillsides, where the necessity to clear land for cultivation resulted in widespread deforestation and soil erosion. In lowland areas, as in the Sula valley, river channeling and river drainage over the course of decades to bring land into banana production also led to occupation of low-lying and flood-prone areas by plantation workforces. Drained wetlands and diked rivers for extensive plantations of export crops in lowland valleys attracted thousands—now homeless and jobless after Mitch—to work for agribusiness in areas acutely vulnerable to flooding.

Because the rural poor were and are increasingly dependent on off-farm income, mostly from agricultural labor, they suffered from elevated risk regarding appropriate procedures for handling and using pesticides. Research in the early 1990s indicated that excessive use of pesticides was not only placing farm workers in jeopardy, but also contaminating soils, surface water, and groundwater. In addition, resistance to pesticides was increasing, and diseases once thought to be declining—such as malaria—were increasing in incidence. Furthermore, access to safe and adequate water supply, sewage, health care, and adequate housing had diminished (Stonich 1992:5).

Rural poverty and land concentration produced waves of migration to cities like Tegucigalpa, where people occupied unsafe structures on hillsides made unstable by deforestation. People seeking homes around the cities cleared steeply sloped lands around highland cities, exacerbating the instability of soils, erosion, and landslides. In the lowland cities, the migrants occupied areas prone to flood, creating extremely densely populated urban neighborhoods along riverbanks and in floodplains. These urban settlements, as in squatter settlements in many areas, suffered from a lack of potable water and sanitation as well as public health facilities, resulting in diarrhea and parasitic diseases that produce high levels of mortality and morbidity among vulnerable populations such as the poor, women, and children (Stonich 1992:4).

After Mitch devastated Honduras, even a glance at the relevant documents reveals that both Hondurans and the international community apprehended and appreciated the complex relationships between environmental, economic, and social factors that led to the destruction that happened in the hurricane. Nonetheless, the outcome of reconstruction plans was a set of policies and practices that addressed symptoms, even effectively in some cases, but did little to address causes, thus condemning themselves to repeat the exercise at some future point as both causes and symptoms evolve with attempts to address them.

The government initially framed the issue in terms of simple reconstruction, but development nongovernmental organizations, trade unions, farmers' organizations, and other elements of civil society saw the disaster as an opportunity to transform the nation (Jansen 2003:49). The Stockholm Declaration of May 1999, on which the reconstruction plan was based, called for reducing the social and ecological vulnerability of the nation as an overriding goal as well as the reconstruction and transformation of Central America based on an integrated approach to transparency and good governance. Eventually, the government integrated the idea of change into its discourse and the national reconstruction plan was entitled the Master

Plan of Reconstruction and National Transformation (Plan Maestro de Reconstrucción y Transformación Nacional). In this context, reconstruction meant rebuilding the infrastructure and the rapid reconstitution of the economic base. The term transformation referred to "a strategy of accelerated, equitable, sustainable and participatory development" and prioritized three key themes: poverty, environment, and democratic participation (Plan Maestro 1999). The plan contained impressive provisions for prioritizing poverty reduction and human development, economic reactivation with productive employment, sustainable protection of natural resources, and strengthening democratic participation (Plan Maestro 1999:12). Specific recommendations included changes in environmental, economic, social, and political activities.

From an environmental perspective, reforestation efforts should have focused on the most seriously impacted areas, such as both rural and urban hillsides, but incentives for resource-poor farmers needed to be included in these programs to protect and preserve the trees. The shrimp farms along the Pacific coast were economically significant at the national level, but their reconstruction should have included regulations to protect nature reserves, particularly the rich mangrove environments (DeWalt 1998).

Economically, jobs needed to be provided to enable individuals, households, and communities to recover and reconstruct. Incentive programs to encourage larger landowners to plant labor-intensive crops on the most productive lands were also recommended (DeWalt 1998). Suggestions by governmental authorities that production patterns be altered toward basic food grains were also important, particularly to reduce migration into the last remaining rainforest regions, predominately in La Mosquitia.

National transformation required that education, health, and nutrition programs be established to sustain local populations and to enhance their capabilities to recover and to develop any new economic options that became available in the process of reconstruction. The public health situation, as it regards issues of disease vector control, sanitation, potable water, and nutrition in particular, were also priorities as forms of vulnerability reduction and reconstruction efforts.

From a political standpoint, the ongoing tension in Honduras and Central America between administrative decentralization (democratization and local decision-making) and traditional centralist tendencies of national governments as it affected long-term reconstruction had also to be resolved in a way that favored local and municipal initiative. Ultimately, the interrelated practices of post-disaster reconstruction, environmental management, and sustainable development had to be grounded at the local level and would then succeed in a context of local-level community mobilization and support. In effect, post-hurricane Honduras provided an important opportunity for economic forms and practices to be altered toward more sustainable forms of use. Furthermore, reconstruction after Mitch could have helped to establish the range of possibilities for sustainability as an achievable goal of specific directed policies within the limitations established by current economic practice.

However, the crucial question remains as to whether the political and economic structures of the nation can, even with the necessity and the incentive of reconstruction, truly come to grips with a set of endemic conditions that are so deeply embedded by both national and international forces in their own forms and practices. In other words, the rhetoric surrounding reconstruction reflected the ambivalence between continuity and change at the ideological level, but also the tension between reestablishing pre-disaster stability as quickly as possible and the recognition that that system itself was part of the problem. Brown-Jeffy and Kroll-Smith (this volume) refer to this as the recovery of inequality.

According to the human rights organization the Washington Office on Latin America, "the landmark Stockholm Accords have yet to be implemented" (Gass 2002:2). This is not to say that reconstruction efforts accomplished nothing. Roughly five years after the disaster, much of the damaged infrastructure had been replaced or repaired. Housing projects, some more successful than others, had been built. Schools had been constructed. Roads and bridges were repaired. Efforts to relocate communities into more secure environments had begun.

However, the basic relationships that exacerbated the effects of the hurricane so severely have not changed significantly. Agricultural policies based on the theory of comparative advantage and the production of nontraditional agricultural export crops continue to dominate the Honduran countryside. Questioning that fundamental premise has not been possible.

The unquestioned acceptance of this policy by national and international interests was quite clearly demonstrated to me personally when I became involved in discussions to prepare a major research and application proposal to be headed by an economist. Based upon my experience and work in Honduras after Mitch, I was invited to participate in what was to be a very large proposal to advise and consult on the recovery process. At one juncture, I suggested that continuing the policies of comparative advantage and nontraditional agricultural export crops might simply reconstruct the previous set of vulnerable conditions. A short discussion took place as to whether Honduras had any options in the process. My opinion was noted and I was omitted from any further planning sessions (the project was not funded anyway). In other words, despite the ideological shift that the emphasis on vulnerability and transformation represented in the national plan, it was not sufficient to disrupt even remotely the global hegemonic model of economic-environmental relations that oriented Honduran realities.

Hurricane Katrina

In the case of Katrina, the research of Shirley Laska et al. (2005) and Robert Gramling and Ronald Hagelman (2005), among others, tells us that the vulnerability profile of New Orleans and the Gulf Coast also combines factors issuing from the use of the natural environment driven in part by political economic factors. The region of the Gulf Coast and particularly surrounding New Orleans was close to or below sea level, leaving it vulnerable to storms and hurricanes. New Orleans is also surrounded by one of the largest rivers in the world, as well as several lakes and bays, swamps, and marshes. Inadequate levees built to control the flow of the Mississippi River and prevent Lake Pontchartrain from flooding the city, added to increased levels of vulnerability. Another measure of vulnerability in New Orleans is due to environmental degradation produced by multiple causes. Louisiana ranks second only to Texas in terms of hazardous industrial waste generated. Much of the waste that is flushed down river leaves through the mouth of the Mississippi, contaminating large areas of the Gulf Coast (Oliver-Smith and Button 2005).

The digging of a Mississippi Gulf outlet has been said to weaken these natural defenses. Coastal erosion has been precipitated by the clear-cutting of cypress for garden mulch and the location of hundreds of gas/oil rigs and pipelines along the coast has further added to environmental degradation. Moreover, the building of canals for the petrochemical industry has contributed to the destruction of the natural protection of coastal marshes as well (Laska et al. 2005). The loss of coastal marshes and land due to these factors and others has accelerated the loss of coastal land and increased the vulnerability of the populace. It is estimated that the Gulf Coast loses about 35 square miles of coast each year due to these factors (Gramling and Hagelman 2005).

The socioeconomic factors contributing to the vulnerability profile of New Orleans and the Gulf Coast are now well known. New Orleans suffers from a condition of endemic economic crisis, compounded by a political system in which corruption also seriously undermined its capacity to cope with an array of serious social problems. Data from the U.S. Census 2000 indicate that approximately one third of the city of 500,000 people was poor. Only 46 percent of the residents owned their own homes. In fact, the poverty rates of the city are more than twice the national average. Louisiana has one of the lowest levels of educational attainment in the nation; one quarter of all adults age 25 and over lack a high school education. Additional vulnerability of the residents is increased by the estimate that approximately 10 percent of the population has special needs, which include being elderly, or having mental or physical disabilities. Further vulnerability is indicated by the fact that in overall health indicators Louisiana ranked last among the 50 states in 2002, and 49th in 2003 (U.S. Census 2000).

Further vulnerability to disasters is evidenced by the fact that approximately 120,000 of the city's residents did not own cars, making evacuation especially problematic in the event of an emergency. The failure of local officials to adequately provide public transportation before Hurricane Katrina struck the city only increased the vulnerability of this population. Indeed, the institutional vulnerability of local, state, and federal agencies was made manifest in the secondary disaster of displacement, subsequent to the failure of the levees after Hurricane Katrina (Button and Oliver-Smith 2008; Oliver-Smith and Button 2005).

The debates and activities unfolding around the reconstruction process in New Orleans and Gulf Coast environs reflect the variety of interests of the different impacted groups. On one hand, we have seen clearly that vulnerability reduction will require reducing the environmental degradation that left the region bereft of natural defenses against a systemic hazard. However, addressing root causes, reversing the Pressure and Release model as Wisner et al. (2004) suggest, will mean taking actions that reduce both the exploitation and efficiency of transport of basic features of the economy, namely petroleum, chemicals, and shipping. Reduction in that sector of the economy will reverberate in the employment sector, eliminating the capital and jobs needed for reestablishing an economic base. If models favoring environmentally sustainable urban planning are implemented, the human rights of the several hundreds of thousands of displaced people and their right to return to their communities will be severely limited, furthering the agenda of certain social classes and their particular vision of a New Orleans of the future (Button and Oliver-Smith 2008).

Globalization and the Political Economy of Disaster

The death and destruction of Hurricane Mitch and Hurricane Katrina can be viewed as due in part to economically and socially inscribed practices associated with the capital and commodity flows that created and sustained them both at both national and international levels. If such is the case, the capacity of either the Honduran or the U.S. system to address the root causes of the disaster was and is limited. At the local level, powerful interests emerge quite quickly to reestablish the political economic structures that supported their dominance. At the international level—both disaster sites were important to, and dependent upon, international markets. The capacity of any single society in the current global economic system to address such root causes is open to question. However, we err, I think seriously, if we attribute to globalization the strength of natural law. Globalization is not gravity and market logics are not physics (Collins 2003). Neither have the force of natural law and both are cultural, essentially ideological phenomena, with deep and widespread material manifestations, that nonetheless admit to wide variation in both interpretation and application. Globalized market logics are clearly guiding reconstruction efforts in both Honduras and the U.S. Gulf Coast, but their exponents today must now contend with alternative understandings of environmental and economic relations. These discussions suggest that while vulnerability may be reinstalled in both contexts by overriding economic and political interests, these ideological models, despite their dominance, are no longer universally accepted or uncontested

In contemporary disasters, perhaps our most important task is to discover and implement those aspects of reconstruction that feasibly, within the limits of action permitted by existing political economic structures, can reduce both environmental degradation and vulnerability to hazards. Post-hurricane Honduras and the U.S. Gulf Coast provide important opportunities for economic forms and practices to be altered toward more sustainable forms of use. Furthermore, reconstruction after Mitch and Katrina can help us understand the range of possibilities for sustainability as an achievable goal of specific directed policies within the limitations established by current political economic practice.

As David Harvey suggests (1996), the built environment in which we live is a material instantiation of our social relations (Harvey 1996). It is both expressive of and shapes our social relations. Nowhere does this relationship become more crucial than in the process of community reconstruction.

Material reconstruction can both support and express social reconstitution. Material reconstruction can be a confirmation of social reconstitution. It can also undermine the process severely and very frequently has (Oliver-Smith 2005).

But if disasters and the losses they cause are complex and compounding, then the process of reconstruction also is inherently complex. Borrowing from my colleague Chris de Wet (2006) in displacement and resettlement studies, I would suggest that post-disaster reconstruction is characterized by an array of interrelated factors of different orders—cultural, social, environmental, economic, institutional, and political—all taking place in rapidly changing circumstances. Moreover, these changes are taking place simultaneously in an interlinked and mutually influencing process of transformation. And further, these internal changes from the reconstruction process are also influenced by and respond to external sources of power as well as the goals of local actors. Again following De Wet (2006), I would suggest the inherent complexity of reconstruction tends to limit the degree of control authorities can exercise over the process and creates a space for local people to take greater control over the process. The vulnerability model tells us that disaster losses are in large part the consequence of human action. Both in prevention and in reconstruction, people can address those conditions which bring about disaster by undertaking political action to not only hold those responsible for vulnerable conditions accountable but also to reframe human-environment relations and the entire process of development. The challenge thus becomes the development of policy that supports a genuine participatory and open-ended approach to reconstruction (De Wet 2006).

Conclusion

To conclude, then, the combination of increasing population, population density, increasing poverty, and occupation of hazardous sites has accentuated vulnerability to both natural and technological hazards and increases the probability of severe impacts of disasters. For example, other cities that face threats similar to those that devastated New Orleans are Mumbai, Caracas, Cape Town, Dar es Salaam, and Manila. Technology has also vastly increased the numbers of hazards to which populations are exposed. When socionatural disasters trigger technological disasters, the resulting complex process may render places completely uninhabitable. While many of the changes associated with increasing state and market integration have established more resilient infrastructures in some regions of the world, they have also frequently undermined traditional adaptations of rural populations to natural hazards. Norman Myers (1997) has asserted that recent human-induced environmental change, such as desertification, deforestation, or soil erosion compounded by natural and man-made disasters, could force as many as 50 million people from their homes by 2010. He sees environmental change and disasters as triggers or detonators that lead to land competition, resource degradation, occupation of fragile regions, and impoverishment that create disastrous conditions and eventually uproot people, turning them into environmental refugees.

Today, because of the globalization of both social and biophysical phenomena, many local problems, including disasters, may have their root causes and triggering agents, and possibly their solutions, on the other side of the globe. Through this globalization process problems have become basically non-linear in causation and discontinuous in both space and time. This renders them inherently unpredictable and substantially less amenable to traditional methods of observation of change and adaptation. Now, people, economies, and nature are in a process of coevolution on a global scale, each influencing the others in unfamiliar ways and at scales that challenge our traditional understandings of structure and organization with serious implications for the adaptive capacities of people and societies (Holling 1994:79–81).

The challenge in this field is to specify the linkages, now on regional and global scales, that generate destructive forces within our local societies and environments. Generally speaking, from my perspective the public must be made aware of how implicated past human actions are in the occurrence of a disaster; that is, how certain kinds of deeply embedded cultural predispositions combine with conscious decision-making to further specific political economic interests in society produce specific kinds of vulnerabilities to disasters. There needs to be a greater understanding about the structural components of vulnerability to disasters and how they contribute as much or more to the unfolding of a disaster as the actual hazard agent itself.

One of the frequent pitfalls of political economic analysis is that people are often subsumed into large aggregates and human agency tends to disappear. Disaster impacts also have a way of diminishing agency, of making human effort appear ineffectual, of rendering people into a passive, victim condition. However, we have also seen that in efforts undertaken in public action to combat vulnerabilities through mitigation as well as through the process of reconstruction, people can reclaim agency and restore a sense of meaning and direction to their lives and communities. By undertaking political action to not only hold those responsible for creating vulnerable conditions accountable, local communities can also work to reframe social and economic relations and press for more sustainable forms of development.

References

Button, Gregory V., and Anthony Oliver-Smith

2008 Disaster, Displacement and Employment: Distortion in Labor Markets in Post-Katrina Reconstruction. *In* Capitalizing on Catastrophe: The Globalization of Disaster Assistance. Nandini Gunewardena and Mark Schuller, eds. Walnut Creek, CA: AltaMira.

Collins, Jane

2003 Threads: Gender, Labor and Power in the Global Apparel Industry. Chicago: University of Chicago Press.

DeWalt, Billie R.

1998 The Human Causes of a Natural Catastrophe, Pittsburgh Post-Gazette, November 22:B1,4.

De Wet, Chris

2006 Risk, Complexity and Local Initiative in Involuntary Resettlement Outcomes. *In* Improving Outcomes in Involuntary Resettlement. Chris De Wet, ed. Pp. 180–202. New York: Berghahn Books.

Gass, Vicki

2002 Democratizing Development: Lessons from Hurricane Mitch Reconstruction. Washington, DC: Washington Office on Latin America.

Gramling, Robert, and Ronald Hagelman

2005 A Working Coast: People in the Louisiana Wetlands. Journal of Coastal Research SI(44):111–133.

Harvey, D.

1996 Justice, Nature and the Geography of Difference. Blackwell: Oxford.

Hewitt, Kenneth

1983 Interpretations of Calamity. Winchester, MA: Allen & Unwin, Inc.

Holling, C. S.

1994 An Ecologist's View of the Malthusian Conflict. *In* Population, Economic Development, and the Environment. K. Lindahl-Kiessling and H. Landberg, eds. Pp. 79–103. Oxford University Press: New York.

Jansen, Kees

1998 Political Ecology, Mountain Agriculture and Knowledge in Honduras. Amsterdam: Thela Publishers.

2003 Crisis Discourses and Technology Regulation in a Weak State: Responses to a Pesticide Disaster in Honduras. Development and Change 34(1):45–66.

Laska, Shirely, George Woodell, Ronald Hagelman, Robert Gramling, and Monica Teets Farris.
2005 At Risk: The Human Community and Infrastructure of Resources in Coastal Louisiana. Journal of Coastal Research SI(44):90–111.

Myers, Norman
1997 Environmental Refugees. Population and Environment 19(2):167–82.

Oliver-Smith, Anthony
2002 Theorizing Disasters: Nature, Culture, Power. *In* Culture and Catastrophe: The Anthropology of Disaster. Susanna M. Hoffman and Anthony Oliver-Smith, eds. Pp. 23–48. Santa Fe, NM: School of American Research Press.
2005 Communities after Catastrophe: Reconstructing the Material, Reconstituting the Social, *In* Building Communities in the 21st Century. Stanley Hyland, ed. Santa Fe: School of American Research Press.

Oliver-Smith, Anthony, and Gregory V. Button
2005 Forced Migration as an Index of Vulnerability in Hurricane Katrina. Paper presented at the 2nd Meeting of the Expert Working Group on Vulnerability of the United Nations University Institute for Environment and Human Security, Bonn, October 14.

Patterson, Thomas.
1994 Toward a Properly Historical Ecology. *In* Historical Ecology: Cultural Knowledge and Changing Landscapes. Carole L. Crumley, ed. Santa Fe, NM: School of American Research.

Plan Maestro de la Reconstrucción y Transformación Nacional.
1999 Reunion del Grupo Consultivo. Stockholm.

Prince, Samuel
1920 Catastrophe and Social Change. New York: Columbia University Press.

Stonich, Susan C.
1992 I Am Destroying the Land: The Political Ecology of Poverty and Environmental Destruction in Honduras. Boulder, CO: Westview.

United States Census Bureau
2000 www.census.gov/

Winograd, Manuel
In press From Natural Events to Natural Disasters: Hurricane Mitch and Environmental Vulnerability in Honduras. *In* The Legacy of Mitch: Lessons from Post-Disaster Reconstruction in Honduras. M. J. Ensor, ed. Tempe: University of Arizona Press.

Wisner, Ben, P. Blaikie, T. Cannon, and I. Davis.
2004 At Risk: Natural Hazards, People's Vulnerability and Disasters. 2nd edition. London: Routledge.

II

CLASS-BASED VULNERABILITY IN DISASTER EXPOSURE, IMPACT, AND RECOVERY

"The Dam is Becoming Dangerous and May Possibly Go"[1]: The Paleodemography and Political Economy of the Johnstown Flood of 1889

3

LESLIE LEA WILLIAMS

THE DEMOGRAPHIC PROFILE OF MASS MORTALITY VICTIMS can tell us about biological and social aspects of risk and vulnerability. Additionally, since disasters have such disruptive and devastating effects on the communities they affect, understanding victim demographics can also provide information relevant to community recovery. The present study examines the paleodemography of the Johnstown Flood of 1889, focusing in particular on the effects of the catastrophe on the community's economic recovery. The Johnstown Flood is interesting because it represents a snapshot view of mass mortality; unlike indiscriminate epidemics or famines, the victims died rapidly in a manner of minutes or hours. This study investigates the age-at-death distributions of those victims using data collected from the victims' list published on July 31, 1890,[2] in the *Johnstown Tribune*, printed as a pamphlet, and reprinted exactly in McGough (2002). The differential mortality of the Johnstown victims, seen in age but especially in sex, speaks to both the risks inherent in natural disasters like floods and the ability of the community to recover.

It is difficult to make statements about the past without knowing the cultural, social, or demographic characteristics of the population being studied. Consequently, reconstructing the past in terms of the age and sex distribution of its populations is an initial concern of archaeologists and physical anthropologists. Although skeletons are the most common data, other sources, such as historical records, are used in paleodemographic research (Chamberlain 2006). While these records may have their own biases, they provide a greater precision in age estimation than skeletal analyses alone.

Although many paleodemographic studies seek simply to describe population structure (e.g., Alesan et al. 1999; Gowland and Chamberlain 2005; Nagaoka et al. 2006), paleodemographic data are frequently used to examine particular biological, social, or economic conditions in past populations (e.g., Gowland and Chamberlain 2002; Sullivan 2004; Van Gerven et al. 1995). In particular, paleodemographers often investigate population age-at-death structures for evidence of mass mortality events like massacres, plagues, and natural disasters. The age structures of mass mortality victims tend to show either an attritional or a catastrophic mortality pattern (Chamberlain 2006). The age-at-death profile of an attritional assemblage shows peaks for the very young and the very old, while those in the middle are relatively less represented (figure 3.1; Chamberlain 2006). In contrast, a catastrophic profile is one in which a mass mortality event strikes indiscriminately and all individuals have an equal chance of dying (figure 3.2).

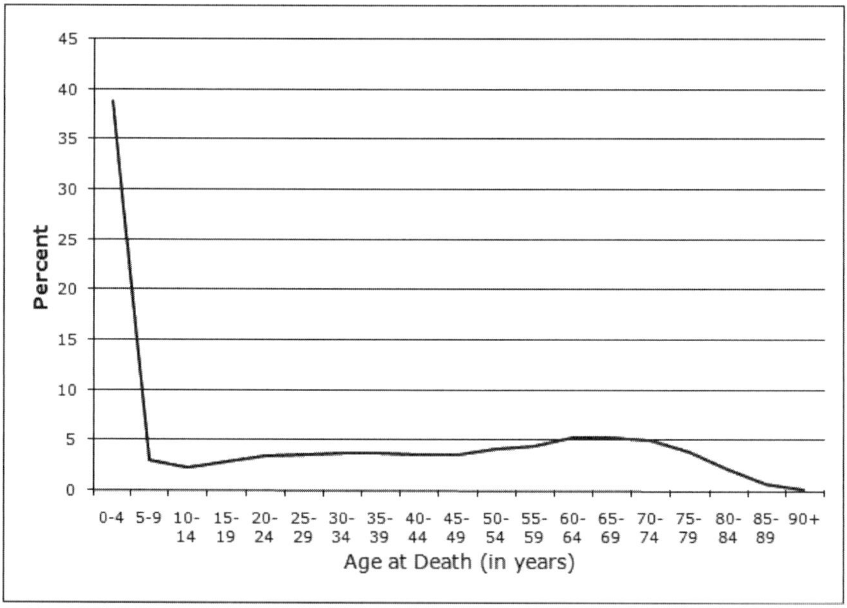

Figure 3.1. Attritional or "Normal" Mortality Profile (Coale and Demeny [1983] Life Tables Adapted from the Female West Level Life Table, with No Population Growth).

The age structure of the catastrophic profile thus matches the age structure of the living population (Chamberlain 2000; Gowland and Chamberlain 2005; Margerison and Knüsel 2002; Paine 2000). Catastrophic events with extreme consequences to both mortality and fertility rates leave signatures on the archaeological record (Paine 2000).

However, not all catastrophic events follow the catastrophic profile, as a population's behavior, culture, and the nature of the disaster can all affect victimization (Chamberlain 2006; Grayson 1990; Grayson 1993; Grayson 1996; Margerison and Knüsel 2002). In particular, these deviations from the expected age-at-death profile can be indicative of cultural or biological buffering of risk. For example, the Donner Party's 1850 starvation deaths show surprisingly high mortality rates in young males (Grayson 1990; Grayson 1993). Grayson concludes that this is likely due to differential mortality risks related not just to age and sex but also to kin group size. An analysis of the victims of the Johnstown Flood provides similar insight into the nature of differential mortality and its resulting socioeconomic consequences.

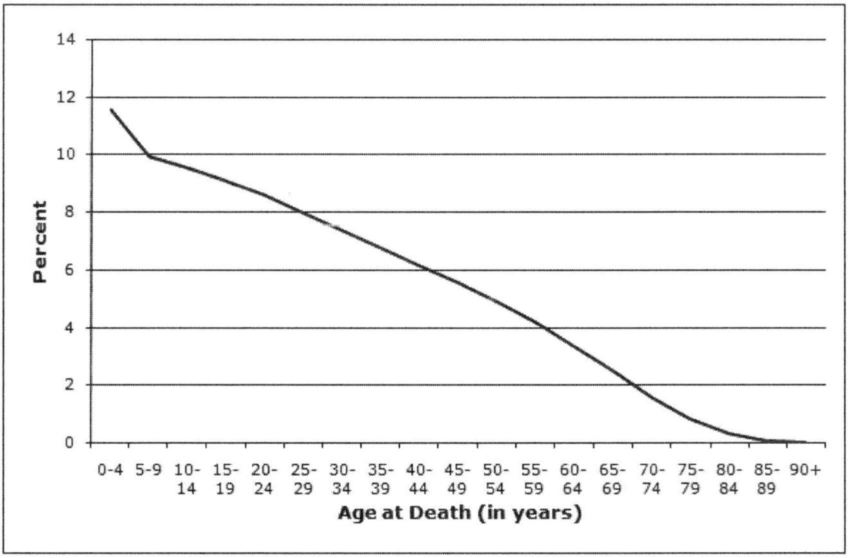

Figure 3.2. Catastrophic Mortality Profile (Coale and Demeny [1983] Life Tables Adapted from the Female West Level 5 Life Table, with No Population Growth).

Johnstown, Pennsylvania: Geographic Vulnerability

The 1889 Johnstown Flood killed approximately 2,200 people in Johnstown, Pennsylvania, and the surrounding areas. In many ways this flood shows the hallmark characteristics of vulnerability in natural disasters, in which human actions exacerbate natural hazards to create increased risk and vulnerability for a group of people (Oliver-Smith 2002). For Johnstown, in Cambria County, the natural hazard is the situation of the town itself. Johnstown is located in the western half of the state, in the Allegheny mountain range. The city was built on a floodplain, encircled by tall hills and situated where two rivers, the Little Conemaugh and the Stony Creek, converge to form the Conemaugh (McCullough 1968). This river is prone to flooding, and its natural geography is unique, dropping down through the mountains surrounding the city at a very steep incline in some places (Shappee 1940).

Human-made risk factors for natural disasters like the Johnstown Flood can include intensification of agriculture, land degradation, restructuring of rural areas, and increased population growth (Oliver-Smith and Goldman 1988). At least three of these risks are seen in the development of Johnstown and the surrounding areas, as the land became degraded, the population exploded, and formerly rural areas were transformed into an industrial metropolis (Shappee 1940). The water of the Conemaugh was used extensively to operate the mills in the Johnstown region, and the rivers in the area were central to the transportation of raw and finished materials. Indeed, the encroachment of the community and industry on the rivers exacerbated the consequences of extended rainfall (Shappee 1940).

Furthermore, during the building of the Pennsylvania Canal in the mid-1800s, the Little Conemaugh River was dammed at the South Fork, regulating the flow of the river. The dam was originally built to help supply the Pennsylvania Canal, and while the construction of the dam was competent, it fell into disrepair and actually broke in 1862 (McGough 2002; Shappee 1940; Unrau 1980). The damage from this break was minimal since the reservoir was not full (Unrau 1980). The dam then went through various states of disrepair before being purchased and renovated by the South Fork Fishing and Hunting Club (Shappee 1940; Storey 1907). This new South Fork Dam held back 20 million tons of water known as Lake Conemaugh (McGough 2002). The lake was five miles in diameter, and during high water in the spring it covered 450 acres and was almost 70 feet deep in several spots (McCullough 1968). A river valley snaked through the hills 16.9 miles from the dam to the stone bridge in Johnstown, dropping in elevation 400 feet (Shappee 1940). Unfortunately, the renovations to the dam made by the South Fork Fishing and Hunting Club were done without the advice of engineers and would prove fatally inadequate (Unrau 1980).

The People of Johnstown: Demographics

Before 1889, Johnstown was surrounded by the boroughs of Millville, Grubbtown, Conemaugh, Woodvale, Prospect, Cambria City, Moxham, Morrellville, Coopersdale, East Conemaugh, and Franklin (Storey 1907). The area was long exploited for its natural resources, first timber, then coal, and then the iron ores used to make the steel for which the city would eventually become known (Shappee 1940). The most recent, pre-flood, United States Census of the Johnstown area was conducted in 1880. The specific age and sex distributions are not given; what is known is that Johnstown had 8,346 individuals. It was a working-class steel and iron community peopled with immigrants from Ireland, Scotland, Cornwall, Wales, Germany, and the Slavic countries, many of whom were employed by the Cambria Iron Company (Camp 1995; Shappee 1940). The total population grew rapidly in the years before the flood,[3] swelling to 20,639 by 1889 (Shappee 1940).

The People of Johnstown: Economics

Beginning in about 1850, Johnstown's industrialization proceeded at an intensive pace, and the character of the community was inextricably linked to its production (Shappee 1940:51):

> . . . the history of the valley is traced in the development of railroads, rolling mills and Bessemer steel. All other phases of community life served these masters. The mills penetrated into the lives of the people as their smoke and dirt entered their homes.

Indeed, the Cambria Iron Company was in many ways a town unto itself, building housing for workers and opening a hospital for its employees when Johnstown could not get one approved (Shappee 1940). Its company store was a large employer and significant economic force within the community, and its managers and leaders exercised immense influence on the shape of Johnstown's infrastructure, conforming the city to the Cambria Iron Company's desires (Shappee 1940). During the 1880s and

1890s, labor in the U.S. was shifting from organization to insurrection, and grassroots labor and populist movements became more prominent (Zinn 1995). In Johnstown, however, Company management had a paternalistic relationship with its workers and strictly opposed labor unions; support for the Company was high among Johnstown's middle classes and press (Shappee 1940). The managers of the Cambria Iron Company formed the "sociopolitical elite" in Johnstown during the 1870s, and maintained this position into the early twentieth century (Morawska 1985:87).

As the main employer in Johnstown, in 1889 the Cambria Iron Company had a payroll of around 7,000 "unskilled" laborers (Shappee 1940). Similar to the general situation in the United States (Zinn 1995), immigration into the Johnstown area following the Civil War was at full force, with many foreign laborers coming into the area looking for work until contract-immigrant labor was outlawed (Shappee 1940). The Cambria Iron Company was also the largest rental agent in Johnstown by 1887, and it struggled to build employee housing at a fast enough pace (Shappee 1940). Workers were lodged in wood and brick tenements and crowded boardinghouses, in various states of repair and often poorly constructed (Shappee 1940). Sanitation was particularly poor, and in combination with crowded conditions led to several epidemic outbreaks of diseases, like smallpox (Shappee 1940). This intensive industrial development of the Johnstown area heightened vulnerability not only to disease but also to natural disaster. The escalation of industrial and population growth in the Conemaugh Valley and the focus on exploiting the land for industrial profits led to encroachment on the rivers and placed a large working-class population directly in harm's way (Shappee 1940).

The South Fork Fishing and Hunting Club

In contrast to the industrial community below, in the hills above Johnstown an idyllic and exclusive summer community built by wealthy steel tycoons from Pittsburgh was nestled around a reservoir. The South Fork Fishing and Hunting Club was composed of some of the wealthiest people in Pittsburgh society, including Andrew Carnegie,[4] Robert Pitcairn, and Andrew Mellon (McGough 2002). The charter of the society lists as its purpose "the protection and propagation of game and game fish, and the enforcement of all laws of this State against the unlawful killing or wounding of the same" (McGough 2002:19). To this end, they purchased the dilapidated South Fork dam in 1879, set about repairing it, and filled the reservoir to create Lake Conemaugh. This reservoir was then stocked with imported fish for the club members' use (McCullough 1968; Unrau 1980). The reconstruction of the dam was a source of concern for some prominent town members, who noted the generally shoddy workmanship and the terrible consequences that would result if the dam were to break (Unrau 1980). Repeated reports over the next decade would express apprehension over the state of the dam and note its weaknesses; nothing would be done to mitigate these concerns (Unrau 1980). By May of 1889, it would be too late.

The Johnstown Flood of 1889

A severe rainstorm on May 31, 1889 caused the Little Conemaugh River to overflow its banks and flood Johnstown, as the water level in Lake Conemaugh reached dangerously close to the top of the dam. By noon on the 31st it was Johnstown's worst flood, with water levels reaching from 2–10 feet (McCullough 1968). This inundation put most of Johnstown's streets underwater, and the swollen rivers washed out several bridges. The severity of the flooding resulted in the mills closing early the morning of the 31st, along with most of the cities' shops and transportation (Storey 1907). The water trapped some people in their homes, while others risked travel to higher ground, a relative's house, or one of the hotels in the town center (McCullough 1968).

At the South Fork Dam high above Johnstown the situation was dire. The mountain had seen torrential rains that caused the lake to rise on its own as the creeks and streams feeding into it doubled in volume. The poorly constructed dam could not take the water pressure and the spillways, which had been covered with grates to prevent the Club's imported fish from escaping the reservoir, became clogged. Although desperate efforts were made to reinforce the dam and create a new spillway, it was too late (McCullough 1968). A last telegraph message was sent from an agent near the reservoir to the town, warning that the dam might break. However, the message did not reach most of the town. Even if it had gotten through to people, the severity of the flooding already affecting the city would have made escape nearly impossible (Storey 1907).

The dam broke at 3:10 p.m., when its middle section was washed away by the force of the water and the river washing away the dam. The water was funneled through the hills, dropping approximately 400 feet to Johnstown in less than an hour, demolishing the communities in its path (McGough 2002). It accumulated debris, including railroad cars, trees, houses, and the bodies of people and animals (McCullough 1968). The towns of South Fork, Mineral Point, East Conemaugh, Woodvale, and Millville were almost entirely destroyed. The wave also hit the railroad yard at East Conemaugh and swept away the Day Express passenger train (McGough 2002).

The wave crashed down onto Johnstown at 4:07 p.m. with more force than anything it had hit up to that point (McCullough 1968). After ravaging a path through the city and its outlying towns, the water was stopped at the stone bridge crossing the Conemaugh River by the accumulation of debris. After dark, the disaster was compounded when the debris gathered at the bridge caught fire with people trapped inside the wreckage; however, it was unknown if anyone perished in the flames (Storey 1907). After the flood "lifeless bodies were lying buried in sand or wreckage at Johnstown and its immediate vicinity, ninety-nine hundredths of whom had yielded their lives within a space of five minutes" (Storey 1907:463).

In the next weeks, the survivors and those not affected by the flood started to collect bodies and tend to the injured. The newly formed Red Cross set up camp and immediately administered aid to the victims, while the story of Johnstown's destruction captivated the nation. Shappee noted that the flood had such emotive power because it "destroyed life when it was still very closely connected to the rural and village roots that supported American ideals and sentiments" (Shappee 1940:373). This sentiment was exacerbated by the sensationalistic reporting of the journalists covering the flood story. Consequently, the outpouring of relief and charity to the Johnstown survivors was immense, as was the demand for flood relics and narrative accounts (Shappee 1940).

The Concept of Vulnerability and the Johnstown Flood

The demographics and economics of the Johnstown Flood highlight the population's vulnerability with respect to a natural and man-made hazard. Vulnerability is the capacity of an individual or group to recover from and deal with the effects of a disaster (Bolin and Stanford 1999; Zaman 1999). This concept is central to disaster studies, since it expresses the broad-ranging social, economic, and physical relationships that combine with natural forces to create a disaster. Patterns of vulnerability must be characterized as the combination of factors like geographic location, pre-existing infrastructure, sociopolitical organization, production systems, distribution structures, and societal ideology (Oliver-Smith and Hoffman 2002). Indeed, different types of vulnerability are inextricably linked:

"physical vulnerability, like exposure to risks and settlement in hazard-prone areas, is a symptom of economic vulnerability," (Zaman 1999:194) and economic vulnerability also influences environmental and social vulnerability.

The economic aspects of vulnerability are starkly seen in the Johnstown Flood. One effect of a production-oriented economy like Johnstown's is environmental degradation and an increased risk of disaster (Oliver-Smith 2002:32). Economic practices that alter the environment are reflections of social systems and are intrinsically linked to the resulting vulnerability of the population (Oliver-Smith 2002). Thus, "social, political, and economic power relations are inscribed through material practices (construction, urban planning, transportation) in the modified and built environments, and one of the many ways they are refracted back into daily living is in the form of conditions of vulnerability" (Oliver-Smith 2002:36). This vulnerability also tends to be less for those with power and access to resources, as they either do not live in zones of higher risk, or have access to technology that allows them to mitigate those risks (Oliver-Smith 2002). Poverty, as a reflection of economic and political forces that create an unequal distribution of resources and power, is especially tied to increased vulnerability (Bolin and Stanford 1999; Torry 1979).

These risks were differentially instantiated for the workers and citizens of Johnstown, trapped in their valley below a fragile dam. In Johnstown, vulnerability was linked to the intensive industrialization and encroachment on the environment by local industry and to the negligence of the most elite members of late nineteenth century Pennsylvania society (Shapee 1940). The combined processes of urbanization and industrialization in Cambria County concentrated residents in areas where environmental and man-made hazards could create greater loss of life. Morawska (1985:82) characterized Johnstown as having:

> an undifferentiated economic base and a low level of horizontal differentiation; a pyramid-shaped social structure with a small elite and a large lower class; a semi caste social organization with sharp ethnic cleavages and a corresponding ascription-based normative system; and an autocratic political system with overlapping political and economic elites.

These political and economic characteristics of Johnstown, in combination with demographic and cultural factors, greatly influenced the mortality of the flood's victims and the recovery of the community. The following analysis examines the differential aspects of vulnerability and mortality by looking specifically at the demographics of the victims.

Methodology of Flood Demography

In a purely demographic sense, the age and sex distribution of those who died can give a good indication of the people most affected by this type of natural disaster. To assess those distributions the names, ages, burial locations, and home street/city were tabulated in a spreadsheet program. Sex was determined based on the first name of the individual given in the written records: individuals with gendered names were assigned a corresponding sex. Individuals with ambiguous names were not assigned a sex and were excluded from the analysis of the age-at-death structure of males and females. Additionally, those individuals who did not have ages listed were excluded from the analysis of the age-at-death structure but were included in the sex distribution analysis. A two-sample Kolmogorov-Smirnov (K-S) test was run to determine if there was a significant difference in the age-at-death distributions of the males and the females killed in the flood. Additionally, a chi-square goodness of fit test with the hypothesis intrinsic to the data was run to determine a possible significance in sex and age distribution of the victims, measuring males, females, and those victims under 15 years old. Both tests were performed in SPSS 11 for PC. The significance level was set at 95 percent ($p < 0.05$). Census data are from the *Statistics of the United States*, the 10th and 11th censuses in 1880 and 1890.[5]

Over 2,200 individuals were known to have died in the flood (McGough 2002), but only 1,641 of them were identified by both age and sex in this analysis. The age distribution by sex is shown in figures 3.3–3.5. The victims included 969 women and 672 men with known ages between 0–99. The K-S test showed that the difference between the age distributions of males and females was significant ($Z = 1.965$, $p = 0.001$). When individuals with unknown age were included in the analysis, it was determined that 857 (39%) males and 1,173 (53%) females were killed. When only individuals of known age are included, almost 16 percent more adult females were killed in the flood than adult males and over 30 percent of those killed were under the age of 15; this is a statistically significant difference ($x^2 = 72.2083$, df = 1, $p < 0.0001$).

Paleodemography and Risk

Populations are not static before mass mortality events (e.g., Thornton et al. 1991), complicating the demographic study of natural disasters. However, the catastrophic and immediate nature of the Johnstown deaths is not skewed by migration into or out of the area. An examination of the age-at-death structure of the Johnstown victims in isolation shows three demographic

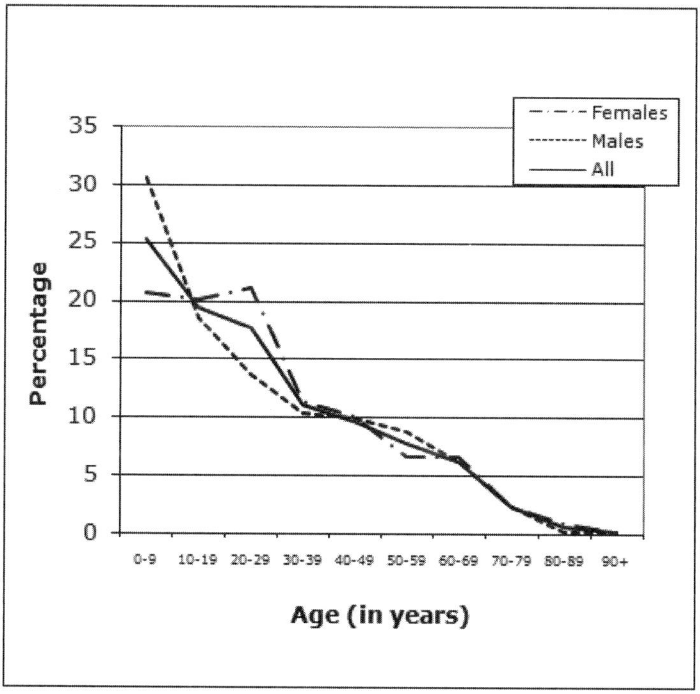

Figure 3.3. Victims of the Johnstown Flood (Data from McGough [2002]).

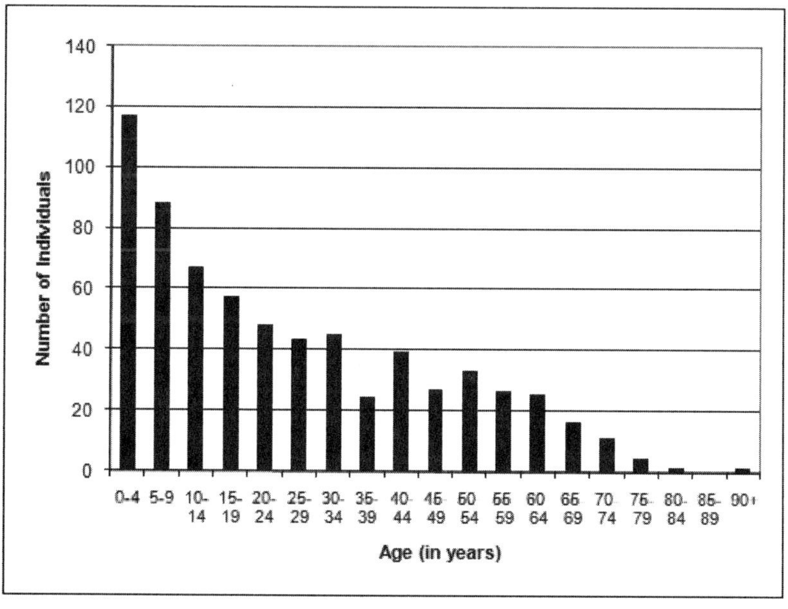

Figure 3.4. Age-at-Death Distribution: Male Victims (Data from McGough [2002]).

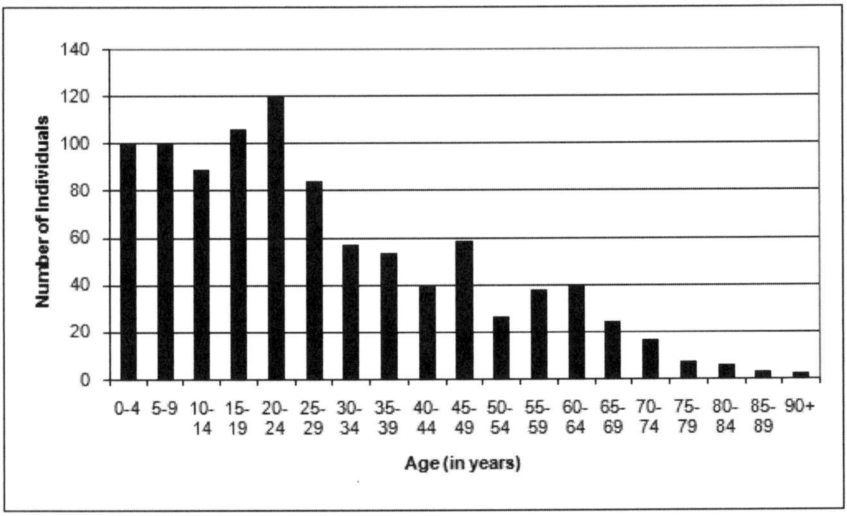

Figure 3.5. Age-at-Death Distribution: Female Victims (Data from McGough [2002]).

patterns. First, while the male curve (figure 3.4) resembles a catastrophic mortality curve, the female curve (figure 3.5) shows considerably more deaths in the younger age ranges than would be expected (Williams 2004). If there were no difference in death rates between males and females, then one might expect the female curve to resemble the catastrophic mortality curve in figure 3.1. Second, significantly more females were killed than males. Third, a high proportion of the victims are children under the age of 15. While some of these patterns may indicate differential mortality due to differential risk, others may be artifacts of the original population.

For example, when the population as a whole is considered, regardless of sex (figure 3.3), the mortality curve is similar to a catastrophic mortality curve (Williams 2004). It is hard to determine if this curve truly represents the age distribution of the pre-flood population of Johnstown without complete demographic data for Johnstown in 1889. The U.S. Census provides information only for 1880 and 1890 and does not give a complete population structure for Johnstown in either year. Modeling Johnstown's population in 1889 is problematic as well, because it is unknown if males, females, or families moved to the area in different proportions. However, the nature of this natural disaster may provide clues to the significance of the age-at-death distributions. Flood mortality resulting from dam breaks differs from that caused by the overtopping of rivers and excess precipita-

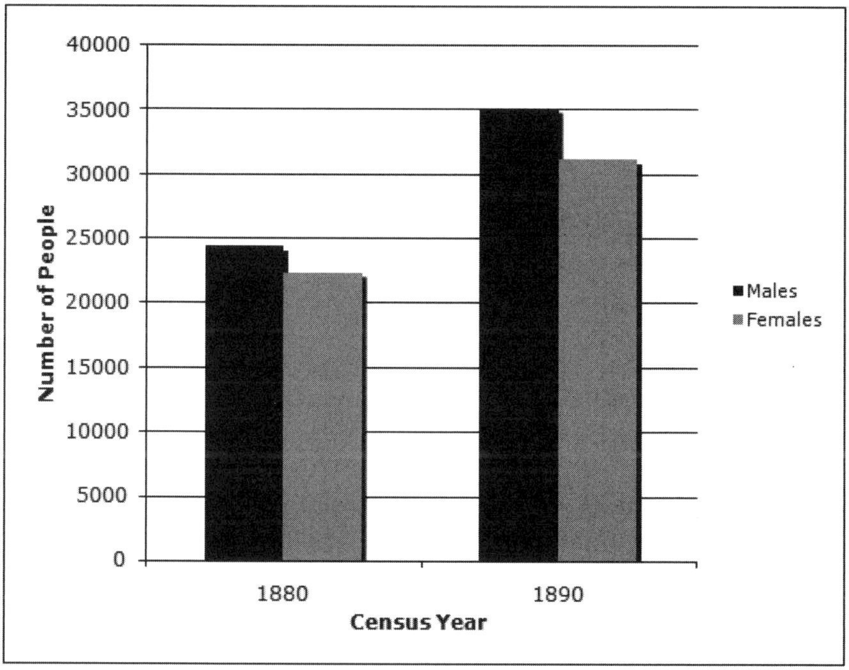

Figure 3.6. Population of Cambria County, Pennsylvania in 1880 and 1890 (Data from U.S. Census 1880 Table XXIII, U.S. Census 1890).

tion; dam breaks are far more likely to cause high mortality since escape is difficult, if not impossible for many individuals (Seaman et al. 1984). Thus, similarities between the flood victim demographic profile and catastrophic mortality curve may not be coincidental.

What is clear is that significantly more women died in the flood than men. This cannot be explained by bias in the original population: Cambria County had more males than females in 1880 and 1890 (figure 3.6), and Johnstown in 1890 also had more males than females. Since it is not an artifact of the pre-flood population, the greater proportion of female flood victims must be explained by geographical, social, or biological vulnerability. Considering that the mills were closed and many men were at home when the dam broke (Storey 1907), it is unlikely that differences in the location of victims within the valley can explain their vulnerability. Sociocultural and biological explanations for increased female victimization provide a better understanding of the high differential mortality of Johnstown's females.

Socially, women were the caretakers of children in the nineteenth century (Matthaei 1982; Ogden 1986). Only around 18.2 percent of women in 1890 were counted by the census as participants in the labor force, making up 17.1 percent of the entire labor force (Matthaei 1982). Most women, therefore, were workers in the home, and married women in particular were occupied in the domestic sphere (Matthaei 1982). Additionally, the cult of domesticity prevalent in nineteenth century conceptions of the family centered on a woman's roles as mother and homemaker. Children were to be educated and raised by the mother, while the father engaged in the capitalist economic sphere to support the family (Matthaei 1982). When the flood hit Johnstown and both men and women were at home, it is likely that the traditional caretakers of the children, the women, were also the ones primarily concerned with their welfare during the chaos of the flood. Consequently, women may have experienced greater mortality when attempting to save their helpless infants or toddlers.

Women were also burdened by heavier, bulkier clothes, decreasing their ability to maneuver in flood conditions. Nineteenth century female clothing was considered so restrictive as to be a health hazard to women; in particular the negative effects of tight-lacing corsets has been noted in both contemporary and modern discussions of women's health (Davies 1982). Other elements of female fashion were also considered dangerous; when the passenger liner *Atlantic* sunk in 1873, social commentators and dress reform advocates partially blamed the volume of the women's skirts for the high female mortality (Mattingly 2002). Women's public fashions in the 1880s were dominated by long skirts, restrictive corsets, and bustles (a gathering of material on the posterior portion of a dress; Cunnington 1990). The working classes emulated these designs and often obtained their clothing secondhand (Burman 2005). Clothing worn around the home and while working was more practical, with smaller bustles and frequently without corsets (Blanchard 1995; Lister 1972), but a complex, layered grouping of undergarments often supplemented these outer garments (Cunnington 1990). In contrast, male fashions were simpler, comprising trousers, jackets, and waistcoats made from durable material meant to provide a measure of convenience for manual laborers (Burman 2005; Lister 1972). Thus while Johnstown's men wore practical clothing for movement and maneuverability, wealthy, middle-class, and working-class women were clothed in long skirts made of much material. These clothes would have restricted the ability to swim and navigate to safety.

The higher female mortality rate may also be due to biological differences between males and females. A global analysis of natural disasters, ranging from earthquakes to hurricanes, showed that women and young children have higher mortality rates, while economically privileged adult men survive more often (Seaman et al. 1984). In natural disasters where survival may depend on clinging to safety, women experience higher mortality than men (e.g., the 1970 East Bengal cyclone and storm surge, see Sommer and Mosley 1972). Flood disasters that mimic a catastrophic dam break, like tsunamis and cyclonic storm surges, also support these findings. Recent studies of the 2005 South Asian tsunami show that the death toll of women was significantly higher than that of men (Nishikiori et al. 2006). An examination of the 1991 Bangladesh cyclone (Bern et al. 1993), notes that this disaster posed a higher risk for women than men, likely due to differences in strength and endurance. These studies show that biological factors may have influenced male survival and female mortality in the Johnstown Flood.

Another significant pattern in the flood data is seen in the age-at-death distributions, which indicate that more young women than young men perished in the flood. This differential mortality may be the result of several factors. It could be an artifact of the population from which it was sampled, potentially due to a surplus of younger women in the population. Indeed, a comparison of the proportion of females aged 5–20 to the total female population in Cambria County in 1880, in Johnstown in 1890, and in the flood victims shows that across the board a greater proportion of the females are young compared to males (figure 3.7). The 1890 Johnstown Census data indicate that women also outnumber men by 23 individuals in this age category. When the high rate of overall female mortality from the flood is considered, this difference becomes more extreme. In 1890, approximately 30 percent of women aged 15–19 and 20–24 were members of the labor force (Matthaei 1982). These are exactly the age categories that are overrepresented in figure 3.5. Although specific data on women's employment in Johnstown is sparse (Brown, 1989), given Johnstown's nature as a company town, younger women might make up a larger proportion of the female population due to their participation in the local economy.[6] It is therefore likely that a bias in the pre-flood population might account for the unusually high frequency of *young* females in the Johnstown Flood victims.

The last pattern seen in the Johnstown data is the high mortality of individuals under the age of 15. Teasing out the significance of this result leads to contradictory conclusions. The 1880 Census non-adult data for Cambria County reports information for the 5–17 age cohort, without providing more specific subdivisions. These records indicate that approximately 33 percent of the pre-flood population of Johnstown's county was

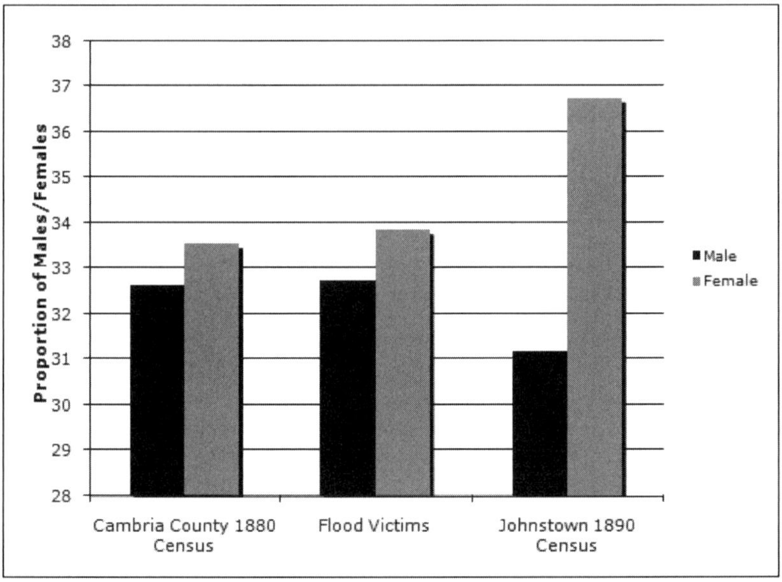

Figure 3.7. Individuals, Age 5–20, as Percent of Total Males and Total Females (Data from U.S. Census 1880 Table XXIII, U.S. Census 1890 Table 80; Data from McGough [2002]. 1880 data only includes individuals from 5–17).

between the ages of 5–17.[7] The later 1890 Johnstown census,[8] which also does not disaggregate the child and young adult information, shows that 34 percent of the Johnstown population was between 5–20 years old. In comparison, Johnstown victims between 5–17 years old and 5–20 years old make up 27 percent and 33 percent of the total aged victims, respectively. Thus, the Johnstown Flood victims under the age of 20 appear to be either a representative sample of the 1890 population, or an under sampling of the 1880 Cambria County population. However, inclusion of young adults (i.e., 15–20 years), who have the lowest risk of death in a flood disaster (Chamberlain 2006), in the comparative census age cohorts may skew the comparison, masking the higher vulnerability of individuals under 15.

Thus in order to understand why such a high percentage of the Johnstown Flood victims are under the age of 15, the data for the youngest victims must be examined. Children under five years old make up 40 percent of the Johnstown Flood victims under the age of 15. The vulnerability of this age group must be explored by looking at the male data in particular.[9] The Johnstown 1890 census reported 11.6 percent of the male population under age five, compared to 17.4 percent of the male Johnstown flood victims. Thus flood victims have a higher percentage of males under five years old than the 1890 Census data; this holds even when those data are corrected for the loss of the flood victims by adding the flood victims to the census data. While these data cannot be extrapolated to the female data or all victims under five, they do show a differential mortality for the youngest children in a natural disaster. This result is not surprising, since children are particularly vulnerable in natural disasters due to their small size, relative lack of stamina and strength (Glass et al. 1977; Seaman et al. 1984; Sommer and Mosley 1972), and dependence upon adults.

These analyses have shown that the demographic patterns in the Johnstown data are due to both pre-flood population bias and differential vulnerability. Though probably exaggerated, contemporary reports also noted the dearth of women and children after the flood (McGough 2002) and are consistent with the mortality profile of the flood victims. The high proportion of young females is likely an artifact of the pre-flood population. The greater proportion of female and young victims highlights the differential vulnerability of women and children when caught in the wave's path.

Economic Recovery

The demographics above indicate that the Johnstown flood was a catastrophic mortality event that impacted women significantly more than men, young children more than adults, but most age classes equally. Beyond these numbers, the devastation to individual families and the entire community was tremendous. The flood created 124 widows, 198 widowers, and 565 orphans and half-orphans; in addition to this, 99 entire families were killed (Gable 1926). The social and economic impact of the flood was extreme, as the town struggled to recover and rebuild.

Surprisingly, however, Johnstown's economy was able to rebound quickly. The city's largest employer, the Cambria Iron Company, resumed operations by the end of the summer, with employees beginning to clear and repair the mill the week after the flood (Shappee 1940). Merchants and banks were back in business soon after, either selling salvageable material or washing and drying the money from the bank vault (Shappee 1940). Within a year after the flood, steel mills were being rebuilt and improved, the railroad was repaired, and buildings were reconstructed: "Johnstown, the largest of the boroughs showed the greatest recovery of all even though some of the new buildings looked as if they had sprouted from the Flood's debris that still lay scattered about" (Shappee 1940:526).

This rapid economic revival of the devastated region was centered on three economic aspects of the flood: 1) the differential survival of the labor force; 2) the tremendous outpouring of aid and relief from the rest of the country; and 3) the preservation of the Cambria Iron Company's infrastructure. The differential survival of the male labor force meant that the manpower needed to run the mills and factories was still available. Indeed, in Pennsylvania in 1880, only 17 percent of manufacturing, mining, and mechanical jobs were held by women.[10] After the flood, the woolen mill, which did employ women, was not rebuilt. The surviving female workers lost their jobs while the male mill workers were hired at the steel mill (Shappee 1940). While it is likely that the death of so many young women had an effect on the industries with which they were associated and the homes they were in charge of running, it seems that the loss did not significantly impact the male-dominated industries of the mills and mines. Indeed, women did not play a large role in the labor force after the flood, as the steel town remained male-dominated in its workforce (Morawska 1985). By 1890, almost 5,000 men were employed by Johnstown's iron and steel industries (Shappee 1940); the company was back to 70 percent of its unskilled labor force within a year of the devastation. Thus, these businesses were able to quickly resume production and the Johnstown economy that depended on them rapidly began to heal.

The aid payments made to survivors of the flood also affected the economic recovery of Johnstown. The outpouring of relief to Johnstown after the flood was remarkable. Donations were sent from around the world, totaling $3,742,818 (Shappee 1940). It was the province of the Flood Commission to manage at least $2,394,000 of this money (Storey 1907). While donations to flood survivors were abundant, the board of inquiry was directed by the Flood Commission to distribute relief money by prioritizing necessity (Benshoff 1890). As seen in table 3.1, this relief was directed primarily at women with dependents and orphans.

Additional aid was dispensed to compensate for destroyed property. The damages from lost property estimated on July 9, 1889 were $8,655,114 (Storey 1907). The vast proportion of these property losses was in Class 6; however, these individuals were the last to be prioritized in terms of compensation and thus did not receive aid money (Storey 1907). In this way aid was given to the people in the economy least able to fend for themselves after the disaster destroyed their resources. Indeed, all applicants for aid had to file sworn statements as to their physical condition and the amount of property lost (McGough 2002). Healthy adult men were unlikely to receive much aid and thus had to return to work as soon as possible. The distribution of aid money thus followed the lines of vulnerability

Table 3.1. Aid Distributions to Johnstown Flood Survivors (from Benshoff 1890 and Storey 1907)

Class One (Received $600 initial aid)	Widows who lost husbands and who had 4+ children under 16	205 cases
Class Two (Received $400 initial aid)	Widows who lost husbands and who had less than 4 children under 16	237 cases
Class Three (Received $200 initial aid)	Widows not in Class One or Class Two, the elderly, and the injured of both sexes	372 cases
Class Four (Received $125 initial aid)	Individuals who lost all property but could still work.	1,168 cases
Class Five (Received $80 initial aid)	Individuals who were not destitute before the flood and lost part of their property, and those who lost all their property but were not immediately in need of aid because of continuous employment.	1,698 cases
Class Six (Did not receive initial aid)	Individuals who lost a great deal but were in relatively stable financial condition.	

seen in the demographic data: women and children were the most likely to perish in the disaster, but were prioritized during relief efforts.

Repair and cleanup proceeded quickly in Johnstown, and the mills were operating within a week of the flood. The Cambria Iron Company's facilities, while damaged, were not unsalvageable. At least 2,000 laborers were brought in from Pittsburgh to begin debris cleanup and repair work, and their above-average wages ensured that up to 10,000 men were soon working in Johnstown (Shappee 1940). While approximately 1,600 people left the Johnstown area in the week following the flood (Shappee 1940), additional labor was provided in the form of immigrants who flocked to the area (McGough 2002). These workers helped speed along the city's economic recovery, which was aided by the prosperity in the rest of the nation (Shappee 1940). In 1907, Johnstown's economy had rebounded and contemporaries noted that "with the unlimited quantity of cheap fuel and other natural advantages, it remains one of the leading steel manufacturing cities of the world" (Storey 1907:346).

Political Economic Consequences of the Johnstown Flood

Not unlike current-day natural disasters, the aftermath of the Johnstown Flood included not only rebuilding but also recrimination. Local communities blamed the South Fork Fishing and Hunting Club for their improper maintenance of the dam, as well as for installing screens to keep their imported fish from going over the spillway (McCullough 1968; McGough 2002). The local newspaper, the *Johnstown Tribune*, wrote scathing editorials against the Club, noting that the rich members wanted "an exclusive resort where, in all their spotlessness and glory, they might while away the summer days" while the citizens of working-class Johnstown were as helpless as "rat[s] caught in a trap and placed in a bucket of water" (*Johnstown Tribune* story "THE CAUSE" cited in McGough 2002:126). Indeed, only one member of the Club lived in Johnstown, and he was the only member to lose family in the flood (McGough 2002).

Nationwide, newspapers carried editorials denouncing the Club for its carelessness and blaming its members for the disaster (Shappee 1940). A jury empanelled by the coroner immediately after the flood found the owners of the dam, through negligence in its maintenance, responsible for flood deaths and damages (Storey 1907). However, further review indicated that it was unlikely that the members of the Club could be held liable (Shappee 1940; Storey 1907). Despite many attempts, no legal action against the South Fork Fishing and Hunting Club was successful; with the reservoir gone, the clubhouse deserted, and public antipathy for the Club high, the Club slowly disbanded until it was dissolved in 1904 (McGough 2002).

A major theme of anthropological research on disasters is the role of the catastrophe in affecting mobilization within the local political system (Oliver-Smith 1996). Disasters are "the contexts for creation of political solidarity, activism, new agendas, and developing new power relations" (Oliver-Smith 1996:309). This is certainly the case in the aftermath of the Johnstown Flood of 1889. One of the results of the disaster at Johnstown was a new sense of *communitas*. Before the flood, the separate towns had resisted consolidation, maintaining their own community pride (McGough 2002). However, after the flood, the residents of the entire valley were treated as though they were one community by the relief workers, and movements toward the consolidation of the separate boroughs of Johnstown grew stronger (Shappee 1940). In late autumn 1889, the boroughs of Johnstown, Cambria, Conemaugh, Grubbtown, Millville, Prospect, and Woodvale were incorporated into Johnstown City (Storey 1907). This was just one visible way in which the flood reshaped the politics of the Johnstown area.

While the 1889 flood did not destroy Johnstown's economy, its being a company town would eventually prove an economic burden. In the second half of the twentieth century, Johnstown's reliance on steel caused rampant unemployment and economic depression with the railroad's decline, rising standards for coal and iron ores, and postindustrial steel manufacturing decline (Wallace 1989). Though not dam related, two other fatal floods struck Johnstown in 1936 and 1977, underlining the fragile relationship between the city and its environment.

Conclusion

Johnstown provides an interesting case study of the effects of natural disaster on the political economy of a nineteenth century mining town. If political economy can be understood as elite strategies for production and distribution, internal versus external foci of elite politics, and elite involvement of public participation in ideology (after Blanton et al. 1996), then we can see clearly the development of the social construction of vulnerability in the Johnstown Flood and its aftermath. The construction of a natural hazard by economic and social elites above the working-class community of Johnstown, where the majority of the population had little political and economic control over their lives, heightened that vulnerability. In this way, the 1889 Johnstown Flood parallels modern disasters such as Hurricane Katrina, in which the transformation of a man-made hazard into a natural disaster differentially affects economically and socially vulnerable individuals. The specific economic and demographic aspects of this vulnerability are seen in the situation of working-class Johnstown and differential male survival. Overall the age structures seen in the Johnstown Flood confirm the data from many other natural disasters: young and female individuals are more likely to die when hazard turns into disaster. Economic recovery after the Johnstown Flood was startlingly fast, a fact that can be attributed to the differential survival of its labor force, the generosity of citizens from around the world, and the good fortune that the main industry was not destroyed. Paleodemographic data thus adds another facet of insight into a population's vulnerability to and recovery from natural disasters.

Acknowledgments

I would like to express my appreciation to Clark Larsen and Jeffrey Cohen for their guidance in this research and their helpful comments on drafts of this manuscript. I am grateful to Andrew Chamberlain for his guidance on the Master's Thesis of which this case study was originally a part. I would also like to thank Paul Sciulli for statistical advice, and Jennifer Spence, Britney Kyle, and Amelia Hubbard for assistance in revising early drafts. Finally, I want to extend special thanks to the staff at the Johnstown Flood Museum in Johnstown, Pennsylvania, especially Robin Rummel for her help in accessing the archives. All errors or omissions are my own.

Notes

1. The quote from the title was the last warning sent by telegraph to Johnstown, alerting the residents of the danger (Degen and Degen 2000). This warning went largely unheeded.

2. The victims' list is compiled from official records (identification of the deceased), and the reports of family and friends. Thus, it is possible that some of the most vulnerable people in Johnstown, such as social outcasts and the homeless, are not counted in these lists.

3. It is not certain what this population estimate is based on, but it likely includes the boroughs surrounding Johnstown Borough itself, but not necessarily the communities in the hills around Johnstown. Other population estimates, like that by McCullough (1968) put the population of the valley at 30,000 people.

4. Evidence for Carnegie's inclusion in the club is ambiguous, with some documents claiming him as a member and others not.

5. For critical appraisal of historical U.S. census data see King and Magnuson (1995).

6. Brown (1989) notes that young daughters and widows of Cambria Iron Company workers were given preference in employment at the woolen mills, the only mills in Johnstown known to employ women. She also notes the need for further research into the roles of women in Johnstown's labor force.

7. U.S. Census 1880 Table XXIII—Population by School, Military, and Voting Ages, by Counties: 1880 Pennsylvania.

8. U.S. Census 1890 Table 80—Persons of School, Militia, and Voting Ages, by sec, general nativity, and color, of places having 2,000 inhabitants or more.

9. Census data is not available for females under 5, nor is it specific for children under 15. Data taken from U.S. Census 1890, Table 80.

10. This number is from U.S. Census 1880, Table XXX "The Number of Persons in the United States Engaged in Each Class of Occupations, with distinction of age and sex and of nativity, by states and territories: Manufacturing, Mechanical, and Mining industries."

References

Alesan, Alicia, Assumpció Malgosa, and Carles Simó
1999 Looking Into the Demography of an Iron Age Population in the Western Mediterranean. I. Mortality. American Journal of Physical Anthropology 110:285–301.

Benshoff, Harry M.
1890 The World's Charity to the Conemaugh Valley Sufferers and Who Received It. Johnstown: Harry M. Benshoff.

Bern, C., J. Sniezek, G. M. Mathbor, M. S. Siddiqi, C. Ronsmans, A. M. R. Chowdhury, A. E. Choudhury, K. Islam, M. Bennish, E. Noji, and R. I. Glass
1993 Risk Factors for Mortality in the Bangladesh Cyclone of 1991. Bulletin of the World Health Organization 71(1):73–78.

Blanchard, Mary W.
1995 Boundaries and the Victorian Body: Aesthetic Fashion in Gilded Age America. The American Historical Review 100(1):21–50.

Blanton, Richard E., Gary M. Feinman, Stephen A. Kowalewski, and Peter N. Peregrine
1996 A Dual-Processual Theory for the Evolution of Mesoamerican Civilization. Current Anthropology 37(1):1–14.

Bolin, Robert, and Lois Stanford
1999 Constructing Vulnerability in the First World: The Northridge Earthquake in Southern California, 1994. *In* The Angry Earth: Disaster in Anthropological Perspective. A. Oliver-Smith and S. M. Hoffman, eds. Pp. 89–112. New York: Routledge.

Brown, Sharon A.
1989 Cambria Iron Company: America's Industrial Heritage Project, Pennsylvania. Historic Resource Study: National Park Service.

Burman, Barbara
2005 Working Class. *In* Encyclopedia of Clothing and Fashion, Vol. 3, Occult Dress to Zoran, Index. V. Steele, ed. Detroit: Thomson Gale.

Camp, Scott D.
1995 Work Response to Plant Closings: Steelworkers in Johnstown and Youngstown. New York: Garland Publishing Group.

Chamberlain, Andrew
2000 Problems and Prospects in Palaeodemography. *In* Human Osteology in Archaeology and Forensic Science. M. Cox and S. Mays, eds. Pp. 101–115. London: Greenwich Medical Media.
2006 Demography in Archaeology. Cambridge: Cambridge University Press.

Coale, A. J., and P. Demeny
1983 Regional Model Life Tables and Stable Populations. Princeton: Princeton University Press.

Cunnington, C. Willett
1990 English Women's Clothing in the Nineteenth Century: A Comprehensive Guide with 1,117 Illustrations. New York: Dover Publications.

Davies, Mel
1982 Corsets and Conception: Fashion and Demographic Trends in the Nineteenth Century. Comparative Studies in Society and History 24(4):611–641.

Degen, Paula, and Carl Degen
2000 The Johnstown Flood of 1889: The Tragedy of the Conemaugh. New York: Eastern National.

Gable, John E.
1926 History of Cambria County, Volume 1. Topeka, KS: Historical Publishing Company.

Glass, Roger I., Juan J. Urrutia, Simon Sibony, Harry Smith, Bertha Garcia, and Luis Rizzo
1977 Earthquake Injuries Related to Housing in a Guatemalan Village. Science 197(4304):638–643.

Gowland, Rebecca, and Andrew Chamberlain
2002 A Bayesian Approach to Ageing Perinatal Skeletal Material from Archaeological Sites: Implications for the Evidence for Infanticide in Roman-Britain. Journal of Archaeological Science 29:677–685.
2005 Detecting Plague: Palaeodemographic Characterisation of a Catastrophic Death Assemblage. Antiquity 79(303):146–157.

Grayson, Donald K.
1990 Donner Party Deaths: A Demographic Assessment. Journal of Anthropological Research 46:223–242.
1993 Differential Mortality and the Donner Party Disaster. Evolutionary Anthropology 2(5):151–159.
1996 Human Mortality in a Natural Disaster: The Willie Handcart Company. Journal of Anthropological Research 52:185–205.

King, Miriam L., and Diana L. Magnuson
1995 Perspectives on Historical U.S. Census Undercounts. Social Science History 19(4):455–466.

Lister, Margot
1972 Costume of Everyday Life: An Illustrated History of Working Clothes. Boston: Plays.

Margerison, B. J., and C. Knüsel

2002 Paleodemographic Comparison of a Catastrophic and an Attritional Death Assemblage. American Journal of Physical Anthropology 119:134–143.

Matthaei, Julie A.

1982 An Economic History of Women in America: Women's Work, the Sexual Division of Labor, and the Development of Capitalism. New York: Schocken Books and The Harvester Press.

Mattingly, Carol

2002 Appropriate[ing] Dress: Women's Rhetorical Style in Nineteenth-Century America. Carbondale: Southern Illinois University Press.

McCullough, David G.

1968 The Johnstown Flood. London: Hutchinson and Co.

McGough, Michael R.

2002 The 1889 Flood in Johnstown, Pennsylvania. Gettysburg: Thomas.

Morawska, Ewa

1985 For Bread with Butter: The Life-Worlds of East Central Europeans in Johnstown Pennsylvania, 1890–1940. Cambridge: Cambridge University Press.

Nagaoka, Tomohito, Kazuaki Hirata, Emi Yokota, and Shuji Matsu'ura

2006 Paleodemography of a Medieval Population in Japan: Analysis of Human Skeletal Remains from the Yuigahama-Miniami Site. American Journal of Physical Anthropology 131:1–14.

Nishikiori, Nobuyuki, Tomoko Abe, Dehiwala G. M. Costa, Samath D. Dharmaratne, Osamu Kunii, and Kazuhiko Moji

2006 Who Died as a Result of the Tsunami?—Risk Factors of Mortality among Internally Displaced Persons in Sri Lanka: A Retrospective Cohort Analysis. BMC Public Health 6:73–80.

Ogden, Annegret S.

1986 The Great American Housewife: From Helpmate to Wage Earner, 1776–1986. Westport, CT: Greenwood Press.

Oliver-Smith, Anthony

1996 Anthropological Research on Hazards and Disasters. Annual Review of Anthropology 25:303–328.

2002 Theorizing Disasters: Nature, Power, Culture. *In* Catastrophe and Culture: The Anthropology of Disaster. S. M. Hoffman and A. Oliver-Smith, eds. Santa Fe, NM: School of American Research Press and James Currey Ltd.

Oliver-Smith, Anthony, and R. E. Goldman
1988 Planning Goals and Urban Realities: Post Disaster Reconstruction in a Third World City. City and Society 2:105–126.

Oliver-Smith, Anthony, and Susanna M. Hoffman
2002 Introduction: Why Anthropologists Should Study Disasters. *In* Catastrophe and Culture: The Anthropology of Disaster. A. Oliver-Smith and S. M. Hoffman, eds. Pp. 1–22. Santa Fe, NM: School of American Research Press.

Paine, Robert
2000 If a Population Crashes in Prehistory and There Is No Paleodemographer There to Hear It, Does It Make a Sound? American Journal of Physical Anthropology 112:181–190.

Seaman, J., S. Leivesley, and C. Hogg
1984 Epidemiology of Natural Disasters. Basel: S. Karger.

Shappee, Nathan Daniel
1940 A History of Johnstown and the Great Flood of 1889: A Study of Disaster and Rehabilitation. PhD Dissertation, University of Pittsburgh.

Sommer, A., and W. H. Mosley
1972 East Bengal Cyclone of November, 1970: Epidemiological Approach to Disaster Assessment. The Lancet i:1029–1036.

Storey, Henry Wilson
1907 History of Cambria County Pennsylvania with Genealogical Memoirs, Volume 1. New York: The Lewis Publishing Company.

Sullivan, Amy
2004 Reconstructing Relationships among Mortality, Status, and Gender at The Medieval Gilbertine Priory of St. Andrew, Fishergate, York. American Journal of Physical Anthropology 124:330–345.

Thornton, Russell, Tim Miller, and Jonathan Warren
1991 American Indian Population Recovery Following Smallpox Epidemics. American Anthropologist 93(1):28–45.

Torry, William I.
1979 Anthropological Studies in Hazardous Environments: Past Trends and New Horizons. Current Anthropology 20(3):517–540.

Unrau, Harlan D.
1980 Historic Structure Report: The South Fork Dam Historical Data, Johnstown National Flood Memorial, Pennsylvania. Denver: National Park Service, U.S. Dept of the Interior.

Van Gerven, Dennis P., Susan Guise Sheridan, and William Y. Adams
1995 The Health and Nutrition of a Medieval Nubian Population: The Impact of Political and Economic Change. American Anthropologist 97(3):468–480.

Wallace, Kim E.
1989 The Character of a Steel Mill City: Four Historic Neighborhoods of Johnstown, Pennsylvania. Historical American Buildings Survey/Historic American Engineering Record. Washington, DC: National Park Service.

Williams, Leslie Lea
2004 Pursuing the Course of the Four Horsemen: The Paleodemography of Mass Mortality. Unpublished Master's Thesis: University of Sheffield.

Zaman, Mohammad
1999 Vulnerability, Disaster, and Survival in Bangladesh: Three Case Studies. *In* The Angry Earth: Disaster in Anthropological Perspective. A. Oliver-Smith and S. M. Hoffman, eds. Pp. 192–212. New York: Routledge.

Zinn, Howard
1995 A People's History of the United States: 1492–Present. New York: HarperPerennial.

The Invisible Toll of Katrina: How Social and Economic Resources are Altering the Recovery Experience among Katrina Evacuees in Colorado

4

MEGAN UNDERHILL

AMERICANS WILL NEVER FORGET the hot summer day in 2005 when the strength of Hurricane Katrina's winds and floodwaters wreaked havoc on over 90,000 square miles of Gulf Coast homes, businesses, and communities (Cutter et al. 2006). In the days following the storm, area residents sat in their hotel rooms or the homes of family and friends, glued to the television in horror, as images of a region drowned and destroyed flitted across their screens. The reality of Hurricane Katrina was experienced in a much more intimate and terrifying way for those individuals who could not or did not evacuate to safety before disaster stuck on August 29, 2005. Those left behind worried about the safety of their person, kin, and community and wondered if and how they would be rescued from the besieged region. From the moment of the storm's inception, issues of economic inequality came to the fore, determining who would be able to evacuate and where. Immediately following Katrina, news stories detailed the unequal ways in which issues of class and race influenced the evacuation and recovery experiences of Gulf Coast residents. Unfortunately, media attention and public debate waned when it became apparent that Gulf Coast recovery would not be swift or decisive. After the media turned their attention to other events in the nation and abroad, what happened to Gulf Coast residents in the months and years following Katrina? How were people's recovery experiences impacted by issues of class and the pervasive but little discussed reality that not all Gulf Coast residents had equal access to resources? In what ways was people's ability to financially recover from the disaster enabled or constrained by their access to different economic and social resources?

Stories and reports abound regarding the U.S. government's handling of the disaster relief effort and its struggle to adequately evacuate or assist the region's most vulnerable residents. However, what has garnered far less attention or public scrutiny is the issue of the wide-scale displacement and resettlement of hundreds of thousands of Gulf Coast residents following Hurricane Katrina. Few historical precedents exist with which to compare or understand the magnitude of the Katrina displacement; however, some have likened the event to mass migrations that followed the Dust Bowl or the Civil War (e.g., Egan 2005; Grier 2005).

In the case of Hurricane Katrina, many Gulf Coast communities suffered economic hardships from the sudden loss of large sectors of their tax base. A dramatic decrease in the local tax base made it difficult for Gulf Coast communities to finance recovery efforts or afford basic public services. Similarly, cities that welcomed tens of thousands of disaster-affected individuals grew worried when it became apparent that many evacuees could not or would not return home. A number of social and economic factors prevented the return migration of Katrina's poorest evacuees, resulting in what was described as the "permanent transfer of a large number of poor people from one city to the next" (Peterson et al. 2006:652). Poor residents were more likely to become permanently displaced than middle-class residents. Many were the recipients of institutional evacuations and were resettled in communities farther from home. For countless people the cost to return home was prohibitive. The few who could afford the return journey discovered that their homes were destroyed and that rental prices had doubled (Peterson et al. 2006). A year after the disaster, many host cities, such as Houston, Baton Rouge, and Atlanta, complained of feeling burdened by the significant increase in the number of financially and emotionally vulnerable residents, and city managers expressed concern about how their local tax base could continue to afford the increased costs of public services and assistance (Axtman 2006; Campo-Flores 2006; Sullivan 2006).

Although the approximate number of those displaced by the hurricane remains unclear, recent FEMA (2007) statistics indicate that more than 650,000 households participated in their housing assistance program. Displaced people found refuge in cities and states throughout the South; however, a significant number of people were flung farther afield to regions of the country that bore little resemblance culturally or geographically to the places they called home. Fourteen thousand Katrina evacuees moved to Colorado after the disaster. The manner in which Gulf Coast residents evacuated to the Rocky Mountain state depended a great deal upon the economic and social resources evacuees had at their disposal. Evacuees with economic and/or social resources came to Colorado of their own initiative while resource-poor individuals exerted less control in regard to their relocation and were often evacuated by the U.S. government. Initially evacuees assumed that their residence in Colorado would be temporary; however, as the days turned into months, evacuees were faced with the startling realization that their residence might be indefinite or permanent.

As discussed in detail below, individuals who had the least degree of difficulty recovering after the disaster were typically white, middle-class non-natives who had relatively small family and friend networks in Louisiana upon whom they relied for little if any economic assistance. The possession of economic savings, nationally recognized educational and occupational credentials, and ties to national businesses helped these economically self-sufficient individuals reconstitute their lives in Colorado and recover from the disaster. In contrast, Hurricane Katrina proved most devastating for native New Orleanians who possessed few economic resources in Louisiana, but who were able to transcend the realities of low-wage work through membership in dense and highly cooperative kin and friend networks. Many were deeply embedded within the local economy of New Orleans and depended upon extensive network ties to help them negotiate an assortment of income earning opportunities in both the formal and informal economy. The rupture of important social networks made resettlement in Colorado difficult, especially for those individuals who lacked economic capital and did not have family members in Colorado to help them negotiate their resettlement experience. Predominantly African American, these individuals arrived in Colorado without a job or the promise of one and few possessed much in the way of savings. Quickly, they learned they would have to start their lives anew in a state where the cost of living was almost twice as high as the one they hailed from.

All of the evacuees in Colorado experienced economic loss; however, the severity of their loss depended in great part upon both the economic and social resources people possessed before and after the disaster.

Economic Impact Vulnerability

In the book *Unequal Childhoods*, sociologist Annette Lareau (2003:236) examines how a person's class position in society influences the type of skills, behaviors, and attitudes that American parents transmit to their children. According to Lareau, a person's class status informs "critical aspects of family life" including but not limited to "time use, language skills, and kin ties." Lareau posits that middle-class children learn a distinctively different set of skills and behaviors than their working-class or poor peers. She argues middle-class parents stress the importance of "concerted cultivation" and make strategic investments of both time and money in the education of their children. As a result, middle-class children develop excellent language and reasoning skills that enable them to effectively communicate with a variety of people in different institutional and interpersonal settings. Interestingly, middle-class parents do not teach their children to make similar investments in the extended family and most grow up geographically and emotionally estranged from their relatives.

In contrast, working-class and poor parents adhere to a childrearing philosophy that stresses the "accomplishments of natural growth." With the help of extended family members (most of who live in close proximity to one another), working-class and poor parents are able to ensure their children are well cared for, fed, clothed, and housed in a safe home in a secure environment. Among the working class and poor, family is the organizing principle of people's lives, and kin members regularly provide one another with instrumental services (Lareau 2003:87). Significant economic constraints make it impossible for working-class or poor parents to make similar monetary or temporal investments in the cultivation of their children's individual talents. Nor are working-class or poor parents able to endow their children with the vocabulary or language skills that will help to effectively communicate with institutional authorities. Lack of financial resources and educational skills causes many working-class and poor individuals to feel powerless and constrained when interacting with institutions or people of different class backgrounds.

In the disaster literature, research has shown that disasters negatively impact the economic situation of the affected population, causing many to experience economic loss (Bolin 1986; Elliot and Pais 2006; Erikson 1976; Fothergill 2004; Fothergill and Peek 2004; Morrow 1997; Peterson et al. 2006). Following a disaster, it is common for people to incur significant debt in an effort to fix salvageable items or replace those that have been destroyed (Fothergill 2004; Morrow 1997). Mounting disaster expenses may exacerbate people's already vulnerable economic situation, making it difficult for some to keep up with pre-existing bills and newly emergent expenses. In a research study conducted after the 1997 Grand Forks Flood, Fothergill (2004:54) notes that disaster-associated expenses increased people's financial burden and caused many area residents to experience "severe and sudden" downward mobility.

The relationship between a person's class and their ability to economically reconstitute their lives after a disaster or resettlement has been well documented (Elliot and Pais 2006; Fothergill 2004; Fothergill and Peek 2004; Morrow 1997; Scudder and Colson 1982). Not surprisingly, disaster researchers note that people of higher socioeconomic status fare better after a disaster than lower income people (Bolin 1986; Fothergill 2004; Morrow 1997; Peterson et al. 2006). Individuals of higher socioeconomic status are less likely to encounter job loss and experience fewer fluctuations in their income following a disaster (Fothergill 2004; Peterson et al. 2006). Indeed, many have salaried positions that provide them with regular and steady paychecks in the harried days and weeks following the event (Fothergill 2004). In addition, individuals of higher socioeconomic status are more likely to possess savings, investments, or lines of credit they can draw on to finance a variety of disaster-associated expenses (Fothergill 2004; Fothergill and Peek 2004; Peterson et al. 2006). In the case of Hurricane Katrina, middle-class residents evacuated closer to their homes and were among the first residents to return to their communities and begin rebuilding their residences (Peterson et al. 2006:650).

Individuals of lower socioeconomic status have a much more difficult time recovering their pre-impact economic status than similarly affected middle-class residents (Fothergill and Peek 2004). Many individuals experience job loss while others lose weeks of critical wages when they are unable to attend their hourly paid jobs for days or weeks after the event (Elliot and Pais 2006; Fothergill 2004). People of lower socioeconomic status are also more reliant upon disaster assistance programs to help them regain their pre-impact economic status. Unfortunately, many lower income people have a difficult time qualifying for, accessing, or negotiating disaster assistance programs. After Hurricane Andrew many individuals were prohibited from seeking government assistance due to the fact that aid offices were located in distant areas of the city that were inaccessible via public transportation (Morrow 1997:153; Oliver-Smith 1991). In addition, researchers have documented instances in which need-based individuals were denied recovery assistance by institutions whose eligibility requirements were based on a nuclear family model such as the U.S. Federal Emergency Management Agency's head-of-household rule (Fothergill 2004; Morrow 1997; Phillips 1993; Scudder and Colson 1982). Inability to provide officials with proof of residency effectively prohibited people from accessing disaster assistance in the Katrina case and others.

Several disaster and resettlement scholars have observed how the presence or absence of social networks impacts an individual's ability to prepare for, respond to, and recover from a disaster or resettlement experience (Barnshaw 2006; Bolin 1986; Fothergill 2004; Hulbert et al. 2000; Moore et al. 2004). Scholars such as Betty Morrow and Anthony Oliver-Smith examine the ways in which social networks provide people with direct and proximate benefits during disaster recovery and resettlement. Morrow (1997:169) argues that disaster recovery is easier for close-knit families and she warns against instituting recovery efforts that might jeopardize or weaken kin assistance. In an extensive review of the resettlement literature, Oliver-Smith (1991:13) concludes that resettlement projects are likely to fail if individuals are separated from their social networks and removed from their communities of origin. Other researchers have also remarked upon the indirect and often invisible forms of assistance social networks can yield after a disaster or resettlement event. After Hurricane Andrew, Hulbert et al. (2000:615) found that family members played an important role in linking their disaster-affected relatives to social service providers involved in the relief effort.

Discussions of network assistance, especially among kin members, have received far greater attention in anthropological and sociological studies of lower income Americans (Halperin 1994; Lareau 2003; Stack 1970). In the ethnography *All Our Kin*, Carol Stack found that lower income African Americans regularly exchanged goods and services with other network members in order to secure daily necessities such as food, clothing, childcare, and transportation. Stack suggests that economic scarcity compelled network members to engage in relationships of reciprocity whereby people exchanged goods and services in an effort to obligate or indebt other network members. She argues that network-provisioned assistance functioned as a "profoundly creative adaptation to poverty" and helped to foster relationships of obligation and trust among network members (Stack 1970:43). Halperin (1994) found that similar relationships of reciprocity and exchange operated among poor, white extended families in an Appalachian region of Kentucky. However, the habitual acts of reciprocity among kin members did not help them to get ahead. Indeed, Halperin observes that in poor and highly interdependent kin groups, "the priority is not profit or even upward mobility, but rather the maintenance of the extended family" (Halperin 1994:32). Similarly, Portes (1998:15) warns that downward leveling norms are more likely to develop in dense and highly interconnected social networks than in networks characterized by a greater degree of openness and diversity.

Finally, scholars like Granovetter (1973) argue it is important to cultivate relationships with people who lie outside one's immediate circle of family and close friends. In contrast to the stability imparted by dense and interdependent kin networks, weak ties broaden both the number and diversity of a person's social network, thus enabling an individual to access information, resources, and even jobs.

Methodology

In this study I used a variety of ethnographic methods including informal interviews, in-depth formal interviews, structured interviews, and participant observation. My research began in September 2005 and concluded 15 months later in December 2006. Over the course of the research project, I interviewed 30 Katrina evacuees who resettled in Colorado following the disaster, of whom all but one were from the city of New Orleans or surrounding towns such as Slidell or Mandeville. My sample consisted of 18 black and 12 white participants, of which 17 were female and 13 were male. In addition, I coded 9 of the individuals as "self-sufficient" and 21 as "interdependent" individuals in my sample. The two groups of people had little in common with one another demographically, economically, or socially, as will be demonstrated in the findings sections of this paper. Stated briefly, self-sufficient individuals were usually middle-to-upper middle class non-natives who had small social networks they relied upon for affective services only. In contrast, interdependent individuals were usually working-class or poor Louisiana natives who possessed large social networks they relied upon for affective and instrumental services.

During the first five weeks of the research project I volunteered at a U.S. Air Force base in Colorado where many of the poorest evacuees were initially housed. As a volunteer, I was able to gather a wealth of observational data about people's immediate experiences with resettlement in Colorado. I began to interview evacuees in October of 2005. In each interview, I asked people 53 questions about their life before and after their resettlement in Colorado. I was particularly interested in learning how disaster resettlement impacted people's financial situations and I wanted to learn if people were able to draw upon pre-existing economic and social resources to aid them with their resettlement in Colorado. I asked evacuees the following kinds of questions: What did you do for a living in New Orleans? How do you support yourself in Colorado? Did you have any side jobs in New Orleans and do you have any in Colorado, why or why not? Have you received any assistance in Colorado, if so what kind and from who? What, if anything, was difficult about your life in New Orleans? What kind of difficulties have you experienced in Colorado?

I also designed and administered a survey wherein I asked people specific questions about the size, composition, and function of their kin networks before and after disaster resettlement. During the 15-month period I attempted to maintain contact with all of the evacuees in my sample, and I spoke to most evacuees two to four times during the course of the project. These conversations made it possible for me to observe how their lives had or had not changed since our first interview.

The Economic Situation Before and After Katrina

Everyone experienced economic loss after Katrina; however, the magnitude of a person's loss depended upon the economic and social resources they possessed before and after the disaster. In order to illustrate how the economic context of people's lives changed following Katrina, I have organized my findings into two separate sections entitled: life before the disaster, and disaster resettlement in Colorado.

In addition, I present my findings in terms of two populations of people; those who I identified as self-sufficient versus those who I saw as interdependent. Both groups can be distinguished from one another in terms of marked economic differences; however, they also differed from one another in regard to their region of origin, values, and norms and the size, composition, and function of their social networks.

Self-sufficient Individuals before the Disaster

Self-sufficient residents were usually middle-to-upper middle class individuals who either worked in, or were retired from, white-collar industries where they received an annual salary and health benefits. Of those who were employed, many worked for national businesses that had offices throughout the country while others worked in well-paying industries such as the healthcare sector where jobs were easy to come by if one possessed the requisite training, skills, and experience. Self-sufficient individuals rarely possessed more than one job in the formal economy and I heard of no instances in which a person tried to increase their income through informal side businesses. Higher annual incomes afforded many of these individuals the opportunity to save money, purchase a home, or invest in other financial ventures.

Table 4.1. Median and Mean Sizes of Family Networks in Louisiana

	Median Size	*Average Size*
Self-sufficient	4	5.7
Interdependent African Americans	37.5	46
Interdependent Whites	20	21

In addition, the majority of self-sufficient individuals were white non-natives who moved to New Orleans with their nuclear family when they were adolescents or adults. As a result of their relatively recent migration to the area, few of these individuals possessed kin networks that exceeded more than five family members (table 4.1). Over half of all self-sufficient individuals in the sample had no children and of those individuals who were parents, the average family size was limited to one or two children.[1]

Similarly, their non-kin networks were also small, consisting of a few close friends and a larger number of associates who were of a similar class background and whom they encountered at work or social events throughout the year. These individuals used their social networks in Louisiana for affective services only. None of these participants relied upon their network members for instrumental services. Indeed, all of these participants assured me that they were "self-sufficient," "economically secure" individuals who need not rely upon network members for services that they were well positioned to purchase. One evacuee, Donald, elaborates upon this sentiment:

> Well, I did not really need help in New Orleans. I was pretty self-sufficient. When I was young and in school my parents helped me out financially and let me borrow some money—but honestly I really do not need that much help. My wife and I do just fine as nurses.

Self-sufficient people were proud to be able to take care of themselves and they viewed assistance from family or friends as unnecessary. They were quick to distinguish themselves from Louisiana natives who they said operated within a network system in which interests were advanced or opportunities accessed based upon one's social connections rather than one's merit, experience, or educational qualifications. One of my informants described the importance of social connections among locals in the following way:

> I don't know what the deal is—what is going on with people but it is just a slower pace of life. There is no hurry down there. There are a lot of

> kickbacks and good old boys down there—you know brother, uncle type deal. In a lot of ways it was all about who you knew there.

Non-native, predominantly white self-sufficient individuals felt as if they had little in common with many of the values or norms that were espoused by native residents. Self-sufficient individuals lived and worked in Louisiana but they did not really see themselves as being part of the area. Armed with economic resources and nationally recognized credentials, these individuals had greater ties and connections to national businesses and interests. They were not embedded within the local cultural economy of New Orleans and few felt it necessary to rely upon local connections to access opportunities or services.

Interdependent Individuals Before the Disaster

In contrast, the financial situation of interdependent individuals was often more precarious and less secure. Most interdependent individuals were working-class or poor African Americans who had small annual incomes derived from hourly positions in the service sector of the economy. The majority of interdependent individuals in my sample were African American; however, a much smaller number of white residents also engaged in interdependent behavior, albeit to a much lesser scale. Indeed, lower income white residents engaged in everyday forms of reciprocity with their nuclear families only—for things like household assistance, childcare, and financial assistance. Like self-sufficient individuals, none of the interdependent whites were natives to the area and they had small family networks composed entirely of nuclear family members. In addition, whites possessed small friend networks that they relied upon mainly for affective services.

Almost no one in my interdependent sample was a homeowner and very few individuals possessed savings. As a result, the majority of people found it necessary to work two or three jobs in order to make ends meet. Close to one half of all lower income people in my sample had more than one job and I heard of many instances in which a person worked 15 or 16 hours a day, six days a week. A small number of individuals also tried to earn extra money working under the table. Only a few people in my sample openly discussed their side business with me. However, one of my informants who worked exclusively in the informal economy noted that many people in New Orleans had side jobs. He states:

> That is kind of what New Orleans teaches you how to do—things that are under the table. People would pay cash for whatever they got and they would let the government worry about themselves.

Informal employment was an attractive option for many because it provided people with an opportunity to exert greater control over their schedule and earnings. The most common side businesses people discussed with me included: babysitting, cooking, repairing cars, cleaning, or helping people with various household repairs. Most of their business clientele consisted of people they knew fairly well; however, many also attracted new and unfamiliar customers through word-of-mouth advertising.

Few lower income people would have been able to survive financially had it not been for the assistance that their social networks provided. The majority of people I have labeled as interdependent were New Orleans natives who were deeply embedded in kin networks that consisted of 40 or more individuals with whom they engaged in frequent acts of reciprocity. Ronald, a 38-year-old African American man who was born and raised in the Ninth Ward, describes the value of reciprocity that existed among network members and New Orleans natives:

> People in New Orleans are giving people. They share and help one another out. I may not know someone real well but they will still let you into their house and fix you something to eat. In New Orleans, many people act as if they be knowing you their whole lives. They help you out and when you can you help them out too. People just treat you like family there.

Interdependent people had an abundance of family, both biological and fictive, whom they could rely upon for assistance, help, or information. Many people lived in extended family households that were composed of three of four generations of relatives, sharing the cost of rent, food, or transportation. In addition, many parents—most often single women—received reliable and affordable childcare assistance from their female kin members. Violet discusses the problems her daughter encountered with many non-kin childcare providers:

> My daughter had a hard time finding babysitters. She would have to take what she could get like night hours—late hours. She didn't have dependable resources so we helped her. I sent my son Ted down there to help her. Then I moved down there to help her. I got part-time work at the Shell station because I needed money. I worked the 3–11 shift and my daughter worked 11–7 and when we were gone my son would be there with the kids so we was just rotating. We just all rotated.

Family members also served as important conduits of information, however; they were by no means the only people whom individuals turned to in order to gather or disseminate information. People relied upon all of their network members to learn about employment opportunities or to advertise their own businesses.

Network members adhered to norms and values that stressed the importance of sharing and reciprocity. People understood that they could not financially survive on their own. As a result, many people gave with the knowledge that they would receive and they understood that sharing was a productive action that all network members benefited from.

Although interdependent social networks helped people to stay financially afloat, they could not help them to get ahead. Evacuees commonly complained about the lack of opportunities in New Orleans and noted how the majority of their family and friends were stuck in low-wage jobs that offered few opportunities for advancement. Many individuals thought it difficult if not impossible to improve their lives in Louisiana. In conversations people stressed the importance of investing in family; however, few interdependent evacuees made similar comments about the benefits that could arise from an investment in education. Evacuees like Jaime, a 19-year-old high school graduate understood the situation as follows:

> In New Orleans education wasn't a big deal. They didn't push it on anyone. The attitude was if you are not doing good in school—well okay—whatever. In New Orleans I was still in contact with my family almost every single day. There was not much else other than that.

For Jaime and other low-income interdependent residents education may not have been considered a big deal because their families could ill afford to extend their education beyond high school. Most interdependent families were living paycheck to paycheck and could not make additional investments of their time or energy in the institutional education of themselves or their relatives. Few individuals I spoke with appeared to be worried about their lack of advanced educational qualifications. People did not perceive that their ability to attain employment in New Orleans hinged upon their possession of an advanced degree. More often, becoming gainfully employed depended upon the resourcefulness of the individual or a word-of-mouth referral from a network member.

Self-sufficient Individuals and Resettlement

Everyone experienced economic loss after the disaster including higher income, self-sufficient residents whose homes, properties, and possessions were badly damaged by Katrina. The financial loss sustained by self-sufficient individuals was significant; however, it never spelled their financial ruin. At most, people experienced a financial setback, a setback they could recover from because they possessed savings, owned insured homes, and were able to resume employment at new or old jobs weeks after the storm.

Savings helped self-sufficient individuals to financially stay afloat in the weeks or months after the storm when many contemplated where they would go and what they would do. Individuals with savings claimed that they "did not really need any help" from the government or nonprofit organizations after the disaster. The institutional assistance that was received by higher income evacuees was usually limited to the procurement of clothes or furniture. Daniel describes his experience with assistance after the disaster:

> After Katrina, there was a great outpouring of generosity and goodwill. People and organizations gave us clothes to wear. I had never worn used clothes before Katrina but we had lost everything so I had to. We also got a lot of free furniture and we were able to furnish our two-bedroom apartment. We were aware that there was a rental assistance program but we did not want to go through all of the paperwork. We didn't need that kind of help. We had enough money saved to live off of for a while.

In addition to savings, many people were able to live off the insurance monies they received from their damaged homes. Insurance payments were an important source of capital that helped self-sufficient individuals to finance their new lives in Colorado and many used their insurance money to make a down payment on a home or to pay for additional start-up costs. The possession of savings and insurance money made people feel comfortable and confident about beginning their lives anew in Colorado, as one of my informants stated:

> We were not scared to move here. We both felt very confident about getting a job and we had savings so we never once felt desperate after the hurricane. We never feared that we were going to become homeless or anything.

Self-sufficient individuals were able to further stretch the largesse of their savings and insurance monies because very few had dependent children for whom they were financially responsible. With but a few exceptions, self-sufficient individuals did not have the added worry or financial stress of having to support a family after their resettlement in Colorado.

Of those individuals who were employed before the disaster, all but one was able to find comparable employment in Colorado shortly after they moved.[2] With relatively little difficulty, salaried employees at national companies were able to transfer their jobs to a local office in Colorado. Employment at a national company yielded other financial benefits as well. Many national companies provided their employees with generous relocation packages that paid for a portion, if not the entirety, of their moving expenses. Joanne describes the ease with which she was able to transfer her job to Colorado:

> So I came up here and I was lucky to be able to transfer with my company because they had an office in Denver. I am just lucky to have fallen into a company with the resources to let me do that. It was pretty easy to get my job transferred. It was a phone call.

Self-sufficient individuals who were not affiliated with a national company also found it relatively easy to attain employment in Colorado. These individuals knew where to look for jobs and felt comfortable submitting applications and résumés as well as being interviewed by potential employers. Local businesses were especially likely to hire individuals who possessed specialized training or skills. Individuals with less specialized skills found jobs through temporary employment agencies that placed them with companies that had "temp to hire" policies whereby a person was permanently hired by an employer after a few months.

The need for kin assistance was a consideration for few self-sufficient evacuees who moved to Colorado. Indeed, only two self-sufficient individuals evacuated to Colorado because they had relatives who lived in the state pre-Katrina (table 4.2).

Both of these individuals' Colorado relatives were financially secure homeowners who possessed the physical space and financial resources to enable them to house their relatives for weeks or months on end. These two evacuees felt extremely fortunate they had relatives who were willing and able to offer them long-term shelter. Living with their family spared them the hardship and frustration of having to find temporary housing

Table 4.2. Median and Mean Sizes of Family Networks in Colorado

	Family who left with them (median size)	*Family who left with them (average size)*	*Family in CO before Katrina: (median size)*	*Family in CO before Katrina: (average size)*
Self-sufficient	1	1.3	0	.33
Interdependent African Americans	4	3.125	1	2.3
Interdependent Whites	11	12.4	1.5	1.4

elsewhere. In addition, family provisioned shelter provided them with several weeks of free accommodation. Despite their gratitude, neither person felt comfortable relying upon their relatives for instrumental services. The following quote from Janet typifies the discomfort they felt about accessing kin assistance:

> I didn't want to have to depend on my relatives here. I didn't want to mooch off anyone and basically I didn't feel like I had to. I am paying them rent and I think that benefits everyone.

Interdependent Individuals and Resettlement

Lower income, interdependent people were more profoundly affected by the hurricane and their resettlement than were higher income self-sufficient individuals. Whereas most self-sufficient individuals described their disaster resettlement experience as an "inconvenience" or a "hardship," lower income, interdependent individuals spoke of their experience in much more dramatic and weighty terms. Like most residents who lived in disaster-affected areas, interdependent individuals experienced huge material loss after the disaster. Katrina's violent winds and swelling waters drowned people's neighborhoods and communities and destroyed their homes and material possessions. However, in many ways resettlement in Colorado proved to be more detrimental for interdependent people than the wreckage wrought by Hurricane Katrina.

Unlike self-sufficient individuals, interdependent evacuees did not have the luxury of savings or insurance payments to help them rebuild their lives in Colorado. A dearth of financial capital made it necessary for them to rely upon governmental or nongovernmental assistance programs to help them get back on their feet. Ninety-five percent of all interdependent evacuees participated in the Federal Emergency Management Agency's housing assistance program, which promised to provide people with 12 to 18 months of rental assistance in Colorado. The housing program made it possible for people to live in Colorado despite rent prices that were prohibitively higher than they were used to paying in Louisiana. Housing assistance was one of the most beneficial sources of aid people received, although it was by no means perfectly instituted in Colorado. Many people were placed in homes or apartments that they would not be able to afford once the Federal Emergency Management Agency program ended. In addition, residences were often distant from town centers, social services providers, or places of employment, which proved particularly problematic for those individuals who did not own cars or live within walking distance of a bus stop. Geographic isolation and a lack of transportation impeded people's ability to find jobs or access alternative sources of assistance.

Job loss and long-term unemployment were two additional factors that contributed to people's increased economic insecurity. Rates of job loss were particularly high among interdependent evacuees. The majority of evacuees lost multiple jobs per person and less than 10 percent of all interdependent evacuees worked for national companies that would have enabled them to transfer their jobs to Colorado. The two individuals who had the least degree of difficulty attaining employment in Colorado were those who were employed at Wal-Mart in Louisiana. Employment ties to a national company like Wal-Mart helped both people to obtain comparable employment in Colorado rapidly.[3] In contrast, it took the other five individuals in my sample four to six months to find a job in Colorado; however, none of them were able to find comparable employment. Indeed, all of the non-Wal-Mart employees complained they were making less money in Colorado than they had in Louisiana. Jennifer, a massage therapist, describes her feelings of frustration in regard to her reduced wages:

> The new place I work pays about half of what I had made in New Orleans and I have to do more massages in order to make less than I did there. I can't work many more days then I already do. I would go into the salon in New Orleans and do three massages but I made over a hundred dollars. Here I have to do 6 massages to make a hundred dollars.

Despite people's frustrations, employed evacuees were well aware they were better off than those evacuees who remained chronically unemployed. Working-class and poor evacuees understood they needed to find employment in Colorado. Most received weekly or monthly reminders from aid agencies regarding the status of their rental assistance, unemployment benefits, or other types of aid. Unfortunately, messages from aid agencies were often contradictory and misleading. Few people had a clear understanding of how long they would be able to access disaster assistance and many feared their benefits would end abruptly. Fears of homelessness were never far from the forefronts of people's minds. Evacuees like Diane, a former nurse's assistant in her mid-fifties, understood assistance would not last forever:

> I need to be on some person's job—where I can be making me some money because when the pay from FEMA runs out I need to be able to pay for my rent, my utilities, and some transportation. If I can't start working and banking up on that now—I am going to be in a pickle. Who will come to my aid then?

Regardless of people's awareness that they needed to begin earning money, a number of factors prevented people from immediately seeking work. Most were unsure how long they would stay in Colorado and they were hesitant to become employed if they planned to return home. Others argued that trying to negotiate disaster assistance programs and social service providers was a full-time job that left few hours in the day to look for work. A smaller number of individuals cited family problems or concerns as yet another reason why they had to postpone their job search. Evacuees like Cheryl, who relocated with her ailing mother, her two dependent children, and her sister's family, postponed her job search in Colorado when her mother became deathly ill and was hospitalized. Cheryl did not feel it was appropriate to begin looking for work until after her mother was released from the hospital and began to show signs of physical recovery.

By mid-November, the majority of interdependent evacuees felt as if they had their lives well enough in order to begin looking for work. In general, people perceived there to be a greater number of employment opportunities in Colorado than Louisiana and felt relatively confident about finding a job. Despite people's optimism, two-thirds of all working-class and poor evacuees were unable to find work a year after their resettlement. A small number of evacuees attributed their inability to find employment in Colorado to issues of racial and gender discrimination while others cited a lack of formal work qualifications as their greatest impediment to finding work. More commonly, I heard people mention that they did not know *how* to go about getting a job in an unfamiliar place where they knew few people. For many evacuees, this was the first time in ten or twenty years they had to look for a new job. Most were used to relying upon friends or family for employment help and were unaccustomed to the formal hiring process in Colorado whereby one filled out an application (often online), submitted a résumé, and attended one or more job interviews. Individuals like Elton were well aware that social connections could help him to acquire a job or learn about alternative ways to earn an income:

> Maybe by the first of the year I will have met enough people to decide and think about if I can financially make it here. I don't know enough people here yet so that means that the job that I decide to pick is not only going to be for money but also for contacts. So I can use those people to help me with my business plan.

Bereft of either financial or social resources, such individuals had few alternatives but to rely upon disaster assistance programs, social service providers, or the charity of local residents, all of which offered evacuees short-term remedies for complex and potentially long-term problems.

Many evacuees tried to circumvent their dependence on aid by engaging in relationships of reciprocity with local volunteers or other evacuees. Frequently unemployed or underemployed evacuees were hired by local volunteers or other evacuees for help with short-term side projects such as house painting, furniture assemblage, or yard work. Evacuees were able to earn extra money through side work; however, side projects were too infrequent to serve as a viable source of income. More commonly, evacuees swapped goods and services with other evacuees who they were friends with or who lived in close proximity to them. Unfortunately, geographic isolation and a lack of transportation prevented many people from initiating or maintaining reciprocal relationships with other Katrina evacuees in Colorado.

Individuals who were able to reinstitute relationships of interdependency and reciprocity with their kin in Colorado fared better than individuals who arrived in the state by themselves.[4] Many interdependent individuals evacuated and resettled in Colorado because they had family that lived in the state pre-Katrina. During evacuation, people strove to keep their family members together and many evacuated to Colorado with 15 or 20 family members from home. Diana, one of seven siblings and a mother of two daughters, explained why many of her family members moved to Colorado following the hurricane:

> We all stay close together because we all want to make sure that the other is taken care of. My daughters are grown and I did not even ask them if they wanted to come up—they just followed me. Many of my sisters and their kids followed me here too. We all just tried to stick together so that we can help one another out.

The presence of kin in Colorado made it possible for people to reestablish many of the forms of everyday reciprocity that characterized their lives in Louisiana. Unfortunately, disaster resettlement in Colorado severely impaired the size, strength, and capacity of many family networks. Single parents, most often women, were especially disadvantaged to be separated from network members who had offered them free and reliable childcare assistance in Louisiana. Parents without family in Colorado found it difficult if not impossible to pay for childcare assistance. The following quote from Annika helps to demonstrate her anger and frustration over the high cost of care in Colorado:

> In New Orleans I didn't have to do everything by myself. If I needed to go to work or go out I knew I could call their father or one of my family members for help. Here I can't do that and I have to do everything by myself. I have tried to put my daughter into kindergarten here but they told me that a full day kindergarten is $285 and I told them that I didn't have that kind of money. I applied for child care assistance but I have yet to hear from the lady who worked for that program. I have been living from paycheck to paycheck and I am over in two separate accounts. I don't know what I am going to do.

Evacuees derived many benefits from their Gulf Coast relatives; however, as a whole their kin networks were in a state of crisis. Few family members were doing well emotionally or financially, making it difficult for people to rely upon their kin members for the same variety of services they had regularly accessed at home. In addition, many evacuees associated exclusively with other disaster affected relatives or evacuees who were privy to similar sorts of information and services as they.

Evacuees with Colorado relatives were lucky to have a native family member to assist them. Evacuees like Troy, a 39-year-old New Orleans native, relied upon his sister and cousin to introduce him to their friends and acquaintances, provide him with useful information in regard to the workings of the city, and help him to find a centrally located and affordable place to live. In addition, Troy was able to secure a job through one of his sister's social acquaintances who informed Troy of an employment opening at her work and encouraged her boss to hire him. However, evacuees with Coloradoan kin were quick to remind me that there were limits to the types of services their relatives could provide or that they felt comfortable requesting. Evacuees did not want their "native" relatives to become overwhelmed by the frequency of their requests, and several individuals admitted they refrained from asking for assistance until their need for help was pressing.

Conclusion

The disaster resettlement experience of Katrina evacuees in Colorado clearly illustrates that the consequences of disasters are not the same for all groups of people, nor is recovery equally likely or equally possible for all groups of people. Therefore, we must investigate *why* such investments are not equally likely or possible for all people to attain. Echoing the words of Annette Lareau (2003:84), this chapter seeks to demonstrate how "families' social structural location gives them a different pool of resources to address similar life problems." Self-sufficient and interdependent residents both experienced material loss after Katrina and both were resettled in new communities they had little familiarity with; however, in few other ways were their experiences comparable.

The financial resources and educational skills possessed by self-sufficient evacuees made it easier for them to recover from Katrina. Their higher annual incomes afforded them the ability to develop and protect their financial assets before Katrina and their national connections and advanced academic qualifications made them feel confident they could attain a secure and well-paid employment position anywhere in the world they should choose to reside, including Colorado. Their financial survival did not depend upon their being rooted in a particular environment, nor was their ability to economically reconstitute their lives dependent upon their ability to exchange goods and services with family or friends. Although not relying upon families for day-to-day survival, they did benefit from long-term financial, educational, and social investments their families made in them. For self-sufficient evacuees, disaster resettlement in Colorado was an inconvenience. It was not an event that spelled their financial ruin or one that led to downward mobility.

Economic factors alone cannot account for what made some evacuees more or less vulnerable in Colorado. For interdependent evacuees, their economic situation was exacerbated by the loss of their social world, or more specifically, the loss of the social institutions or networks that provided them with instrumental services and assisted them in their day-to-day survival. In Louisiana, network assistance, especially among kin members, functioned as an adaptation to poverty and provided working-class and poor residents with a built-in safety net. In contrast to self-sufficient evacuees, interdependent evacuees arrived in Colorado with limited or no financial resources in tow. For those individuals who were lucky enough to resettle in Colorado with family or friends, many were able to reinstitute some of the forms of everyday reciprocity that characterized their lives pre-Katrina. Such assistance helped evacuees; however, a significant decrease in network size and a lack of weak ties to local community members made it difficult for evacuees to access the same number or diversity of services they received at home. A year after their resettlement, few interdependent evacuees were employed and most were still in a state of financial and emotional crisis.

The two groups of individuals experience recovery via the broader political economy in different ways: the poorer interdependent individuals with government assistance programs, and the more economically self-sufficient individuals with insurance companies. After years of having their wealth appropriated by government taxes or insurance policy plans, disaster-affected residents assumed they would be able to rely upon the government and other private institutions to help them finance their recovery. For example, housing insurance companies tried to avoid claims despite being able to pay them. Thus, middle-class Americans were tied to the larger political economy not only through their transferable professional skills/jobs but also through the insurance industry that protects their major form of investment. But disasters alter political economies, and the insurance industry has reacted by putting in stronger clauses about flooding, thus altering a major political economic link between the middle class and private capital.

Similarly for the lower class, the government began to absolve itself—months after Katrina—of its responsibility to assist the disaster-afflicted population. Thus, the poor are not only tied to the larger political economy through replaceable service and unskilled labor, but also via a variety of public housing programs in disaster and/or normal situations. These housing programs are costly to governments, especially if they involve financing, and thus governments tend to use them to the extent necessary to decrease outcries from either those in need of housing, or from people of higher socioeconomic status that find the neediness of others annoying. The definition of necessary, then, is what is politically economically expedient. As government assistance began to wane after Katrina, Gulf Coast residents were encouraged to assume complete responsibility for their recovery regardless of whether they were necessarily ready or able to take on such an endeavor.

Disasters help to make visible pervasive and often ignored social issues (Fritz 1968:658). Hurricane Katrina is no exception to this rule. In the aftermath of Katrina, the American public once again became aware of the fundamental ways in which issues of class and race influence people's life chances in routine and nonroutine circumstances. Although Katrina did not foment broad-scale social change in the United States, it did provide Americans with a lens with which to critically evaluate the political economy of a finance-driven economy.

Notes

1. It is important to note that only one self-sufficient individual had dependent children in Louisiana for whom they were financially responsible. All of the other parents had children who had reached adulthood and were financially responsible for themselves.

2. The one instance of job loss among self-sufficients occurred with a male home inspector in Louisiana. He could not attain equivalent employment in Colorado because he was unfamiliar with Colorado's building codes and architectural styles and would have to obtain new training.

3. The other five people who found employment in Colorado possessed a technical skill, worked in the restaurant industry, or found a job through a family member or a family friend.

4. Seven interdependent individuals were forcibly evacuated to Colorado by the U.S. government. They arrived in Colorado by themselves and did not have family or friends who lived in the state pre-Katrina.

References

Axtman, Kris

2006 A Year after the Hurricane, Houston Complains about their Persistent Joblessness. State Officials Plea for Federal Help. The Christian Science Monitor. Electronic document, csmonitor.com/2006/0822/p04s01-usec.htm, accessed August 22.

Barnshaw, John

2006 Beyond Disaster: Locating Katrina within an Inequality Setting. *In* Learning From Catastrophe: Quick Response Research in the Wake of Hurricane Katrina. Natural Hazards Center, ed. Pp. 47–70. Boulder, CO: Natural Hazards Center.

Bolin, Robert

1986 Disaster Impact and Recovery: A Comparison of Black and White Victims. International Journal of Mass Emergencies and Disasters 4(1):35–50.

Campo-Flores, Arian

2006 Houston Wants Katrina Evacuees to Move On. Newsweek, March 13. Electronic document, www.msnbc.msn.com/id/11677333/site/newsweek/page/0/print/1/displaymode/1098/, accessed August 30, 2007.

Cutter, Susan L., Christopher T. Emrich, Jerry T. Mitchell, Bryan J. Boruff, Melanie Gall, Mathew C. Schmidtlein, Christopher G. Burton, and Ginni Melton

2006 The Long Road Home: Race, Class, and Recovery from Hurricane Katrina. Environment 48(2)11–20.

Egan, Timothy

2005 Uprooted and Scattered Far From Familiar. New York Times, September 11. Electronic document, www.nytimes.com/2005/09/11/national/nationalspecial/11diaspora.html, accessed August 30, 2007.

Elliot, James R., and Jeremy Pais

2006 Race, Class, and Hurricane Katrina: Social Differences in Human Responses to Disaster. Social Science Research 35(2):295–321.

Erikson, Kai T.
1976 Everything in Its Path: Destruction of Community in the Buffalo Creek Flood. New York: Simon and Schuster.

Federal Emergency Management Agency
2007 Hurricane Katrina Information. Electronic document, www.fema.gov/hazard/hurricane/2005katrina/, accessed August 30, 2007.

Fothergill, Alice
2004 Heads Above Water: Gender, Class, and Family in the Grand Forks Flood. Albany, NY: State University of New York Press.

Fothergill, Alice, and Lori Peek
2004 Poverty and Disasters in the United States: A Review of Recent Sociological Findings. Natural Hazards 32:89–110.

Fritz, Charles
1968 Disaster. *In* Contemporary Social Problems. Robert K. Merton and Robert A. Nisbet, eds. Pgs. 651–694. New York: Harcourt, Brace & World.

Granovetter, Mark
1973 The Strength of Weak Ties. The American Journal of Sociology 78(6):1360–1380.

Grier, Peter
2005 The Great Katrina Migration. The Christian Science Monitor, September 12. Electronic document, www.csmonitor.com/2005/0912/p01s01-ussc.html, accessed August 30, 2007.

Halperin, Rhoda
1994 Cultural Economies Past and Present. Austin, TX: University of Texas Press.

Hulbert, Jeanne S., Valerie A. Haines, and John J. Beggs
2000 Core Networks and Tie Activation: What Kinds of Routine Networks Allocate Resources in Nonroutine Situations. American Sociological Review 65(4):598–618.

Lareau, Annette
2003 Unequal Childhoods: Class, Race, and Family Life. Berkeley, CA: University of California Press.

Moore, Spencer, Mark Daniel, Laura Linnan, Marci Campbell, Salli Benedict, and Andrea Meier
2004 After Hurricane Floyd Passed: Investigating the Social Determinants of Disaster Preparedness and Recovery. Family and Community Health 27(3):204–217.

Morrow, Betty Hearn
1997 Stretching the Bonds: The Families of Andrew. *In* Hurricane Andrew: Ethnicity, Gender, and the Sociology of Disasters. Walter Gillis Peacock,

Betty Hearn Morrow, and Hugh Gladwin, eds. Pp. 141–170. New York: Routledge.

Oliver-Smith, Anthony
1991 Successes and Failures in Post-Disaster Resettlement. Disasters 15(1):12–23.

Peterson, John S., Laura D. Stanley, Edward Glazier, and James Phillip
2006 A Preliminary Assessment of Social and Economic Impacts Associated with Hurricane Katrina. American Anthropologist 108(4):643–670.

Phillips, Brenda D.
1993 Cultural Diversity in Disaster: Sheltering, Housing, and Long Term Recovery. International Journal of Mass Emergencies and Disasters 11(1):99–110.

Portes, Alejandro
1998 Social Capital: Its Origins and Applications in Modern Sociology. Annual Review of Sociology 24:1–24.

Scudder, Thayer, and Elizabeth Colson
1982 From Welfare to Development: A Conceptual Framework for the Analysis of Dislocated People. *In* Involuntary Migration and Resettlement: The Problems and Responses of Dislocated People. Art Hansen and Anthony Oliver-Smith, eds. Pp. 267–287. Boulder, CO: Westview Press.

Stack, Carol
1970 All Our Kin: Strategies for Survival in a Black Community. New York: Harper and Row Publishers.

Sullivan, Bob
2006 Katrina Exodus Felt Far and Wide: Many Communities Wrestling with Explosive Growth, Long Term Problems. MSNBC.COM, August 28. Electronic document, www.msnbc.msn.com/id/14542913/print/1/displaymode/1098/, accessed August 30, 2007.

Recovering Inequality: Democracy, the Market Economy, and the 1906 San Francisco Earthquake and Fire

5

SHELLY BROWN-JEFFY AND STEVE KROLL-SMITH

In every city the people are divided into three sorts:
the very rich, the very poor, and those who are between them

(ARISTOTLE 350 BC BOOK IV, CH XI)

NATURE DIRECTS ITS FURY SEEMINGLY AT RANDOM. A small town in Kansas is leveled by a tornado; an earthquake swallows a highway; floodwaters ravage a town in Iowa. The scale of disaster is measured not only in the strength or invasive power of an aversive agent but also in the amount of human and environmental damage it causes. The shock of disaster is routinely followed by the more prosaic, organizational, and inevitably political economic process of recovery.

Recovery is a term used by government agencies, corporations, nongovernmental organizations, neighborhoods, and scholars. More importantly, for our purposes, it is used almost always the same way by all parties. In other words, researchers who write about recovery are apt to draw from the same vocabularies and perspectives as those who work in government and other arenas of the real politic. Brawny words like "renewal," "rebirth," and "revitalization"—the vocabulary of disaster recovery—are found in both the civic booster response to calamity and in scholarly papers and books (see for example, ACORN 2006; Daniels et al. 2006). Indeed, anthropologists and sociologists have added to this lexicon, coining terms like "amplified rebound" and the "Phoenix Effect" (see for example, Dyer

1999; Krausse and Dyer 2000; Quarantelli and Dynes 1977). Among their many messages, these sanguine concepts signal the "can do" ethos of modernity.

This chapter revisits the idea of recovery but does so from a critical stance. It accepts the idea that post-disaster recovery is an essential human pursuit but adds an additional idea, to wit, that not only do communities labor to restore housing, social services, businesses, and other essential collective needs, but they also repair, and often amplify, previous patterns of social inequality. Recovering inequality is always a part of human disaster response, but it is a particularly thorny practice in societies with a firm belief in citizenship, human rights, and the fair conduct of the state.

Disasters, we propose, provide an uncommon occasion to inspect the dynamics of social inequality inherent in capital-based economies and normally concealed in the mundane praxis of democracy. A disaster ruptures that always precarious arrangement that hides the dilemma of free-market democracies: structured inequality cloaked in the language of human parity. Our approach is to shift the idea of recovery from a standardized term used in a similar fashion by both scholars and private and public practitioners to a decisive appraisal of the inherent contradictions within any political economy, in this case a free-market democracy.

Scope and Methodology of Study

A particularly vivid example of the synergy between scholarly insight and civic pride is found in the academic concept "Phoenix Effect" (Krausse and Dyer 2000) and the phoenix emblazoned on the official seal of San Francisco, a city once leveled by an earthquake and fire. We draw from this spectacular catastrophe to fashion an argument for a political economic approach to disaster studies.

Political economy analysis is concerned with the interaction of political and economic processes in a society. We are interested in how the distribution of power and wealth between different groups and individuals, and the processes that create, sustain, and transform these relationships, are reconstructed during the disaster recovery process. In this chapter, we distill primary and secondary sources on San Francisco's ruinous earthquake and fire in 1906 into a narrative account of the first several months of recovery or, as we see it, two different recoveries. First, volunteer, municipal, and state groups and organizations, assisted by the U.S. Army, acted with a single purpose: to bring San Francisco back from almost total destruction to a shining example of America's progressive urban spirit. Second, restoring the city also required mending the fabric of class and race distinction rent by the cataclysm. The two recoveries—the planned, reflexive renewal of the city's infrastructure and the unplanned but inevitable revival of social and material inequality—as we will see, required one another. Our thesis is that the process of recovery was one that depended on the political and economic structures and tendencies of that time.

Narrative accounts from multiple sources, dissertations, and secondary scholarly sources are pieced together in a chain of evidence that discloses the recovery process. Our primary sources include any type of data that was originally collected, reported, or documented about the San Francisco earthquake and fire in 1906. Careful review of the literature revealed a body of research that also examined some of this firsthand data. The bibliographies of these documents proved to be invaluable to the data collection process. The narrative accounts presented in these documents are the data we analyze to illustrate the system of inequality in the immediate and delayed recovery experience.

This investigation ends with a postscript: a short annotation on the recovery of New Orleans following its devastating flood in 2005, one hundred years after the destruction of San Francisco. The differences and similarities between those two urban catastrophes beg additional work for understanding the political economy of disaster recovery in a comparative light. In particular, we discuss how the entire political economy of the United States has coherently shifted enough in the past century to produce a different kind of vulnerability and a different kind of strategy for disaster recovery.

Human Inequality, Market Economics, and Democracy

Well before the appearance of capital-oriented market societies, Aristotle observed the seemingly human need to sort people into groups based on their possessions. Dumont's (1981) "homo hierarchicus," it appears, is not limited to the Indian caste system. Whether or not social inequality is a species necessity, it is found everywhere humans gather. In some cultures, hierarchy and inequality are explicit and conscious principles. For others, social and economic disparity are more implicit and less mindful, but no less potent than their more deliberate and unconcealed variation. Massey (2007) recently updated Aristotle, referring to the ubiquitous presence of inequality in American communities. Inequality, discrimination, favoritism, lineage, and other means to perpetuate social distinctions are, it appears, found in all societies. So, what we are proposing should come as no surprise to someone who thinks about such issues.

In the first sentence of his "Gospel of Wealth," published in 1900, Andrew Carnegie (1900:14) acknowledged the dilemma created by the absence of parity in a market society founded on the Enlightenment ideas of universal human worth:

> The problem of our age is the proper administration of wealth, that the ties of brotherhood may still bind together the rich and poor in harmonious relationship.

Inequality was inescapable. Carnegie's dilemma, and the dilemma of the political economic elite of the time, was how to govern wealth in such a way as to keep the poor relatively contented with this arrangement. For Carnegie the answer was philanthropy—a political economic strategy that involves an ideology of paternalism toward those who are worthy, access to political influence via gifting, and an economic focus on production of elite goods that can be occasionally shared with commoners (e.g., visual arts, theater, music, jewelry). However, the fashioning of consent among commoners also relied heavily on the splayed fingers of democracy reaching into and molding a moral temper of hard work and individual responsibility consistent with the goal of capital accumulation.

This political economic orientation of society produces specific kinds of vulnerability to disasters, as well as particular opportunities and constraints for recovery. For example, following a disaster such as the San Francisco earthquake of 1906, the recovery of the status quo for the political economic elite depends, in part, on returning to predictability in terms of what commoners believe in, how commoners interact with decision-making, and how the labor of commoners is managed. In 1906, this predictability required decisive yet careful responses on the part of the governing class. We argue that such responses were constrained by one of the tenets of Carnegie's political economy—that you must respond in a paternalistic fashion to help those who work hard.

Much social science work on disasters proceeds as if cities and towns existed in a cultural and moral vacuum, largely detached from these cultural, social, political, and economic forces that act, to paraphrase Kurt Lewin (1951), as a dynamic field shaping the life-space of a population. Most studies of emergent groups, evacuation behavior, first-responders, and even the transparently political and economic work of reconstruction proceed with little or no mention of the seemingly inescapable force of the conscience collective or the peculiar shape of a nation-state economy.

In short, we have, only by virtue of its neglect, a modest motive for this type of inquiry.

But there is another reason to probe the connection between epoch, city, and disaster. Kenneth Hewitt (1983:23) reminds us that "Most natural disasters are characteristic rather than accidental features of places and societies where they occur." The idea that Mother Nature and human history are inextricably intertwined is surely an invitation to consider disasters, cities, and their intersections with a more inclusive, but also more obscure collage of potent social, cultural, and economic forces.

An Urban Cataclysm

On April 18, 1906, at 5:12 a.m., the San Andreas Fault opened its craggy face and fired a potent shock of seismic energy toward sleepy San Francisco. At 5:13 a.m. the earthquake made landfall directly beneath the lighthouse on Point Arena, roughly 90 miles north of San Francisco. Its speed as it shot underground is estimated to have been seven thousand miles an hour (Thomas and Witts 1971:66). Moments later a quake ripped through the city with a force greater than all of the explosives used in World War II. Buildings swayed like tall grass in the wind; some collapsed. Panicked, livestock ran amok on city streets, adding appreciably to the chaotic spectacle.

Called by many the greatest U.S. catastrophe of the twentieth century, the story of the San Francisco earthquake and fire has been told in dozens of books and hundreds of newspaper and magazine articles. Ironically, the quake itself caused comparatively little damage to the city. Resultant fires and efforts to extinguish the fires proved far more devastating to the city's infrastructure and population. When the last fires were extinguished on April 21, a total of 514 city blocks (4 square miles) were destroyed by incineration. Over 28,000 buildings burned to the ground. There were approximately $500 million of property damage. More than 250,000 or roughly 60 percent of residents were homeless (Strupp 2006:7–8). Close to 90 percent of the city's 400,000 residents evacuated, many carrying only what they could throw in a bag or trunk. Between 700 to 3,000 or more people died (Bolton 1997:9). Eighty percent of the city infrastructure was in ruins (Strupp 2006:8).

San Francisco

In 1906, at the western edge of the United States, a disaster, one part natural and two parts human error, laid waste a unique and irreplaceable American city. It would take another 99 years before another cataclysm, the 2005 flooding of New Orleans, wrought similar damage to an urban landscape in the United States.

San Francisco at the turn of the twentieth century was a thriving port city with a promising future. It boasted a population of 375,000 registered residents (Strupp 2006:9). Like all cities in the early 1900s, the city by the bay was in the grip of progressivism and social reform. By the turn of the twentieth century, San Francisco was considered to be the most unionized city in America (Douty 1977; Rozario 1997; Strupp 2006). It is worth noting that in July 1906, just two months after the disaster, unionized plumbers struck for higher wages (Douty 1977:174).

The America of the early 1900s was shaped by an emerging middle class, many of whom thought their destinies were tied to moderating the influence of labor, on the one hand, and on the other, persuading the industrial elite, the "upper ten" or "The Leisure Class" as Veblen (1899) famously called them, to act in a manner that would benefit all of society and not just the privileged few. Not surprisingly, while it worried about the "upper-ten," which comprised about 1 or 2 percent of a U.S. population of 76 million, the middle class fretted more about the working class and the poor.

In his essay "What Makes Life Significant" William James (1899/1992:51) pokes a little fun at progressivism. "You have culture," he writes, "you have kindness, you have cheapness, you have equality (you have, in short, a) middle-class paradise." While paradise might be a stretch, it is true that middle-class America at the turn of the twentieth century was organizing, finding its voice, becoming, as it were, conscious of itself and its importance to a market-oriented society. By the turn of the century, the middle class numbered about one in five Americans (Sullivan et al. 2001).

From clerical workers to small business owners, the middle class was not monolithic. The concept of an "upper-middle class" emerged in the 1890s, yet it would take another several decades for the idea of a "lower-middle class" to enter the vernacular (Coser 2004). Though a diverse lot, this emerging middle class was aware of itself as straddling the boundary between the obscenely affluent and what were tellingly called "the deserving" and "the undeserving" poor. San Francisco's middle class lived in the city. It was their city. They walked its streets, shopped its stores, their children attended its schools; it was, in short, home. Their stake in restoration is a key variable in making sense of the recovery of inequality following the earthquake and fire.

There are no definitive or even dependable figures on the number of working-class and poor that lived in San Francisco at the time of the disaster. The range of manufacturing jobs, from agricultural processing to textiles, suggests there was a substantial working-class population (Bolton 1997:14). In addition, seasonal work and part-time work along the docks contributed to a substantial number of underemployed and unemployed. In the early twentieth century, one was not poor if he/she was working; but work was often contingent and short-lived. One became poor when he/she was out of work, a frequent enough occurrence to blur the line between employment and poverty (Polanyi 1944:250).

Importantly the working class and poor in early twentieth century San Francisco lived apart from the middle and upper classes. Sequestered in neighborhood clusters with colorful names like the Mission District and the Barbary Coast, labor and the poor were spatially segregated from the middle and upper classes. Distinction, the public coin of capital accumulation, was a prominent feature of the city's social topography.

In contrast to the difficulty in determining accurate numbers of the middle, working, and lower classes living in San Francisco, the number of Chinese and Japanese was well documented. In 1906, 4.6 percent of the city's population was either Chinese or Japanese (Bolton 1997:9). In 1900, 13,954 Chinese lived in a ghetto. Most were forbidden from working outside what was called Chinatown (Bolton 1997:10; Pan 1995). For all practical purposes, Chinatown was a self-governing municipality within San Francisco (Pan 1995:5). The outside-inside relationship Asians endured in the city would prove critical to what would become their more atypical recovery, which we discuss later.

The Great Derangement

> So at the last moment I picked up some of my little trinkets and filed my dress case, what victuals I had I put into a little telescope and with that slung over my shoulder like all the rest—rich and poor alike (Bertha 1906).

The 1906 earthquake and fire radically altered the physical and social boundaries that worked in normal times to fashion a market-based society composed of groups with markedly different life chances. Those whose life chances were many and varied not only experienced the shock of material loss, perhaps injury and even death, they were often brought face-to-face with a broad swath of people whose lives rarely if ever intersected with theirs. The seismic shock and fires commencing April 18 exposed, however briefly, the fragile arrangements fashioned by tradition, law, and routine that worked together to conceal the poor and dispossessed. There was, in short, a collapse of social distinction.

Victor Turner was not thinking of disasters when he refashioned van Gennep's idea of liminality to help us see that hierarchy and structured inequality might on occasion be temporarily replaced by "communitas" (Turner 1967, 1969). The immediate aftermath of most disasters approximates a liminal moment, a space betwixt and between the routines of inequality and distinction. It is a space alive with a profound feeling of togetherness, a fellow feeling of social equivalence. Consider an account from a well-educated survivor in San Francisco:

> Men and women who had lived in wealth and elegance stood in the bread line with Chinese and Negroes, with street-sweepers and paupers. Rich and poor struggled along, side-by-side . . . On the winding drives in the Presidio reservation the spectacle was indescribable. Destiny had thrown together in that spot all sorts and conditions of human beings. (Hooker Family Papers 1906).

The earthquake and fire shoved people into streets where the ordinary markers of segregation were no longer discernable. There, in public spaces, the fragile skein of civility that routinely acted as a flimsy, but needed, protocol for life in a class-conscious society was momentarily shredded. A witness offers a thick description of this unruly moment:

> Men whom their friends counted wealthy foraged in the morning for a cup of coffee and a roll to stay the hunger of their dear ones. In doing so they rubbed elbows with sodden beggars and with the outcast women of Chinatown. Golden Gate Park was the Mecca of the destitute. . . . here the common destitution and suffering had wiped out all social, financial and racial distinctions. The man who was a prosperous merchant was occupying with his family a little plot of ground that adjoined the open-air home of a laborer. The white man of California had forgotten his antipathy to the Asiatic race and was maintaining friendly relations with his new Chinese and Japanese neighbors. The society girl who on Tuesday night was a butterfly of fashion at the grand opera performance was assisting some factory girl in the preparation of her crude breakfast (Wilson, page number unavailable, 1906).

Disasters often disrupt the physical and social distance that preserves the hierarchical arrangement of socioeconomic classes. Muddled class relationships were on the minds of each narrator as he or she recounted different scenes from this urban maelstrom. For a period of time, however brief, personal survival trumped physical and social distinction as people with widely varying life chances were thrown together in the common desire to live to tell the tale.

The Therapeutic Community and the Need for Class Restoration

Students of disaster often refer to this unusual collage of people from all social and cultural distinctions thrown together in a mutual effort to survive a "therapeutic community" (Barton 1970; Gurney 1977; for a somewhat different take on this idea see Kroll-Smith and Couch 1990). The somewhat romantic idea of an emergent society that disregards social and cultural inequality while engaging in mutual aid, sympathetic identification, and emotional solidarity is not without empirical evidence. The narratives just presented strongly suggest the emergence of precisely such a community.

Granting the emergent therapeutic community, step to the side for a moment and examine this idea from we might call a *class restoration* perspective. Communal sharing, personal identification, and collective feelings of solidarity violate the areal and figurative features of a class-based social order. From this vantage point, the emergence of the therapeutic community—an unstructured, protean violation of the order of things—requires that a measure of material and symbolic work is employed to secure the restoration of those hierarchical arrangements that reflect and assemble class power and also provide order and meaning to virtually all members of society. The question at the heart of this chapter is how, in fact, that restoration takes place.

Recovery is a key concept in disaster research. The United Nations' International Strategy for Disaster Reduction offers a widely accepted definition of recovery as the:

> decisions and actions taken after a disaster with a view to restoring or improving the pre-disaster living conditions of the stricken community, while encouraging and facilitating necessary adjustments to reduce disaster risk (UN-ISDR 2007).

Although the use of the conjunctive "or" signals more than the author of this definition intended, we find this conjunctive an apt usage. To restore is one goal, to improve is another. To labor to restore distinction based on wealth is a quite different goal than to work to improve the life chances of the disposed and underemployed. This official definition of recovery also calls for "necessary adjustments to reduce disaster risk" that would require, among other actions, a substantial increase in the standard of living among those populations most likely to be in harm's way. However, for a city organized around the invisible hand of the market, restoring the living conditions of a community damaged by disaster requires repairing the ladder of inequality—the chain of class differences—that organizes and regulates social and cultural life.

Recovering San Francisco

On the minds of many middle-class San Franciscans was the restoration of the society they inhabited prior to the early morning of April 18, 1906. No one in early twentieth century America expected the federal government to provide mandated assistance. The first Federal Disaster Relief Act would not be passed until 1950 and FEMA was not organized until 1979. The sole federal response to the destruction of San Francisco was the assignment of more than five thousand Army troops to the stricken city (Thomas and Witts 1971:154–155).

It was the middle class that welcomed the arrival of the military shortly after the earthquake. Writing in June 1906, Mary Austin (1907:409) observed, "The will of the people was toward authority, and everywhere the tread of soldiery brought a relieved sense of things orderly and secure." Among the sense of things orderly on the minds of many middle-class residents was the perceived need to restore the pre-disaster classification of the laboring classes, to realign the "worthy" and the "unworthy" poor. The historical origin of these classes is found in a robust manufacture economy that emerged in the waning years of the nineteenth century—by the end of the century, the U.S. exceeded the combined production output of both Great Britain and Germany (Cashman 1993:8).

A country with the most concentrated production output in the world requires a viable, stable labor force. A principal strategy for identifying and sustaining a stable pool of labor was a nomenclature created from an unalloyed faith in the democratic ideal of the individual, the citizen, who used his or her freedom to fashion a secure place in a market-oriented society. Despite a progressive emphasis on environmental versus biological causes of human misery, it was the yardstick of morality that served to measure the good citizen (Bolton 1997; McGerr 2003; Rozario 1997). Labor was located in a taxonomy that included the "dangerous classes," the "unworthy" or "undeserving poor," and the "worthy poor." If a typical middle-class family in 1900 earned close to $1,200 per year, a comparable working-class family—employed in mines, factories, construction, and semiskilled trades like tailoring—earned somewhere between $240 and

$435 annually, thus the title the "worthy poor" (McGerr 2003:16; see also Painter 1987). It was not, however, the deserving or worthy poor the middle class feared and hoped the military would control, but the "brigands of civilization" (Morris 1906:78).

Armed with a taxonomy derived from a democratic belief in the primacy of the citizen and the utility of the moral measure, the first task for relief workers in San Francisco was to match all those individuals and families who were not middle and upper class with the categories working poor, unworthy poor, dangerous classes, and a fourth group, the Chinese (Bolton 1997:4). The earthquake and fire had a leveling effect that created the need to actively identify the worthy from the unworthy. More visible than the poor and working class, the Chinese were easier to resegregate, though in the end, as we will see, this group proved impossible to control.

To reimpose a classificatory scheme, social agencies responsible for relief quickly concocted an impromptu worthiness scale to determine who would receive aid, what kind, and how much (Bolton 1997; Rozario 1997). In fact, several worthiness measures were deployed at various times to make these determinations, among them, the simple distinction between "able-bodied" and "disabled" (Bolton 1997:103–108; Russell Sage Foundation 1913) and "moral" and "immoral" (Russell Sage Foundation 1913:293). Indicators of immorality included "laziness," "vicious habits," "drunkenness," and "unemployment" (Russell Sage Foundation 1913:80). A third measure of worthiness allowed that most disaster survivors were at least nominally dependent and sorted them into three categories:

1. Dependency because of abnormal conditions.
2. Dependency because disaster had converted semi-dependency into complete dependency.
3. Dependency because character and circumstance, irrespective of abnormal conditions, induced dependency (Russell Sage Foundation 1913:297).

Category 1, those people whose dependency was created by the earthquake and fire, were generally understood as the middle and upper classes. Category 2, those people who lived marginally before the disaster, comprised the working poor. And category 3 corresponded with the group generally called the "unworthy poor."

Unpacking the nomenclature of social kinds that quickly emerged to re-sort the city into some facsimile of its pre-impact structure of inequality sheds light into how the civility of difference is recovered. Recall from a few pages past, the social parity and egalitarianism that followed in the immediate wake of the earthquake and fire: "Men and women who had lived in wealth and elegance stood in the bread line with . . . street-sweepers and paupers. Rich and poor struggled along, side by side." Referred to as a therapeutic community, that equal opportunity moment could not last.

"Demixing"

If San Francisco was to recover its economic niche as a port, trade, and manufacturing center it would require a socially stable and cooperative workforce. Far from disposable, labor was critical to rebuilding the city. By May 13, 1906, city relief organizations established over 100 separate housing camps (Bolton 1997:55) with over 8,000 "refugee houses," wooden structures capable of accommodating several families (Strupp 2006:32). One prominent goal of the housing movement was "demixing" San Francisco's citizens. An awkward term, "demixing" conveyed the need to resegregate the city according to class and race distinctions (Strupp 2006:32).

Together these dozens of camps reflected middle-class concerns with the recovery of proper civil conduct, spatial segregation and, ultimately, economic viability. A quick distinction was made between unofficial and official camps. Unofficial camps were inhabited by people who had either been forced out of the city's authorized areas or knew beforehand they would not be welcomed in these more "proper areas." A quick correspondence emerged between the unofficial camps and the unworthy poor, and the official camps and the worthy poor (Russell Sage Foundation 1913:216). While not an exact equivalence, the correlation between unofficial-unworthy, on the one hand, and official-worthy on the other was striking.

The unofficial camps were considered both unsanitary and immoral, two terms intimately tied together in the progressive era (Bolton 1997; Strupp 2006). The official camps, on the other hand, were regulated and something close to a martial order emerged, bringing in the words of one middle-class reformer: "a certain calm and order to the working people" (Russell Sage Foundation 1913:211). Calm and order were enforced by both the military and the readiness of the relief agencies to withhold resources from those who did not comply with the rules of the camps.

Demixing, it appears, applied to more than temporary housing. A telling anecdote signaling the recovery of that civic skein of inequality is found in the *San Francisco Relief Survey* (Russell Sage Foundation 1913). Describing the "community kitchens" that sprang up throughout the stricken city, the report draws a distinction between the solvent and the debtor: "The disaster kitchens [were opened] to serve hot meals both to refugees and to persons able to pay for their food" (Russell Sage Foundation 1913:49). The distinction between refugees and able persons was further drawn in this description: persons able to pay were afforded "the privilege of sitting at separate tables and of ordering a better quality of food than that furnished at the free tables" (Russell Sage Foundation 1913:49). In this seemingly innocuous practice we begin to see the recovery of inequality. The class politics of food is illustrated on a broader, policy level with the decision by the Red Cross, just days after the last fire was extinguished, to sell food rather than give it away. The aim was to encourage personal discipline among survivors and ensure a timely return to regular political economic processes (Strupp 2006:24; see also Henderson 2005).

Disaster Assistance and the Political Economy of Housing Segregation

The San Francisco earthquake and fire was the first major catastrophe to be photographed and filmed. Dozens of photos create a visual portrayal of the blending and combining of disparate social and ethnic groups into tent cities. With approximately two-thirds of the city's housing stock destroyed, over 200,000 residents required some type of shelter (Russell Sage Foundation 1913:217).

Tent camps were made available on a first come first serve basis. By May 13, 1906, over 100 separate camps were providing refuge for survivors. While separate camps were provided for the Chinese and Japanese, the other camps housed side-by-side San Francisco's once spatially segregated social classes. The restoration of areal inequality began with two separate initiatives.

First, in a matter of months, small, green, "earthquake cottages" were built and placed in various open spaces throughout the city. In all, 5,610 cottages were constructed, some with two, some with three bedrooms, plus 885 "bonus houses" and 19 tenement houses of one or two bedrooms (Russell Sage Foundation 1913:219). The city's leadership and relief organizations had one group in mind when they built this housing: "the worthy" or "working poor":

> Spokesmen for the Relief and Aid society complained . . . about the difficulties of identifying the honorable. In the case of housing, officers applied a material calculus of virtue, recognizing the moral credentials of any per-

> son or family that had purchased or leased a home before the fire—those who had rented in tenement buildings were presumed to be unworthy (Rozario 1997:73).

Behind the parsing out of dwellings via a material calculus of virtue was the need to get working people into decent housing and back at their jobs. Disaster cottages solved, at least for the moment, the problem of housing for the worthy poor while simultaneously segregating them from the unworthy population at the one end, and from the middle and upper classes on the other.

Second, by summer's end in 1906, stable or upwardly mobile middle- and upper-class families were being offered generous subsidies and loans to rebuild their homes or purchase new ones (Bolton 1997:108; Strupp 2006:32). By 1908, most middle- and upper-class families were living as they had before the earthquake and fire (Bolton 1997:61). Disaster relief had fulfilled its role, unwittingly described by Deacon (1918:137) in his report to the Russell Sage Foundation on "The American Red Cross in Disaster Relief," this way: "The object of . . . relief is to assist families to recover . . . and to regain their accustomed social and economic status."

In a cruel twist, however, a series of nationwide recessions reduced the need for labor in the city; accordingly, privileged families became less tolerant of the painful reminders of the 1906 calamity. A concerted movement quickly emerged to evict laborers from their cottages and move them to less visible parts of the city (Bolton 1997; Rozario 1997; Strupp 2006). In the recovery of San Francisco, social service, government, and business interests worked unwittingly via a democratic ethos to repair the damaged boundaries of economic and social class distinctions. The significance of a democratic ethos in securing the recovery of inequality is nicely illustrated by the curious case of the Chinese.

The Privilege of Alien Status

Chinatown was destroyed by the fire. What was left was looted by white residents under the blinking eye of the military (Strupp 2006). The inescapable presence of hundreds of homeless Chinese and Japanese wandering about the ruined city was too much to bear for many white residents. Several prominent businessmen beseeched the army to keep "the yellows" in groups and place them at the far edges of the city (Pan 1995:63). Camps were created at some distance from white enclaves. Nevertheless, white property owners continued to complain of the smells coming from the Chinese camps (Pan 1995:44). "[W]itnesses recounted a pitiful procession of elderly Chinese women, hungry, tired, and barely able to walk . . . as they struggled" to move from one camp to another (Bolton 1997:57).

Adding to their misery, city government officials worked tirelessly to keep the Chinese from moving back to Chinatown. Viewed as a prime piece of urban real estate, it made good business sense to rezone the land and make it available to more lucrative mainstream business interests. The Sub-committee for Permanent Location of Chinatown was created to secure the rezoning. Losing all or most of their material possessions, wandering refugees in their own city, now "the Orientals" were to lose the right to rebuild their homes.

"The Orientals," however, were not without representation. Critical to their well-being and chances for recovery was the simple fact that they were *not* American citizens. China had been watching the events in San Francisco from afar. The city's port was the export and import center of trade between the U.S. and China. In May of 1906, the Chinese ambassador to Washington complained to President Roosevelt about the treatment of his countrymen. Having succeeded once in boycotting American goods in response to a race-based Exclusion Act passed by Congress in 1904, Roosevelt was well aware of the economic troubles China might create if he did not act quickly. He instructed Secretary of State Howard Taft to act immediately to ensure fair treatment of the Chinese:

> . . . suffering and destitution are peculiarly great among the Chinese. I need hardly say that (relief) work must be done wholly without regard to persons and be as much for Chinese as for any others" (quoted in Pan 1995:78).

The Chinese would regain their right to live in the old Chinatown. Indeed, the new Chinatown would be both more artistic and architecturally sound than the pre-disaster settlement. Ironically, the Chinese experienced a legitimate version of the amplified rebound, with many families acquiring better housing and increased financial security (Bolton 1997; Pan 1995; Strupp 2006). The moral yardstick deployed to distinguish the worthy from the unworthy could not be so readily applied to a citizen of another country. The political economy of early twentieth century San Francisco did not depend only upon patrimonial responsibility of worker and governing elite as elsewhere in the United States; it also depended relatively heavily on the international economic ties that were important for the city's economy and necessary for the development of its middle class.

To Conclude

The fault lines of cultures and societies are often exposed when disasters upend the routines of order and hierarchy. But more than fire, brimstone, and cracking earth are necessary for us to see and identify just what is revealed when disasters strike. Drawing from the inherent tensions between the ethos of democracy and the dynamics of a free market economy at a particular historical moment, we amend the idea of recovery. We account, at least in part, for how San Francisco restored more than a semblance of its pre-disaster way of life but also its inherent patterns of social and economic inequality. Perched on our critical branch we make a case for the need to expand the idea of recovery.

Postscript: Recovering New Orleans

It would take another 99 years before another cataclysm wrought similar damage to an American city. In 2005, at a precarious sliver of land between a mighty river and a massive lake, a quite different disaster—again, more human made than natural—engulfed another matchless urban landscape. The 2005 flooding of New Orleans submerged 80 percent of this languid southern city under 3 to 12 feet of brown, fetid water. Two-thirds of the city evacuated and as of February 2008, almost three years later, 71.5 percent of city's pre-disaster population has been reestablished, though it is not clear if these are the same returning residents (The Brookings Institute 2008). The recovery of New Orleans differs significantly from the recovery we've described of its sister city 100 years ago. Many of those differences are closely tied to the changing character of corporate capital and its peculiar relationship with labor.

Early twentieth century America valorized the working man. Upon his back was placed the country's economic hopes and dreams. While the economy of San Francisco was based in part on finance, city revenue and taxes were based primarily on manual jobs including skilled building trades and unskilled trades such as cannery workers and transit operators (Douty 1977:65). To recover from its catastrophe, to re-create its skilled and unskilled economic base, San Francisco required the labor of those who lived in the city. The working poor were viewed as worthy of assistance because they had something immediate and tangible to contribute to the economy.

If 1906 San Francisco needed able-bodied residents—the worthy poor—to rebuild the city, New Orleans in 2005 did not. By the early twenty-first century "the figure of the 'disposable worker' emerges as prototypical upon the world stage" (Harvey 2005:169). When the floodwaters topped the levees on August 29, 2005, America had already shifted from a muscle-based production economy to a cerebral-based finance economy (Baxter and Kroll-Smith 2005). In this economy, wealth is more important than wages and what happens on Wall Street is far more significant than what happens on Main Street.

The economy during the rebuild of New Orleans was much different from the economy during the rebuild of San Francisco. New Orleans' economy did not depend on skilled and unskilled labor. It depended rather on service labor in tourism, and revenue from the port authority. There was an immediate need for labor to rebuild its infrastructure, but most of this need was met by outsourcing and undocumented workers.

Prior to the flood, approximately 25 percent of New Orleans residents lived below the poverty line. Another 40 percent lived just above it (Brookings Institute 2008). The storm wiped out most of the high-poverty census tracts (Brookings Institute 2005). Thus there was no incentive for poor non-homeowners to return to New Orleans, especially those with little or no hope for employment in the recovery phase. There was nothing to *pull* them back. The delayed rebuilding of some parts of the city was enough to *push* the poor population to get out and resettle somewhere else. However, it was those with a vested interest in the city (business and homeowners) who returned to the city. Those who were the first to return are estimated to be those with dwellings in the high-elevation areas, which consisted of mostly upper- and middle-class whites.

In closing, it is worth noting that tied closely to the diminishing role of labor in a finance driven economy—where money, not sweat, makes money—is the truncated role of society. The attenuation of society, or at least that part of society that took responsibility for human welfare and well-being, is starkly captured in Margaret Thatcher's now famous obituary to its demise:

> I think we've been through a period where too many people have been given to understand that if they have a problem it's the government's job to cope with it. 'I have a problem, I'll get a grant.' 'I'm homeless, the government must house me.' They're casting their problem on society. *And, you know, there is no such thing as society.* There are individual men and women, and there are families (Margaret Thatcher, talking to *Women's Own* magazine, October 31, 1987).

If the culture of early twentieth century America needed society, in the form of a viable population of "worthy" labor coupled with a reformist-minded middle class, America in the twenty-first century can dispense with both. It is here, in the crucible of a wounded New Deal and the ascendancy of finance capital, that we might well begin an inquiry into the recovery of New Orleans. Recovery, as we will likely see, assumes a quite different shape when it is configured around the absolute need for labor and wages than when it is organized around finance and wealth.

References

ACORN

2006 New Orleans: ACORN Katrina Recovery and Rebuilding Campaign. ACORN, HIC-GS Case study #17. Electronic document, www.acorn.org, accessed August 30, 2007.

Austin, Mary

1907 The Tremblor: A Personal Narration. *In* The California Earthquake of 1906. David Starr Jordan, ed. Pp. 341–360. San Francisco: A. M. Robertson.

Barton, Allen H.

1970 Communities in Disaster: A Sociological Analysis of Collective Stress Situations. Garden City: Anchor Books.

Baxter, Vern and Steve Kroll-Smith

2005 Napping at Work: Shifting Boundaries between Public and Private Time. Current Sociology 53(1):33–55.

Bertha

1906 Letter from Bertha in Berkeley, California to Ellsa Billerbeck in Milwaukee, Wisconsin. The Bancroft Library, Berkeley, CA.

Bolton, Marie

1997 Recovery for Whom?: Social Conflict After the San Francisco Earthquake and Fire, 1906–1915. PhD Dissertation, Department of History, University of California Davis.

The Brookings Institution Metropolitan Policy Program

2005 New Orleans After the Storm: Lessons from the Past and a Plan for The Future. Washington DC: The Brookings Institution.

The Brookings Institute Metropolitan Policy Program

2008 The New Orleans Index: Tracking Recovery of New Orleans the Metro Area. The Brookings Institution Metropolitan Policy Program & Greater New Orleans Community Data Center

Carnegie, Andrew

1962[1900] The Gospel of Wealth. Cambridge: Harvard University Press.

Cashman, Sean Dennis

1993 America in the Gilded Age. New York: NYU Press.

Coser, Lewis
2004 Class. The Dictionary of the History of Ideas, Electronic Text Center, Charlottesville, Virginia: 1–21.

Daniels, Ronald J., Donald F. Kettl, and Howard Kunreuther, eds.
2006 On Risk and Disaster: Lessons from Hurricane Katrina. Philadelphia: University of Pennsylvania Press.

Deacon, J. B.
1918 Disasters and the American Red Cross in Disaster Relief. New York: Russell Sage Foundation.

Douty, Christopher M.
1977 The Economics of Localized Disasters: The 1906 San Francisco Catastrophe. NY: Arno Press.

Dumont, Louis
1981 Homo Hierarchicus: The Caste System and Its Implications. Chicago: University of Chicago Press.

Dyer, Christopher
1999 The Phoenix Effect in Post-Disaster Recovery. *In* The Angry Earth: Disaster in Anthropological Perspective. Anthony Oliver Smith, ed. Pp. 278–300. New York: Routledge.

Gurney, Patrick
1977 The Therapeutic Community Revisited. Preliminary Paper, 39. Newark: Delaware Disaster Research Center.

Harvey, David
2005 A Brief History of Neoliberalism. New York: Oxford

Henderson, Andrea Davies
2005 Reconstructing Home: Gender, Disaster Relief, and Social Life after the San Francisco Earthquake and Fire, 1906–1915. PhD Dissertation, Department of History, Stanford University.

Hewitt, Kenneth
1983 The Idea of Calamity in a Technocratic Age. *In* Interpretations of Calamity from the Viewpoint of Human Ecology. Kenneth Hewitt, ed. Pp. 3–32. Boston, MA: Allen and Unwin.

Hooker Family Papers
1906 Manuscripts and Archives, Yale University Library.

James, William
1899 What Makes a Life Significant? *In* On Some of Life's Ideals. William James, ed. Pp. 49–94. New York: Henry Holt and Company.

Krausse, Gerald, and Christopher L. Dyer
2000 There's A Big Wind A Comin': A Profile Of Survival And the Culture of Response after Hurricane Mitch, on Isla Guanaja, Honduras. Quick

Response Report, 123. Natural Hazards Research and Applications Information Center, University of Colorado.

Kroll-Smith, J. Stephen, and Stephen Couch
1990 The Real Disaster Is Above Ground: A Mine Fire and Social Conflict. Lexington: University Press Kentucky.

Lewin, Kurt, ed.
1951 Field Theory in Social Science: Selected Theoretical Papers. New York: Harper & Row.

Massey, Douglas S.
2007 Categorically Unequal: The American Stratification System. New York: Russell Sage Foundation.

McGerr, Michael E.
2003 A Fierce Discontent: The Rise and Fall of the Progressive Movement in America, 1870–1920. New York: Free Press.

Morris, Charles
1906 The San Francisco Calamity by Earthquake and Fire. Urbana: University of Illinois Press.

Painter, Nell
1987 Standing at Armageddon: The United States, 1877–1919. New York: W. W. Norton.

Pan, Erica Y. Z.
1995 The Impact of the 1906 Earthquake on San Francisco's Chinatown. New York: P. Lang.

Polanyi, Karl
1944 The Great Transformation: The Political and Economic Origins of Our Time. New York: Rinehart & Company.

Quarantelli, E. L. and Russell R. Dynes
1977 Response to Social Crisis and Disaster. Annual Review of Sociology 3:23–49.

Rozario, Kevin L.
1997 Nature's Evil Dreams: Disaster and America, 1871–1906. PhD dissertation, Department of History, Yale University.

Russell Sage Foundation
1913 San Francisco Relief Survey: The Organization and Methods of Relief Used After the Earthquake and Fire of April 18, 1906. New York: Survey Associates, Inc.

Strupp, Christoph
2006 Dealing with Disaster: The San Francisco Earthquake of 1906. Paper #060322. Institute of European Studies, University of California Berkeley.

Sullivan, Teresa A., Elizabeth Warren and Jay Westbrook
2001 The Fragile Middle Class in Debt. New Haven: Yale University Press.

Thomas, Gordon, and Max M. Witts
1971 The San Francisco Earthquake. New York: Stein and Day.

Turner, Victor
1967 The Forest of Symbols: Aspects of Ndembu Ritual. Ithaca, NY: Cornell University Press.
1969 The Ritual Process: Structure and Anti-Structure. Chicago: Aldine Publishing Co.

United Nations' International Strategy for Disaster Reduction (UN-ISDR)
2007 Terminology on Disaster Risk Reduction, Introduction: Basic Terms of Disaster Risk Reduction. Electronic document, www.adrc.or.jp/publications/terminology/top.htm, accessed August 30, 2007.

Veblen, Thorstein
1899 The Theory of the Leisure Class: An Economic Study in the Evolution of Institutions. New York: MacMillan.

Wilson, James R.
1906 San Francisco's Horror of Earthquake and Fire: Terrible Devastation and Heart-Rending Scenes. Philadelphia: Memorial Publishing Company. Percival Supply Company The 1906 San Francisco Earthquake and Fire Digital Collection, The Bancroft Library, University of California, Berkeley.

III

THE LINE BETWEEN HAZARD AND DISASTER FOR PRIMARY PRODUCERS

Weak Winters: Dynamic Decision-Making in the Face of Extended Drought in Ceará, Northeast Brazil

6

DONALD R. NELSON AND TIMOTHY J. FINAN

DROUGHT IS A CONTINUOUS HAZARD in semiarid regions where livelihoods are sensitive to environmental variations. The circumstances under which a hazard becomes a disaster are, however, mostly determined by a set of socioeconomic factors (Comfort et al. 1999; Oliver-Smith 1996). The literature on social and climate vulnerability provides a framework for explaining the transition of a hazard into a disaster and the consequent variable distribution of impacts (Blaikie et al. 1994; Thomalla et al. 2006). In the semiarid environment, hazards are environmental elements that defy time—they are always present, but disasters occur in time-specific situations and are intimately related to non-environmental elements defined in terms of the varying relationships that differentiate society. They are the outcomes of cumulative decisions taken over many years, and often reinforce the conditions that will permit the next disaster (Comfort et al. 1999).

The first historical droughts in northeast Brazil were recorded in the early 1500s, shortly after discovery and settlement by the Portuguese, and official statistics have documented 20 droughts over the last 100 years (Carvalho 1994; Marengo 2007).[1] Although drought-based disasters had occurred previously, the 1877–1879 drought changed the course of history. It is generally accepted that over 500,000 people perished in the northeast from starvation and pestilence, a mortality rate so staggering that it awakened the awareness of the entire nation. In response to this disaster, the federal government assumed the responsibility for solving the problem of drought.

Since this time, a major debate in northeast Brazil has focused on how to minimize the disaster potential of drought, and two distinct camps have emerged. The one, dominant in the political discourse, focuses directly on combating the hazard itself. Thus government has invested consistently in water reservoir and storage systems, cloud seeding, climate forecasting, and a range of measures aimed at improving agricultural technologies. The second camp, dominant in academic circles, focuses on the social relations that characterize the vulnerabilities throughout the region. This position stresses the variability in access to resources, the concentration of wealth and income, and the patron/client relationships that maintain power in the hands of the elite few.

These arguments echo similar debates in the wider academic and practitioner literature. On one hand is the viewpoint that disasters stem from events that are out of human control. On the other hand is the now more widely accepted position that disasters are the result of larger social structures and processes as well as human-environment relations (Burton et al. 1978; Hewitt 1997; Oliver-Smith 1996). This political economic perspective is essentially about the understanding of social vulnerability (Oliver-Smith 1996). It considers vulnerability as a function of the characteristics of the climate event and the population's economic and social standing (Moran et al. 2006; Ribot et al. 1996).

While both camps in northeast Brazil might share a similar goal—that of eliminating the scourge of drought—an integrated, coupled approach has not been achieved. Since the nineteenth century tragedy, public resource allocation has gradually succeeded in reducing the disaster component of drought to the point that even the most severe drought does not result in mortality. On the other hand, the public expenditure in drought mitigation for each episode has risen in direct proportion to the drop in mortality. The need for ongoing and increasing public expenditure reflects the persistent exposure of households to drought and their incapacity to respond, which together transform hazards into disasters (Nelson and Finan 2007).

Scope of Study

It is clear that disaster in northeast Brazil (see figure 6.1) is a function of government intervention and underlying vulnerabilities. In this chapter, we examine the nature of these vulnerabilities and seek to identify variations in the experience of drought which articulate the range of human agency in relation to the participation of government in drought mitigation. We recognize the role of the environment, but also acknowledge the importance of choices and the processes that create and constrain choice (Minnegal and Dwyer 2000). Local populations have developed strategies of adaptation to semiarid conditions which are challenged during climate extremes. The success of such strategies is directly influenced by underlying vulnerabilities; and the government, rather than expand the range of alternative strategies, has traditionally provided the emergency safety net.

In rural Ceará, much of the population is directly dependent on rainfall for basic needs. Rains nurture the crops, which provide food and job opportunities; rains restore the pastures for livestock; and rains fill the reservoirs and cisterns that are the principal repositories of drinking water. Given the direct impacts of rainfall variability on households, people have learned to manage their resources under regimes of high uncertainty.

Figure 6.1. Map of Northeast Brazil. **[AQ1]**

Household livelihood strategies in the region reflect the variable nature of the climate system as well as the specific social and ecological context in which they are practiced. The threshold that marks the transition point from drought to disaster is determined by the interactions between this human agency, the physical environment, and social relations.

To illustrate this argument, this chapter documents a drama that extended over three years in two *municípios* in the state of Ceará.[2] The period of time, 1997–1999, encompassed a period of three years that began with a year of "normal" rainfall followed by a very severe El Niño drought year and then a year of below average rainfall. We use an economic metric to assess the differences in the interaction of human agency and government intervention over a sample of 52 households. In so doing, the study tests the hypothesis that the limits of human agency and the need for government intervention are directly related to underlying vulnerabilities, which themselves are analyzed below. The implications of this argument focus on how government intervention can be directed more effectively toward vulnerability reduction as a means of minimizing the potential for disaster. We begin with a description of the physical and economic characteristics of the research area, followed by basic profiles of the vulnerability of the sampled households. The third section traces the evolution of the drought across the time period in question and documents the response both by households and by government. In the final section, we identify the key findings of the research and interpret what these mean from a policy perspective.

The Context

The physical landscape of Ceará is characterized as part of the unique *caatinga* biome, defined by the constant temperatures, variable rainfall, and high rates of evapotranspiration. The scrub forest vegetation grows in shallow, rocky soils, which are formed over crystalline bedrock, prohibiting the infusion of rain, and even minimal anthropogenic disturbances can lead to significant erosion. The state sits squarely within what is known as the drought polygon; a region of nearly 1,000,000 square kilometers and 9 states, over 86 percent of which is semiarid. While certainly not the only region in the country to experience drought, the frequency, severity, and socioeconomic impacts of the droughts throughout the years serve as the basis for its infamous characterization as an area of extreme suffering.

The arrival of an official drought year is first heralded in the headlines of the state newspapers; only one word is needed to convey the overpowering sense of anguish, suffering, and acute distress—*seca*. The word drought not only describes the physical and social situation in the hinterlands, but it also evokes thoughts of the humanitarian relief mechanism that responds to the needs of the population. In official drought years, vast amounts of resources are infused into the system to stem the human suffering. In 1998, for example, public relief efforts were valued at the equivalent of 10 percent of Ceará's state budget (Governo do Estado do Ceará 2005). However, reflecting the high level of spatial and temporal variation of rainfall, localized droughts occur somewhere within the state almost every year. These localized droughts do not trigger the relief mechanism and, as a result, according to our informants, may have larger impacts on the population than do the officially designated droughts.

The rainfall needs of the rural population are not adequately captured in annual precipitation summaries—although they are used to characterize official droughts—but rather precipitation requirements are best understood in terms of the timing of the rains. The primary staple crops in the region, produced under a rain-fed regime, are corn, beans, and manioc. Each of these crops has specific water needs for each stage in its growth cycle. So, although the inter-annual variation has significance, particularly for water managers, the intra-annual timing of the rain has a very telling impact at the local level. Dry periods are common climate features in Ceará, and dry periods of ten days or more are capable of destroying an entire crop even during years with high rainfall totals.

The state generally receives four to five months of rain per year, and depending on the geographic region within the state, the rains begin in December or January and continue through until April or May. In general, planting begins with the first rains and may continue until March 19, the day of Saint Joseph, which is considered by traditional consensus to be the last viable planting date. The bulk of the rain normally falls in March and April, yet due to the variable nature of the climate this is far from guaranteed. Figure 6.2 shows the amount of rainfall variation that occurred in the município of Parambu during the month of March from 1979 to 2006. Although March is typically one of the wetter months, the graph demonstrates a variation of 15 to 400 millimeters. Successful household livelihood strategies must accommodate this extreme variation.

The social landscape of the state is characterized by intense poverty and, in some rural areas, 75 percent of the population lie beneath the poverty line. This poverty is associated with high levels of concentration of wealth (Gini coefficient of inequality is high at 0.68), low educational achievement, and minimal social mobility (PNUD and IPEA 2003). Nearly one

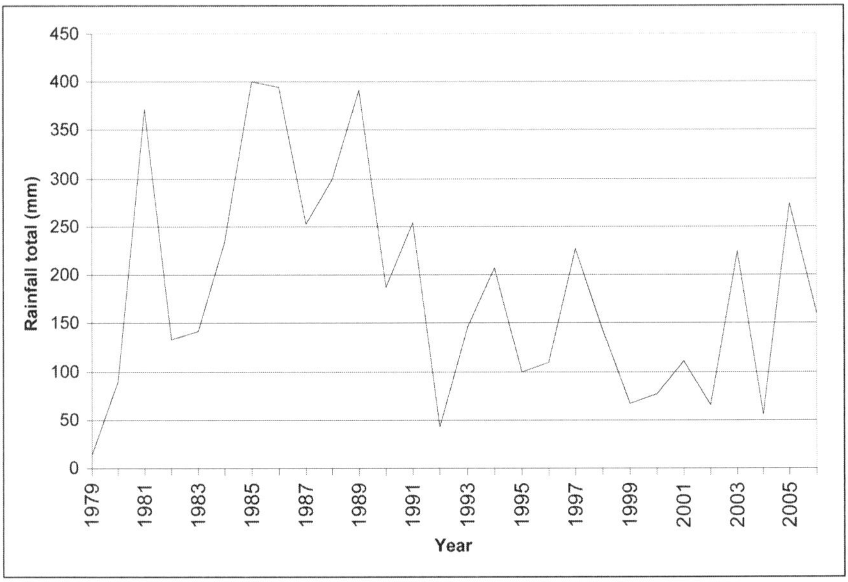

Figure 6.2. Average March Rainfall (mm) for Parambu, 1979–2006.

million people—48 percent of rural workers and 31 percent of the state's entire workforce—are currently involved in the agricultural sector, which contributes only 5 percent of the state gross domestic product (Araújo and Falcão 2004). The vast majority of agricultural production is rain fed, and the 15 percent of cultivated land that is under irrigation is not evenly distributed throughout the state. There is no naturally occurring perennial surface water in the state, but rainfall captured in both small and large reservoirs throughout the state is fed into several valleys to provide water for drinking and irrigation. Large-scale irrigated perimeters and shallow wells are concentrated in small regions along the courses of these river valleys.

Our analysis is based on a series of field visits between 1997 and 2000. During the first year we conducted interviews with individuals in 484 households in 6 municípios throughout the state. The municípios of Limoeiro do Norte and Boa Viagem represented the range of climatic, social, and agricultural variables identified in the entire sample and we revisited these 2 municípios to conduct further interviews with 52 randomly selected households that were part of the original sample.[3] Limoeiro do Norte is located in the Jaguaribe River Valley, an area that permits intensive irrigated agriculture since the Jaguaribe River has become perennialized as the result of investment in large-scale dams and reservoirs. The município is comprised of three distinct sub-regions: irrigated highlands, lowlands located in the alluvial plain, and the caatinga, one of the driest and poorest areas of the entire sample region. Boa Viagem sits in the *sertão-central*, a rolling, dry region that occupies the largest part of Ceará. The hilly, rocky nature of the landscape limits the types of agricultural technologies that can be employed. There is very little irrigation in the region, and livestock plays a larger role than in Limoeiro do Norte.

The year 1997 was officially labeled a drought year by FUNCEME based on average precipitation for the entire state.[4] That same year Limoeiro do Norte received 55 percent of the annual average and Boa Viagem received approximately 20 percent less rainfall than the average. Nevertheless, less than 10 percent of the sample population classified 1997 as a drought year. The discrepancy between the official classification and local perceptions highlights the difficulty in using a purely meteorological definition of drought. For farmers, drought is defined by the impact that the timing of the rains has on crop production rather than quantity of rainfall. Thus, based on farmers' interpretations, the analysis uses 1997 as a baseline year to examine changes in 1998 and 1999, both of which were considered drought years by the respondents.

Household Vulnerability to Drought

In this study, disaster is defined as a function of vulnerability and hazard (Blaikie et al. 1994; Tompkins et al. 2005; disaster = *f* [vulnerability, hazard]), and a situation is considered a disaster when there are large numbers of associated deaths or refugees (IFRC 2002). Hazards are potentially destabilizing events and may be social, economic, or natural perturbances. In this case, the hazard is drought, characterized by the timing and quantity of rainfall as it relates to agricultural production. For its part, vulnerability reflects the sensitivity to a hazard as well as the capacity of people to respond (Finan and Nelson 2001) and represents a measure of the susceptibility of the household to the hazard (Adger 2006; Eakin and Luers 2006). Both sensitivity and the capacity to respond are determined by the relationships of people to their specific environment and to broader political economic forces (Oliver-Smith 2004). We argue that households in Ceará respond differently to the risk and the occurrence of drought depending on their level of exposure, sensitivity, and capacity to respond—that is, their level of vulnerability.

In this analysis, we assign each of the 52 households to one of three vulnerability categories, which are measured in terms of an economic metric. It is important to note that vulnerability is a relative measure among the categories and, as we will show, even the least vulnerable households demonstrate sensitivity to drought events. Thus, we use the value of rainfall-dependent household resources to indicate levels of exposure, while sensitivity is measured by the impact this income loss would have on the household. We discuss the capacity to respond in our discussions of human agency and drought impacts.

We have classified all income sources, not derived from crop and livestock sales, as climate-sensitive, climate-neutral, or social transfers. Climate-sensitive income refers to casual agricultural labor, which is dependent on the quality of the agricultural season. Although drought affects all aspects of the rural economy, agricultural labor represents the front line, so to speak, and is the most directly and severely impacted. Forty-six percent of the households received climate-sensitive income in 1997 (table 6.1). Climate-neutral income refers to a variety of income sources that are not directly impacted by drought, and slightly more than half the sample had access to climate-neutral employment. These sources include public service employment, permanent employment in factories, commerce, and skilled and unskilled (nonagricultural) labor, both full-time and part-time. Social transfers are defined as economic assistance by the state or federal government of Brazil. Social security (pension) payments were the only form of social transfers in 1997, but in subsequent years these transfers also included public work front salaries. In 1997, 27 percent of the households had at least one member receiving social security. Seven of the households (14 percent) did not report any income other than that generated by crop and livestock production and were classified as part of the most vulnerable group.

The most vulnerable category includes those nonirrigating households who would fall beneath the indigence line if they lost all of their climate sensitive income.[5] Those in the middle category would lie below the poverty line with the loss of climate-sensitive income or are irrigators who would otherwise be in the most vulnerable category.[6] The least vulnerable households in the sample are those that have sufficient income to remain above the poverty line even if their climate-sensitive income was reduced to zero.

In 1997, the most vulnerable households have a greater participation in climate-sensitive activities and their average level of annual revenue from these activities is significantly lower than with the other groups (table 6.1). Table 6.1 also demonstrates that the mean revenue value for the other two types of activities is lowest for the most vulnerable households, thus allow-

Table 6.1. 1997 Annual Income Sources by Vulnerability Category

	Climate-Sensitive		Climate-Neutral		Social Transfers	
Categories	Percent of Households in Sample	Mean Income ($R)	Percent of Households in Sample	Mean Income ($R)	Percent of Households in Sample	Mean Income ($R)
Most	54	424	40	794	20	1,766
Medium	27	860	82	1,813	27	2,080
Least	33	1,476	67	5,795	67	3,510
Limoeiro do Norte	45	870	60	1,406	45	2,240
Boa Viagem	47	384	47	2,249	16	2,496
Average	46	566	52	1,874	27	2,331

ing the conclusion not only that more of these households are exposed to drought but also that their livelihoods are the most precarious in general. The presence of a social security pensioner in a household tends to reduce vulnerability (depending upon household size), since it provides a climate-neutral source of income. Only 20 percent of the most vulnerable households are pension recipients.

What distinguishes income levels between Limoeiro do Norte and Boa Viagem is the prevalence of irrigation. So, while the reliance on climate-sensitive revenue is similar in the two municípios, the average revenue stream is significantly higher in the irrigated município of Limoeiro do Norte due to a larger demand for agricultural labor. Limoeiro is also a smaller, more urban município with more climate-neutral employment opportunities—both skilled and unskilled. The greater participation of Limoeiro households in pension income is possibly due to demographic factors or to the presence of an active rural worker union.[7] The presence of irrigation is reflected in the vulnerability distribution across the two municípios (table 6.2). In Limoeiro do Norte, 55 percent of the households are in the most vulnerable category, while in Boa Viagem three-quarters of the households fall into the most vulnerable group.

Table 6.2. 1997 Vulnerability Categories by Municípios

	Limoeiro do Norte		Boa Viagem		Total	
Category	n	Percent	n	Percent	n	Percent
Most	11	55.0	24	75.0	35	67.3
Medium	7	35.0	4	12.5	11	21.2
Least	2	10.0	4	12.5	6	11.5
Total	20	100	32	100	52	100

Table 6.3. Livelihood Assets by Vulnerability Category

Vulnerable	*n*	*HH Size*	*Literacy Rate HH Heads*	*Mean Income Earners*	*Per capita Annual Income (R$)*	*Land-Owner (percent)*	*Irrigated Land (ha)*	*Cattle Owners (percent)*
Most	35	5.1	69	1.3	498	49	0.0	51
Medium	11	4.4	82	1.8	1,094	91	0.8	64
Least	6	4.3	83	2.8	2,096	100	0.9	67
Total	52	4.8	73	1.6	808	63	0.3	56

HH = Households; ha = hectares

Household Livelihoods

Vulnerability, as posited above, varies directly with the characteristics of household livelihoods, particularly resource endowments upon which households lay claim (Adger and Kelly 1999; Blaikie et al. 1994). Those households with greater asset packages will be better positioned to respond to a hazard such as drought. Assets accumulated during good years can be divested during drought and thus can serve as self-insurance policies for drought events (Corbett 1988). They also serve as buffers to shocks, providing more time and flexibility of response (Vásquez-León et al. 2003). In Ceará, as in the rest of the northeast, the major livelihood assets are labor, land, livestock, and income capital. Table 6.3 reveals important differences among the vulnerability categories with regard to these assets. Household size in the most vulnerable households is larger, but the number of contributors to household revenue is smaller. Also, using household head literacy as a proxy indicator suggests that the level of educational achievement in the most vulnerable households is decidedly lower. Land, livestock, and technology (irrigation) assets are also more highly concentrated among the least vulnerable.

Land has traditionally been the symbol of inequality and asset disparity in northeast Brazil. Farmers use a variety of mechanisms for land access, including ownership, sharecropping, or various forms of usufruct rights; however, reflecting historical imbalances in ownership, the primary mechanism is sharecropping. While some families own and work their own land, the farm size is not enough to support the household, and these farmers tend to increase their cultivated area though sharecropping the land of others. The most frequent form of land payment is a portion of production, and the value of agricultural land rent has steadily decreased from approximately 30 percent of production to around 15 percent. The decrease is primarily associated with diminished demand for land as a result of rural-urban migration. The dynamics of land access are also indirectly

Table 6.4. Land Access and Area Cultivated

Categories	*Sharecroppers (Percent of Households)*	*Usufruct (Percent of Households*	*Owner* (Percent of Households*	*Hectares Cultivated (Mean)*
Most	45.7	20.0	48.6	5.1
Medium	27.3	9.1	90.9	4.6
Least	0.0	0.0	100.0	12.6
Limoeiro do Norte	35.0	20.0	80.0	7.4
Boa Viagem	37.5	12.5	53.1	4.9
Average	36.5	15.4	63.5	5.9

*Because many households may have more than one strategy for accessing land, these numbers may not add up to 100.

related to vulnerability. Landowners normally do not allow their tenant sharecroppers to maintain large animals or to use the crop residue to feed their own cattle. Thus, sharecropping, despite low rents, limits the diversity of possible production activities of the sharecropper household. Table 6.4 shows that the most vulnerable households tend to be sharecroppers and cultivate smaller plots of land.

Income-earning strategies are comprised of different combinations of informal and formal employment, small businesses, pension payments, woodcutting, migration, and the sale of crop and livestock products. Informal employment—mostly found in agriculture—is the least reliable and poorest paid source of income. Formal nongovernment employment and small business revenue are much less common. Woodcutting is a common strategy that gains importance during the onset of drought and the corresponding decrease in demand for agricultural labor. The *prefeitura* is, with rare exception, the largest employer in rural areas, and its impact is evident in the fact that it employs individuals from 23 percent of all the households.[8] Such prefeitura employment varies widely from well-paid agency heads to lower paid positions as secretaries, teachers, health agents, and school cooks. While the income from public employment helps buffer families from drought, the jobs are doled out at the whim of the *prefeito* and continue only as long as the administration remains in power. It is common for an incoming prefeito to dismiss all current employees and fill the positions with political supporters. In this sample of households, the more vulnerable households experience half the rate of participation in public employment compared to the others.

Migration has been the traditional solution to severe drought, and stories of *flagelados* (fleeing victims) are seared in the social memory of the region. Today it is difficult to separate the rural-urban exodus from a drought-related demographic shift. One-third of all households in the sample have at least one family member that has emigrated, although the causes are unclear. Some families have members that practice a seasonal migration to participate, for example, in the sugar cane harvest in southern Brazil. Regardless of the motive for migration—either temporary or permanent—the remittances in cash and kind contribute significantly to the household livelihood for households involved.

In the context of human capital, land, and income-earning capacity, the most vulnerable households suffer a clear disadvantage within the sample. Their livelihoods display a structural fragility that increases their level of sensitivity to drought events. Nonetheless, even the most vulnerable household seeks to devise risk management strategies that add some sense of security and certainty to their lives. The potential (and limit) of this human agency is now the focus of our analysis.

Human Agency and Drought

In our analysis, we have defined disaster as a function of hazard and vulnerability, and it is now necessary to explore the dynamic interactions between these two concepts. The intersection of hazard and vulnerability has two components, which we will call "private response" and "public intervention." The former can be thought of as the decision range that is available to households, or the level of human agency that has emerged as part of a broader adaptive process. Over the generations, rain-fed (and irrigated) households in rural Ceará have devised strategies that are effective during most years. These strategies, while not affording a luxury existence, have met the litmus test of survival, and they demonstrate the universal ingenuity of human beings confronting a difficult environment. We assert here, however, the range of human agency is shaped by social conditions and defined by environmental extremes, to which there is little or no response, and beyond which disaster looms. Our analysis seeks to identify these limits, and to relate them to the point in time that the government intervenes to avoid the disasters that occurred in the past—often, as we have stated, at great cost to society. Thus, we argue, vulnerability is ultimately determined by the underlying factors that define human agency and access to the public sector in time of crisis.

The fundamental element of human agency in highly variable environments is the capacity to diversify (Ellis 2000). In Ceará, households employ diversification strategies in terms of their technological choices and in terms of their income earning activities. One such example lies in how farmers plant. All households practice agricultural risk management strategies that demonstrate a rather sophisticated adaptation to drought. These strategies are a combination of experiential knowledge and activities promoted by the government agriculture and extension agencies. Farmers grow corn, beans, a variety of cucurbits, and sometimes cassava in a rain-fed regime, with beans being the principal home consumption crop. The range of human agency in rain-fed agriculture involves decisions of when and where to plant, how much to plant, which seeds to use, how to feed livestock, and how to allocate household labor on and off the farm, including migration.

Both sophisticated forecasting systems and empirically rooted local rain prophets prognosticate the arrival of the *inverno* (winter) rainy season, and rural society engages in a collective angst as January draws near. Public discourse focuses on the prospects for a "weak winter" or "strong winter," and the media dedicates daily coverage to the topic. The capricious nature of the rainfall means that planting dates may change by months, but the standard practice is for crops to go into the ground once the soil is moist to a depth of one *palmo*—a measure from the tip of the thumb to the tip of the little finger of an open hand, which is approximately equivalent to 20 centimeters. Thus, one significant rain will signal the beginning of the agricultural cycle for many households. This decision may reflect such factors as crop and soil type, but it is also related to the level of dependence on agriculture. The more dependent a household, the more likely the farmer is to plant with the first rains. Households that have more neutral income or depend to a greater degree on social transfers are more likely to plant later in the season.[9] Early planting increases the risk of having to replant, but planting later increases the risk of no harvest. Thus, some households plant with the first rains that normally fall in late December or January, while others—citing the likelihood of dry spells in January and early February—tend to wait until February or March to plant. If the rains have not arrived by March 19—Saint Joseph's Day—the year is considered a drought year by the farmers.

Farmers may also vary the planted area and the seed types. One strategy is to plant only a portion of the field each day for several weeks which both spreads out the risk of an entire crop being wiped out by a short dry spell and provides access to fresh production for a longer period of time. It is also common to diversify soil type, planting in both the moister clay soils and in the higher and sandier soils. While the public and private sectors have made available drought-tolerant, high-yield varieties of beans and corn, farmers also do their own seed selection and carefully maintain a stock of locally recognized varieties that have different production and tolerance characteristics. For example, there are highly precocious varieties of the genera *Vigna* and *Phaseolus* that produce under minimal moisture conditions, but whose total productive cycles are short. In contrast, other *Vigna* varieties begin production later in the cycle, but will continue producing as long as there is moisture.

With regard to livestock, most families keep small numbers of goats, pigs, and chickens, mostly for consumption. The wealthier households also raise cattle, which are prized for consumption value (milk) and monetary value. Sharecroppers, however, are usually not permitted to maintain any large livestock, because the landowner also has cattle and reserves access rights to all pasture, including crop fodder. If sharecroppers do maintain cattle, the size of the herd is limited by the amount of fodder that can be manually foraged. This social constraint limits the diversification potential of landless farm families.

Income diversification is critical to most rural rain-fed families, and they seek to allocate household labor in many different ways. We have discussed how employment opportunities might be considered climate-sensitive versus climate-neutral. During a normal winter, the demand for labor is significantly increased in rural regions, as larger farmers seek day laborers for land preparation, planting, and other agricultural tasks. Labor demands for livestock management are also increased. During the dry season, families may seek to cut firewood for sale or, where available, work in some extractive industry (e.g., quicklime or charcoal). Seasonal migration during the off-production part of the year is also common for some households.

These representations of human agency in the diversification of technology and income constitute an adaptation to a semiarid environment on an average year. They do not, however, assure livelihood protection during an extreme year. Furthermore, due to underlying vulnerabilities determined by livelihood assets, human capacities, and access to resources, the limits of human agency vary greatly in terms of the ability to stave off the disaster situation. Households in the most vulnerable category face specific disaster thresholds that are much different than those encountered by the least vulnerable. For example, the most vulnerable tend to be landless, to have less education and employable skills, and to be more confined to climate-sensitive forms of income-earning activities.

It is possible to estimate the level of human agency in monetary terms by adding the implicit value of agricultural and livestock production to reported levels of income from different sources. Table 6.5 shows the total monetary value of household activity for 1997 (considered by most to be an average

Table 6.5. Total Economic Value of Household Production (1997)

Categories	*Total Value (R$)*	*Monthly per capita (R$)*	*Agriculture Production (R$)*	*Agriculture as Percent of Total Value*
Most	1,763	35	563	33
Medium	3,723	81	890	20
Least	8,927	162	1,342	18
Limoeiro do Norte	3,576	68	1,008	29
Boa Viagem	2,895	54	543	28
Total	3,157	59	722	28

year in these municípios). As the table demonstrates, the most vulnerable families have both the lowest monetary value and the highest dependence on agriculture as a source of income. Also of note, the value of agricultural income in the município of Limoeiro do Norte averages twice that of the rain-fed município, indicating the importance of irrigation on family income. On average, the less vulnerable households have much greater access to nonagricultural income and a smaller reliance on agriculture.

Weak Winters Arrive: The Impacts of the 1998–1999 Droughts

Northeast Brazil displays a strong El Niño signal, and 1998 was a particularly severe event throughout the world. In Ceará, there was generally significant rainfall in January of that year, and many farmers planted early. The rains then ceased and restarted in March only to disappear again in April. Crop production was nearly totally compromised, and the natural pastures did not produce much forage. In addition, the critical reservoirs for drinking water did not fill, and many small towns found themselves with precarious supplies of water for human consumption. With reference to table 6.5, predictable changes occurred to the income profiles, with climate-sensitive income and agricultural income dropping precipitously in 1998 and 1999. Livestock income maintained a stable level as many households decided to sell their animals as a coping strategy. Climate-neutral income remained stable, as expected, and pension income actually improved due to an increase in the national minimum salary.

Figure 6.3 draws upon the data from 1997 to offer a conceptual model of the relationship between vulnerability and hazard. As a caveat, we state initially that it is difficult to quantify how an individual household experiences the process of disaster, since it is a dynamic rather than a static condition. In figure 6.3, we posit a disaster threshold realizing that it is a

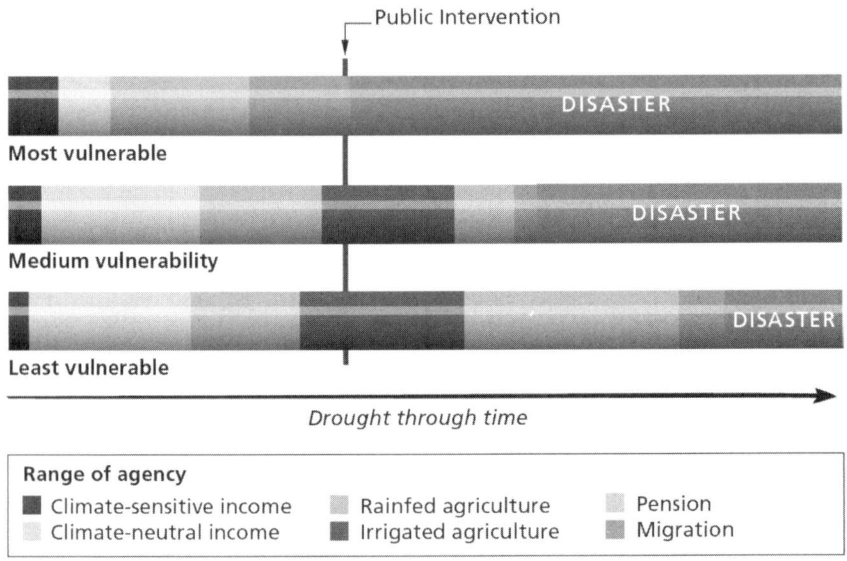

Figure 6.3. Range of Agency, Public Intervention, and Disaster Thresholds through Time by Vulnerable Category (Bands in Order as Shown in List, 1st - Climate Sensitive Income, 2nd - Climate Neutral Income, etc.).

heuristic devise, since people do not wake up one morning in a disaster mode, as they might in the case of a hurricane. In reality, a family without food and inadequate water would likely be taken in by another or move in with a nearby relative in the city thus surviving, but the livelihood has become severely compromised and, in some cases, irreversibly so.

The horizontal bars in figure 6.3 contain the sources of agency for an average household in each of the vulnerability categories. As the drought unfolds the sources become scarcer. For example, climate-sensitive income and agricultural production are most sensitive to drought and will diminish rapidly. Climate-neutral income is less sensitive, and pensions are unchanged during droughts. Nevertheless, the value of these two income sources must be stretched to cover the losses. This pushes the household closer to the disaster threshold. The length of each income-source box reflects the value and relative importance of the income within each vulnerability category. Migration and irrigation are two additional sources that can help lengthen the period of time before disaster becomes inevitable. The lengths of the migration and irrigation boxes represent the relative importance of these strategies between vulnerability categories.

The most vulnerable households, even in 1997, are found perilously close to the disaster threshold, with over two-thirds of their total income dependent on rainfall. In comparison, the less vulnerable households have a much reduced disaster threshold due to a greater set of livelihood options and a more unrestricted level of human agency. They consequently can withstand a hazard over a longer period of time. The area to the right of the solid vertical line in figure 6.3 in effect represents the implicit value of public intervention that prevents disaster from reaping its cruel harvest as it used to do. Access to public emergency relief acts to replace the income lost to the hazard and its impacts. This dynamic between human agency, in the form of allocative and mobilizing choice, and public intervention was dramatically played out during the 1997–1999 time period.

In 1998, by St. John's Day (June 24)—the traditional feast day of the harvest and usually a time of great merriment—it was clear that a crisis situation had spread across the interior of the state. Rural populations began to block off the main federal north-south artery that connects Ceará with the south of the country as well as a federal east-west highway. Merchants in rural towns began to feel threatened, and there was a general clamor for water tanker deliveries—all of this was captured and reported on the national media. As is the case with emergency drought situations, the various levels of government mobilized their emergency programs, which begin with local declarations of emergency, state requests to the federal government for relief funds, federal appropriations, then the channeling of resources through the state government to the prefeituras to the drought victims (Finan and Nelson 2001).

While based on household vulnerabilities we would expect that the drought would wreak havoc with lives and livelihoods, the data presented in tables 6.6a–c demonstrate differently. In effect, the income levels for the most vulnerable increased by nearly 20 percent in 1999, as the structure of household shifted sharply to social transfers, that is, emergency relief.

During 1998, the most vulnerable households were held at the threshold of disaster, and their income levels actually improved once the emergency measures became widely implemented. For the medium vulnerability families, income levels dipped in response to the 1998 drought due to loss of agricultural production, but they recuperated in 1999, due to both climate-neutral income (human agency) and access to emergency relief (public intervention). The least vulnerable households also suffered a loss of agricultural income and have a minimal amount of access to emergency relief; however, through their coping strategies, they are able to maintain their income levels. Thus, in terms of household income, the drought did not bring disastrous consequences and actually increased for

Table 6.6a. Income Shares—Most Vulnerable*

Year	*Income Sources (percent)*				*Per capita (R$)*
	Climate-Sensitive	*Climate-Neutral*	*Agricultural Production*	*Social Transfers*	
1997	21	21	43	15	374
1998	4	19	22	55	376
1999	4	23	26	46	440

*The categories of income sources with statistical significance between vulnerability groups (Kruskal-Wallis test $p < 0.05$) are climate-sensitive 1997, climate-neutral—all years, and social transfers 1998 and 1999.

Table 6.6b. Income Shares—Medium Vulnerable

Year	*Income Sources (percent)*				*Per capita (R$)*
	Climate-Sensitive	*Climate-Neutral*	*Agricultural Production*	*Social Transfers*	
1997	9	43	33	16	976
1998	9	52	27	13	906
1999	9	53	22	16	993

Table 6.6c. Income Shares—Least Vulnerable

Year	*Income Sources (percent)*				*Per capita (R$)*
	Climate-Sensitive	*Climate-Neutral*	*Agricultural Production*	*Social Transfers*	
1997	4	31	23	41	2,188
1998	0	37	16	47	2,003
1999	0	37	12	51	2,018

some of the most vulnerable. It is now opportune to examine the structure of this public intervention.

The emergency response program was comprised of three basic strategies—direct distribution of food, distribution of drinking water by tanker truck, and public work front employment.[10] This intervention package has a long, bitter history in Ceará, one steeped in corruption and abuse, and is widely and wryly referred to as the "drought industry" (Magalhães 1993; Nelson and Finan 2007; Villa 2000). While the more egregious forms of abuse have been eliminated, emergency relief still confers great political advantage upon the local elites (Lemos et al. 2002; Tendler 1997). Although the occurrence of a drought is an expected event,

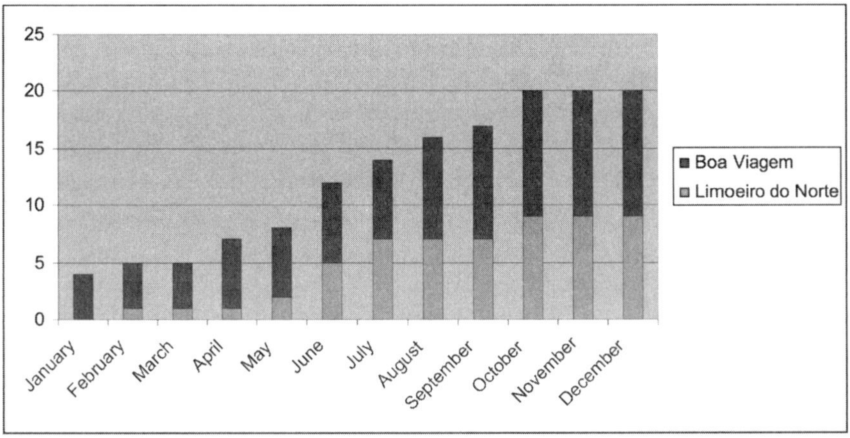

Figure 6.4. Number of Households Enrolled on the 1998 Work Front, by Month and Município.

there is no drought response plan prepared in advance at the government level. Once the first signs of drought begin to appear, the government slowly begins to mobilize. Water distribution is the first help that arrives, followed by work fronts and food baskets (*cesta básica*).

Since the most vulnerable families are at the disaster threshold, even the slightest delays in government responses can have a non-trivial impact. Figure 6.4 portrays the sequence of monthly enrollment on the public work fronts. The work fronts are federally and state-funded efforts that are responsible for a variety of infrastructure projects including drilling wells, building water reservoirs, opening and repairing roads, and constructing public buildings. Individuals work five half days per week and receive the equivalent of one-half the minimum salary, around R$60 per month. The work fronts are designed to target the most vulnerable of the population through the application of six eligibility criteria.[11] The process is far from perfect but has advanced significantly over the last ten years in response to the reform measures. In 1998, only 3 percent of the households enrolled in the program were not from the most vulnerable category; however, due to some of the excluding criteria such as the ownership of a motorcycle, 45 percent of the most vulnerable households were not eligible. Thus although there is little program leakage, the depth of the program does not fully meet the demand.

Food baskets come primarily from the federal government (some non-governmental organizations also provide food) and were distributed monthly, in principle. The monetary value of these baskets was around

R$15 in 1998 and 1999, although it varied by *município* and month to month, making the total contribution hard to quantify. In addition, delivery was inconstant and households rarely received the food donations 12 times a year. The basic food basket contained rice, beans, sugar, oil, cornmeal, manioc flour, and pasta. Even for the smaller families, the quantity of food did not last for a month and for the larger families, not even a week. Reduction in food consumption was reported in more than 70 percent of the most vulnerable households. Nevertheless, many families reported that the food basket was responsible for getting them through the toughest times.

A third of the households reported outside assistance to help with their food needs. The vast majority of this help came from family members, although some households were assisted by non-relative neighbors. This family-based assistance normally flows from parents to children and grandparents to grandchildren. Most of the older residents receive social security and use the money to help their children and grandchildren by providing either food or cash. While not quantified in the analysis, many households confirmed that their local shops extended informal credit, providing basic goods for a future payment.

In sum, all households from the most vulnerable group received some form of public assistance during the crisis. All but one family received food baskets, and 43 percent benefited from both the food basket and public work front programs. All, of course, were supplied with drinking water by tanker delivery. By contrast only one of the least vulnerable households received food. Thus, for the majority of the rural households, we can argue that the difference between drought and disaster was, in effect, social transfers from the public sector in the form of emergency relief.

The impacts of the drought cannot be measured only with our economic metric, for it does not adequately address the long-term consequences and the possibilities for recovery. While there was no massive flight from the region, no major demographic shift, and little, if any, mortality, the drought did compromise livelihoods, increasing the vulnerability of the already vulnerable. For example, with regard to cattle ownership, all the cattle-owning households sold some cattle, either to ensure their own survival or the survival of the remaining cattle. The most vulnerable households, however, were the only ones to sell all their cattle, and it was this group that reported the only cases of animal mortality. The two-year drought also exhausted the seed stocks of the most vulnerable populations. In the year 2000, some of the most vulnerable households reported that they were unable to plant their fields due to lack of seed. Of those that did plant, most planted with the first January rains, which tapered off and caused significant losses. On the other hand, the least vulnerable households were holding off planting until the rainy season "took hold." They were also planning to increase the cattle herds that were depleted in response to the drought. Thus when the rains finally arrived, ending the drought, the most vulnerable households found themselves in an even more precarious situation, but now without government assistance.

Conclusions: Disaster and Sustainability

Rural households in Ceará, northeast Brazil, endured an extended El Niño drought that put great pressure on local livelihoods during 1998–1999. Some of the households, buffered by their own resources, were able to cope using such strategies as selling off some assets, moving their cattle to pastures in the highlands or to other states, using their savings, and switching to climate-neutral employment opportunities. The most vulnerable households also survived, but through the massive flow of emergency aid. The drama of these two years poignantly demonstrates how the dynamic interaction of hazard and vulnerability determines disaster outcomes. In this case, it further illustrates the implicit tension between the private disaster response and the public role in mitigation.

There is an inverse relationship between the loss of life in a drought and the level of public expenditure. In the drought of 1877–1879, government mitigation in the tragedy of the northeast was negligible, although discourses in government did alert the nation to the plight of the region. The drought of 1958, while still marked by high rates of mortality, represents a watermark in the government commitment to eliminate mortality caused by drought. In 1983, over $350 million was spent in the northeast. During the 1998–1999 relief effort, the public government spent the equivalent of $450 million in Ceará alone to avert disaster. As a result of these exorbitant emergency investments, we assert that large-scale, drought-induced disaster has been eliminated in Ceará. Although there is still human suffering related to drought events, particularly during the drawn-out period between the failure of the rains and the first emergency responses, vast out-migrations and high levels of mortality are no longer consequences of weak winters. Nevertheless, government activities have done little to address the underlying vulnerabilities, resulting in a population that is precariously dependent on the relief efforts of the government. Any perturbation in the emergency relief mechanism will have drastic consequences on the population. Current climate projections indicate a drier and hotter sertão, suggesting, all things remaining equal, that the government will be forced to increase the level and frequency of humanitarian responses. From economic and political perspectives it is unclear how long these activities will be sustainable.

While drought is a natural phenomenon that is used to *describe* regional events, it is *experienced* at the individual level. The experience of a weak winter is determined by the range of household options and capabilities, in conjunction with natural physical features and public investments. We must, then, conclude by recalling the false dichotomy of different ways to solve the problem of drought. It is clear by examining the way in which drought unfolds among the population that any long-term, sustainable solutions require an integrated approach that addresses both the lack of water as well as the underlying vulnerabilities of the population. The latter requires public investments to increase the range of human agency during years of "normal" rainfall. It is tempting to imagine how $450 million, almost $300 per capita for all rural residents, might have been invested. Such possibilities are plentiful and include investment in health and education (also, higher education), the creation of an industrial/service sector that can absorb and improve such low-valued labor in climate-neutral activities, and the decentralization of power and decision-making into local communities themselves. Pursuit of these possibilities will require significant political will and it is our hope that changes will come as a result of foresight and planning, rather than as forced responses to changing physical and political conditions in the sertão.

Notes

1. Droughts have occurred more frequently over the last two decades, with six official droughts recorded since 1990.

2. A *município* is about the size of a U.S. county and is the local administrative political unit.

3. A t-test for equality of means indicates that there is no statistical difference in family size, per capita incomes by source ($p < 0.05$), or for the amount of land cultivated ($p < 0.10$) between the 1997 and the panel samples.

4. FUNCEME (Ceará Foundation for Meteorology and Water Resources) is the state entity responsible for providing seasonal climate forecasts and for monitoring daily precipitation throughout the state.

5. These calculations were based on an indigence threshold of less than R$45 per capita/month, and poverty is defined as less than R$90 per capita/month, as determined by the World Bank (2000).

6. The presence of irrigation does not eliminate vulnerability to drought. A common misperception is that irrigators are able to maintain production levels

even during drought years. Our analysis, however, does not support this conclusion, and most irrigators suffered losses during 1998–1999. Production is the result of many factors including the source of irrigation water, the size of the property, and the crops planted. For example, due to consecutive years of diminishing water levels in reservoirs during the drought years, regulators were forced to limit the amount of cultivated land in the irrigation perimeter.

7. The proportion of the sample receiving rural pensions is consistent with the larger sample of households surveyed in 1997.

8. The government of a município is a *prefeitura*, which is led by the local mayor, or *prefeito*.

9. In some years, farmers may replant parts of their fields up to four times or more.

10. It is important to note that there was, for the first time, a significant flow of drought relief through private channels due to the dramatization of the drought on national television. Sons and daughters from drought-stricken communities now resident in faraway places such as São Paulo and Rio de Janeiro organized relief drives.

11. These household level criteria, established by the Civil Defence, include: no one under the same roof can be receiving a social pension—unless there are more than 5 family members; no one can be a public employee—unless they live in the rural area, receive <½ minimum salary and the family size is >5; no one can be participating in a credit program; no one can have permanent employment; no one can own a motorcycle or car; no one can have more than 20 cattle or 60 sheep; and no one can own a commercial establishment.

References

Adger, W. N.
2006 Vulnerability. Global Environmental Change 16:268–261.

Adger, W. N., and P. M. Kelly
1999 Social Vulnerability to Climate Change and the Architecture of Entitlements. Mitigation and Adaptation Strategies for Global Change 4:253–266.

Araújo, A., and F. B. Falcão
2004 Desenvolvimento Regional e Combate a Pobreza. *In* Desenvolvimento Endógeno: Um Novo Paradigma para a Gestão Local e Regional [Regional Development and Combating Poverty. In Endogenous Development: a New Paradigm for Local and Regional Management.]. P. Vergara, ed. Pp. 357–375. Fortaleza: IADH.

Blaikie, P., T. Canon, I. Davis, and B. Wisner
1994 At Risk: Natural Hazards, People's Vulnerability and Disasters. London: Routledge.

Burton, I., R. W. Kates, and G. F. White
1978 The Environment as Hazard. Oxford: Oxford University Press.

Carvalho, O. D.
1994 O Impacto Social da Seca no Nordeste. Paper presented at the Conferência e Seminário Latino-Americano da Desertificação [The Social Impact of Drought in the Northeast. Paper presented at the Latin American Conference and Seminar on Desertification.]. Fortaleza.

Comfort, L., B. Wisner, S. L. Cutter, R. Pulwarty, K. Hewitt, A. Oliver-Smith, J. Wiener, M. Fordham, W. Peacock, and F. Krimgold
1999 Reframing Disaster Policy: The Global Evolution of Vulnerable Communities. Environmental Hazards 1:39–44.

Corbett, J.
1988 Famine and Household Coping Strategies. World Development 16(9):1099–1112.

Eakin, H., and A. L. Luers
2006 Assessing the Vulnerability of Social-Environmental Systems. Annual Review of Environment and Resources 31:365–394.

Ellis, F.
2000 Rural Livelihoods and Diversity in Developing Countries. Oxford: Oxford University Press.

Finan, T. J., and D. R. Nelson
2001 Making Rain, Making Roads, Making Do: Public and Private Adaptations to Drought in Ceará, Northeast Brazil. Climate Research 19:97–108.

Governo do Estado do Ceará
2005 Tableas Sobre o PIB do Ceará [GDP Tables for Ceará.]. Fortaleza, Secretaria do Planejamento e Coordenação, Instituto de Pesquisa e Estratégia Econômica do Ceará:5.

Hewitt, K.
1997 Regions of Risk. Harlow: Addison Wesley Longman Limited.

IFRC
2002 World Disasters Report. Geneva: International Federation of Red Cross Red Crescent Societies.

Lemos, M. C., T. J. Finan, R. W. Fox, D. R. Nelson, and J. Tucker
2002 The Use of Seasonal Climate Forecasting in Policymaking: Lessons from Northeast Brazil. Climatic Change 55(4):479–507.

Magalhães, A. R.
1993 Drought and Policy Responses in the Brazilian Northeast. *In* Drought Assessment, Management, and Planning: Theory and Case studies. D. A. Wilhite, ed. Boston: Kluwer Academic Publishers.

Marengo, J. A.
2007 Caracterização do Clima no Século XX e Cenários no Brasil e na América do Sul para o Século XXI Derivados dos Modelos de Clima do IPCC [Characterization of the 20th Century Climate and 21st Century Scenarios for Brazil and South America derived from the IPCC Climate Models.]. São Paulo: Ministério do Meio Ambiente, Secretaria de Biodiversidade e Florestas, Diretoria de Conservação da Biodiversidade.

Minnegal, M., and P. D. Dwyer
2000 Responses to a Drought in the Interior Lowlands of Papua New Guinea: A Comparison of Bedamuni and Kubo-Konai. Human Ecology 28(4):493–526.

Moran, E. F., R. Adams, B. Bakoyéma, S. Fiorini T., and B. Boucek
2006 Human Strategies for Coping with El Niño Related Drought in Amazônia. Climatic Change 77:343–361.

Nelson, D. R., and T. J. Finan
2007 Persistent Vulnerability: The (False) Security of Ironic Adaptation to Drought in Ceará, Brazil. Paper presented at the Annual Meeting of the Society for Applied Anthropology, Tampa, FL, March 29.

Oliver-Smith, A.
1996 Anthropological Research on Hazards and Disasters. Annual Review of Anthropology 25:303–328.
2004 Theorizing Vulnerability in a Globalized World: A Political Ecological Perspective. *In* Mapping Vulnerability: Disasters, Development and People. G. Bankoff, G. Frerks, and D. Hilhorst, eds. Pp. 10–24. Sterling, VA: Earthscan.

PNUD and IPEA
2003 Atlas do Desenvolvimento Humano no Brasil [Atlas of Human Development in Brazil]. Brasília: Programa das Nações Unidas para Desenvolvimento.

Ribot, J. C., A. R. Magalhães and S. Panagides, eds.
1996 Climate Variability, Climate Change, and Social Vulnerability in the Semi-arid Tropics. Cambridge: Cambridge University Press.

Tendler, J.
1997 Good Government in the Tropics. Baltimore, MD: Johns Hopkins University Press.

Thomalla, F., T. Downing, E. Spanger-Siegfried, G. Han, and J. Rockström
2006 Reducing Hazard Vulnerability: Towards a Common Approach between Disaster Risk Reduction and Climate Adaptation. Disasters 30(1):39–48.

Tompkins, E. L., S. A. Nicholson-Cole, L. A. Hurlston, E. Boyd, G. B. Hodge, J. Clarke, G. Gray, N. Trotz and L. Varlack

2005 Surviving Climate Change in Small Islands. Norwich, UK: Tyndall Centre for Climate Change Research, University of East Anglia.

Vásquez-León, M., C. T. West, and T. J. Finan

2003 A Comparative Assessment of Climate Vulnerability: Agriculture and Ranching on Both Sides of the US–Mexico Border. Global Environmental Change 13(3):159–173.

Villa, M. A.

2000 Vida e Morte no Sertão [Life and Death in the Sertão]. São Paulo: Editora Atica.

World Bank

2000 Brazil Poverty Reduction, Growth, and Fiscal Stability in the State of Ceará, Vol. 1, Policy Report, 19217-BR. Washington, DC: Latin America and the Caribbean Regional Office, World Bank.

The Impact of Volcanic Hazards on the Ancient Olmec and Epi-Olmec Economies in the Los Tuxtlas Region, Veracruz, Mexico

7

OLAF JAIME-RIVERÓN AND CHRISTOPHER POOL

MESOAMERICA IS AMONG THE MOST VOLCANICALLY and tectonically active areas in the world, a fact Eric Wolf (1959) captured eloquently in the title of his book, *Sons of the Shaking Earth*. Archaeological and geological records document the effects of volcanic eruptions on ancient societies in this region (Richardson and Keating 2002). Prominent examples include the eruption of Xitle contributing to the abandonment of Cuicuilco and displacement of its population to the growing city of Teotihuacán in central Mexico (Córdova et al. 1994; Sanders et al. 1979); the Classic period destruction of nearby Tetimpa, Puebla by ash flows from "the smoking mountain" Popocatépetl (Plunkett and Uruñuela 1998); and the third century AD devastation of western El Salvador and southeastern Guatemala by the massive eruption of Ilopango (Sheets 1980, 2002).

Though volcanic activity has more frequently affected the inhabitants of Mesoamerica's highlands, lowland societies like the Olmec of Mexico's southern Gulf Coast also suffered—and benefited—from volcanism. Olmecs in the great centers on the coastal plain of southern Veracruz and Tabasco fashioned their cutting tools from obsidian obtained from volcanoes hundreds of kilometers away. Closer to home, the volcanic Sierra de los Tuxtlas rises from the plain in the Olmec heartland. The Tuxtla volcanoes produce no obsidian, but they were the source of raw materials for the hundreds of monumental sculptures that are the hallmark of Olmec art (Williams and Heizer 1965) and the origin of innumerable grinding stones and other tools that helped Olmecs clear land and process food—making maize-based subsistence possible in this tropical region. In addition, volcanic ash often tempered their pottery (Shepard 1952), and in the Tuxtlas it weathered to some of the richest soil in Mexico.

However, the Tuxtla volcanoes were not entirely beneficent, especially to those who lived on their slopes. Archaeological and environmental research has documented ten ancient eruptions in the southwest and central Tuxtlas, nine of them occurring in a span of 1,650 years (Reinhardt 1992; Santley 2003; Santley et al. 2000). Archaeological survey suggests that societies in the Tuxtlas responded to these eruptions either by moving away from the most severely affected areas or by relying on regional and interregional social and economic networks, depending on their particular social scale, political integration, and subsistence adaptation (Pool and Britt 2000; Santley and Arnold 1996; Santley et al. 2000).

In this chapter, we consider both edges of the volcanic sword—the dangers of eruptions and the benefits of igneous resources. We do this through analysis of shifts in settlement and productive technology of ground stone artifacts across the cultural transition period known as Protoclassic (1650–1950 BP), also a time of intensified volcanic activity in the Tuxtlas.

Theoretical Background

Archaeological discussions of natural disasters have often emphasized macroenvironmental variables without examining the interior dynamics of cultural systems. However, nested political hierarchies have stabilizing or destabilizing effects because they offer the memory of the past and knowledge of distant places that provides models for recovery after change occurs. Such dynamic systems, in which nature and human communities are involved, show three complementary attributes (Walker et al. 2004): *resilience*, *adaptability*, and *transformability*.

Resilience is the capability of a system to continue to operate in spite of stress, and reorganize to recover at the end of a perturbation (Redman and Kinzig 2003). Adaptability is the capacity of the actors within a system to influence or manage resilience. Transformability indicates the capability of creating a system completely anew when the economic, ecological, or social structures cannot maintain the existence of the system. Tres Zapotes and the Tuxtlas exemplify all these instances of interaction diachronically during the cultural history of the Olmec culture.

For the purpose of this chapter, which focuses on ground stone technology, relationships between "small-and-fast" cycles (e.g., changes in economic behaviors observable by participants) and "large-and-slow" cycles (e.g., multigenerational changes in mode of production) are relevant for understanding resilience dynamics and discontinuities in ancient political economic systems in the aftermath of a volcanic event.

Ground Stone Technology and Resilience

Both large and small adaptive cycles are important with respect to understanding the lithic technology of our case. For over 1,000 years in the Early and Middle Formative periods, a settlement pattern of small, autonomous villages and hamlets of relatively impermanent structures in a sparsely occupied landscape (Santley and Arnold 1996) was part of a large-and-slow adaptive cycle in the Tuxtla Mountains that also included the adoption of a subsistence pattern of maize cultivation, exploitation of fruit trees and wild protein sources (VanDerwarker 2006), and a basic technology for making grinding implements from basalt to process maize and other foods. This combination of settlement pattern, subsistence practices, and craft technology allowed for the implementation of small-and-fast adaptive cycles that included emigration from areas affected by volcanic disaster (Santley et al. 2000), variation in the reliance on specific components of the subsistence base (VanDerwarker 2006), and, in the case of basalt working, the implementation of selected manufacturing techniques from the wider repertoire of technological choices stored in social memory. At the end of the Formative period, population grew and the inhabitants were able to survive the aftermath of volcanic eruptions. By studying a key resource from the productive sphere through all the steps of its life cycle to its final consumption and discard, we can shed light on patterns in political economic changes. For our case, volcanic events may produce considerable effects on productive networks and commodity flows.

The Olmecs based their subsistence on agricultural products that needed to be transformed through the use of basalt grinding tools. In the same rock, the Olmecs elaborated monuments and architectural features that reinforced identity, ethnicity, and memory. An igneous rock, basalt was obtained in the volcanic Tuxtla Mountains and transported to Olmec sites where the communities finished their artifacts. The specific case of the transformation of technological choice in a context of volcanic hazard helps reveal the relationship between large-and-slow and small-and-fast adaptive cycles, operationalized as productive techniques.

Geological, Cultural, and Chronological Context

The Tuxtla Mountains are located on the Mexican Gulf Coast in the southern part of the state of Veracruz. They are an active volcanic area today and have been in the recent geological past. Volcanism began 7 million years ago, in the Late Miocene, and continued to recent times with archaeological eruptions since 1400 BC and historical eruptions in AD 1664 and 1793 (Nelson and González-Caver 1992; Reinhardt 1992).

Two types of eruptions have predominated in the Tuxtlas Volcanic Field (Santley et al. 2000:148–149). The most common type is a Strombolian eruption, during which explosive ejections of ash occur at intervals ranging from minutes to hours. Of the approximately 250 satellite vents producing recent eruptions in the Tuxtlas Volcanic Field, about 210 are cinder cones with associated lava flows. Most of these cones were produced by Strombolian eruptions, as evidenced by bedding of ash and scoria exposed on their flanks. Strombolian eruptions do not distribute ash over as large an area as the more explosive Plinian eruptions, and in general, the lasting effects of ash falls from Strombolian eruptions in the Tuxtlas appear to be localized within areas on the order of about 100 square kilometers (Pool and Britt 2000; Reinhardt 1992).

The second most common eruption type in the Tuxtlas Volcanic Field has been a phreatic or phreatomagmatic eruption responsible for producing maars, most of which contain crater lakes (Santley et al. 2000:149). Of the approximately 250 satellite vents in the Tuxtlas Volcanic Field, 40 vents are maars or phreatic explosion craters, 13 of which are young enough to still contain crater lakes. The rest have been partially filled by epiclastic sediment or younger lava flows, but are still recognizable by their surrounding circular or arcuate cliffs. The distribution of maars in the Tuxtlas Volcanic Field is apparently controlled by the depth to the main water table in the underlying Tertiary sediments. Exceptions occur around the Laguna de Catemaco, where the presence of the lake raises the water table. Most deposits produced by the maar-forming eruptions lack juvenile material, indicating that the eruptions were phreatic rather than phreatomagmatic. One example can be seen northeast of Tres Zapotes at Laguna Colorado. Some maars, however, formed in combination with Strombolian activity, such as at Cerro Puntiagudo and Cerro Nixtamalapan, where both maars and cinder cones were produced.

Archaeological Record of Volcanic Hazards

Archaeological investigations have documented at least ten eruptions in the western Tuxtlas in the past 5,300 years, but direct evidence of the consequences of the eruptions for human settlement is scarce before 1400 BC. Archaeological data from excavations at Matacapan, one of the major sites in Classic times, and from two neighboring sites, Bezuapan and La Joya, show that nine of these eruptions occurred between 1400 BC and AD 659. The source of the eruptions was a series of vents near Cerro Mono Blanco on the western shores of Laguna Catemaco, situated only five kilometers east of Matacapan. Reinhardt (1992) has distinguished many of these ash falls based on radiocarbon determinations, archaeological ceramic associations, and trace element analysis.

Interpretations of Volcanic Hazards and Archaeological Sites in Los Tuxtlas

The earlier occupations in Los Tuxtlas date to the Early and Middle Formative period, a time when the region was occupied by small-scale societies. A settlement survey conducted by Santley and colleagues in the early 1990s (Santley and Arnold 1996; Santley et al. 1997; see also Pool and Britt 2000) showed that the Middle Formative population shifted southward out of the upper Río Catemaco Valley, which had been affected by a volcanic eruption at the end of the Early Formative period. As discussed by Santley (2003; Santley et al. 2000) the limited degree of social complexity exhibited in villages and hamlets composed of simple perishable structures, the simple mode of foot transport that integrated them, and the scant craft specialization made movement from the area impacted by volcanism the most viable alternative to catastrophic stress during the Early and Middle Formative periods. Mobility could be an effective option because regional population density was low, which could facilitate emigration to areas farther from the eruption source (Santley and Arnold 1996; Santley et al. 2000). More research and excavations in the future could support more effectively this working hypothesis concerning population growth. Now, we only have data from settlement pattern studies.

By the Late Formative period settlements in the central Tuxtla Mountains were better integrated through the establishment of a regional administrative center in the lower Río Catemaco valley at Chuniapan de Abajo, which was subsequently moved a short distance eastward to Chuniapan de Arriba in the Protoclassic period (Santley and Arnold 1996). Despite this greater regional integration, the Upper Río Catemaco Valley was again abandoned as the result of the combined Plinian and phreatomagmatic eruption of Nixtamalapan, which deposited over a meter of ash over some 100 sq. km. One of the settlements affected by this eruption was the village of Bezuapan. Excavations at Bezuapan revealed the effects of two Protoclassic volcanic events on successive, stratigraphically superimposed house lots, as well as the presence of an earlier Late Formative occupation (Pool and Britt 2000). The first eruption, which appears to have been a minor precursor to the second, deposited a thin ash that resulted in the collapse of a house and adjacent structure. The house lot was reoccupied shortly thereafter, as evidenced by the presence of four bell-shaped storage pits and deposition of middens containing household debris. Interestingly, the inhabitants of this last Protoclassic occupation appear to have increased their reliance on tree fruits and hunting of small game associated with agricultural disturbance (VanDerwarker 2006) at the same time that they expanded their craft activities, adding basalt working and possibly textile manufacture to ceramic production. Pool (nd) views these developments as risk-reduction strategies undertaken in response to increased volcanic activity. The second eruption, however, filled the bell-shaped pits to a depth of two meters with volcanic ash, and deposited over a meter of ash on the ancient ground surface. Following this eruption, Bezuapan, and as indicated by Santley and Arnold's surface survey, the rest of the upper Río Catemaco valley, lay abandoned for as much as a half century, until the Early Classic period (Pool and Britt 2000; Santley and Arnold 1996).

The first inhabitants to fill this population vacuum founded the Classic period site of Matacapan. They brought with them a new cultural tradition that included distinctive forms of cylindrical tripod jars, annular-based hemispherical bowls, ritual objects called *candeleros*, triangular-headed figurines, and temple platforms with stepped façades composed of vertical rectangular *tableros* projecting over sloping *taludes*. All of these elements are associated with the enormous urban site of Teotihuacán located over 500 km away in central Mexico (Santley et al. 1987). Teotihuacán also controlled the source of green obsidian at Pachuca, which appears with greater frequency in the Early and Middle Classic periods at Matacapan (Santley 1994).

Teotihuacán was a major power in Early and Middle Classic Mesoamerica, with economic and political ties that stretched from Western Mexico to the Maya region, as well as the nearer regions of Oaxaca and the Gulf Coast (Braswell 2003; Cowgill 1988, 1997). Interpretations of the Teotihuacan presence at Matacapan have changed from that of way-station for traders (Coe 1965) to a colonial enclave under Teotihuacán hegemony (Santley et al. 1987) to a diaspora community of refugees from political turmoil. Whichever may be the case, it is clear that the inhabitants of Matacapan and surrounding parts of the Tuxtlas maintained close economic and social ties to Teotihuacán through the Early and Middle Classic periods (e.g., Pool 1992), as well as with eastern Puebla, the source of most of their obsidian (Santley et al. 2001).

In addition to expanded interregional interaction, the central Tuxtlas saw an enormous population increase in the Classic period, from a low of some 2,500 people around AD 250 to some 50,000 by AD 550 (Santley and Arnold 1996). In tandem with population growth, more complex polities emerged, integrating large territories within the Tuxtlas region (Santley and Arnold 1996; Stoner 2008). These expanded, more well-integrated populations reacted in very different ways to volcanic eruptions than did their Formative and Protoclassic predecessors (Santley et al. 2000). During the Middle Classic period the central Tuxtlas experienced two major eruptions of the Puntiagudo volcanic cone near Catemaco (Reinhardt 1992). The distribution of volcanic ash indicates that these eruptions were similar in extent and severity to the Nixtamalapan ash fall (Pool and Britt 2000; Reinhardt 1992). Nevertheless, Santley and Arnold's (1996) settlement survey revealed no significant population displacement following these volcanic events. Matacapan, in particular, continued to flourish, with no detectable population decrease. Emigration out of the area most impacted by volcanic disaster would have been a much less effective response because of the dramatic increase in population densities accompanied by the more fixed territorial claims of state-level polities. Instead, Matacapan and its dependents would have been able to rely on the center's enhanced ability to coordinate and deploy substantial amounts of labor on the supra-community level to deal with the volcanic events more quickly than was possible in the Formative period In addition, Matacapan may have been able to sequester capital resources from regions outside the local area most severely impacted by the eruptions (Santley et al. 2000).

Volcanism and Settlement at Tres Zapotes

Archaeological survey and excavation have documented the growth and decline of Tres Zapotes, a major center of Olmec and epi-Olmec culture located at the western edge of the Tuxtlas uplift. This 500-hectare site sprawls across terraces of Tertiary period sediments and the floodplain of the Arroyo Hueyapan between the ancient volcano of Cerro el Vigía and the flat alluvial plain of the Papaloapan river system. First occupied as a village in the Early Formative period (ca. 3200–2950 BP), the site grew to become a regional Olmec center by the close of the Middle Formative period about 2350 BP, then expanded rapidly to its full extent in the Late Formative period (2350–1950 BP; Pool and Ohnersorgen 2003). In the course of the next 300 years, the site declined slightly in extent as it began to abandon the part of the floodplain nearest the Arroyo Hueyapan, then continued to decline as populations became concentrated in pockets separated by areas of lighter occupation within the former bounds of the center. By 950 BP the site lay abandoned, to be lightly reoccupied for a brief time in the Early Postclassic period (Drucker 1943).

Carl Wendt (1998, 2003) encountered deposits measuring up to two meters thick of Drucker's (1943) post-Formative ash fall layer as well as a deeper one below cultural deposits. Although it is not clear whether the ash fall at Tres Zapotes was from the same eruption of Nixtamalapan that buried Bezuapan, radiocarbon dates of underlying deposits and ceramic stratigraphy indicate that it fell during the Protoclassic Nextepetl phase (1950–1650 BP). Temporally sensitive materials from surface collections and auger tests indicate that the ash fall at Tres Zapotes hastened the abandonment of the center of the Arroyo Hueyapan floodplain, although that process had begun earlier in the phase (Wendt 2003). However, people continued to inhabit the alluvial terraces that border the floodplain, where they were out of the reach of flooding and where ash would erode more quickly (Pool and Ohnersorgen 2003).

Just as at Bezuapan, Protoclassic households at Tres Zapotes appear to have expanded their crafting activities. In particular the occupants of a house lot excavated in Operation 3B, whose basalt industry we describe below, expanded their craft activities to include ceramic production and obsidian blade production as well as ground stone manufacture. Pool (n.d.) interprets this multi-crafting as a response to greater household subsistence risk during a documented period of increased volcanic activity. It is particularly illuminating that in the upper levels of the Operation 3B excavations, which date to the very late Protoclassic or very early Classic period, the Operation 3B household reverted to a bipolar technology for reducing obsidian. This technology, which had fallen into relative disuse during the Late Formative period, allows for more material to be extracted from obsidian cores and recycled artifacts, and can be seen as a means of conserving a scarce resource (Knight and Pool 2008).

Coeval with this shift in the pattern of residential occupation, Tres Zapotes also saw significant changes in the construction programs in its four major civic-ceremonial mound groups in the Protoclassic (Pool 2008). Previously, during the Late Formative period, these four plaza groups at Tres Zapotes followed a consistent plan—the plaza oriented east-west, a long elite residential mound lying to the north, and a conical or pyramidal temple mound to the west (other temple mounds and low platforms were added to the basic plan in three of the four groups). Pool (2003, 2008) interprets this redundancy of function and layout among coeval mound groups as reflecting power sharing among allied factions who ruled Tres Zapotes. During the Protoclassic period new construction in mound Group 3 flouted earlier conventions by reorienting the group's principal plaza from east-west to north-south. A carbon sample from within an offering placed before a plain stela (Monument 44) in this plaza yielded a date of 1870 ± 50 BP. Incidentally, a fragment of Stela C, bearing a Long-Count date in 32 BC (1982 BP) was erected on its side about ten meters north of Monument 44, at about the same time, during the dedication of the plaza. Also, distributions of pottery sherds suggest that Group 1 saw changes in the functions of its structures, with the temple and the elite residence effectively switching positions (Pool 2008). In sum, the Protoclassic period saw the diversification of monumental construction programs, which we interpret as a weakening of power sharing as leaders of factions asserted their individuality.

The dissolution of the governing alliance at Tres Zapotes may also have weakened the control of Tres Zapotes over its hinterland. During the Late Formative period, the site of El Mesón, 10 km north of Tres Zapotes, constructed a formal mound group whose layout precisely mimicked the Tres Zapotes Plaza Group (TZPG) plan (Loughlin and Pool 2007). During the Protoclassic period, the population of El Mesón and its hinterland began an expansion that continued into the Classic period as Tres Zapotes was in decline, but architectural layouts of formal mound groups broke free of the Tres Zapotes model. The fragmented dynamics of Protoclassic political interactions also appear to be reflected in the epigraphic record of the time. Found in the Papaloapan delta west of Tres Zapotes, the La Mojarra stela bears Long Count calendrical dates in the mid-second century AD and a

400-character inscription in the epi-Olmec script that Justeson and Kaufman (1993) read as an account of battles and ritual offerings leading up to the accession of the ruler depicted next to the text.

Table 7.1 places the above political, demographic, and economic changes into the categories of small-and-fast and large-and-slow adaptive cycles. Short-and-fast adaptive cycles are underlined. Small-and-fast adaptive changes took from the repertoire of techniques in the social memory ways of adapting rapidly to changing conditions. Large-and-slow changes are continuous in the history of the region. Note that short-and-fast cycles are most numerous and pervasive in the Protoclassic period, which saw increases in volcanic activity and political factionalism, as well as greater flexibility in complex political-economic strategies. This flexibility enhanced resilience in coping with natural disasters.

Although the causes of the Protoclassic political changes and demographic shifts in the Tuxtlas and the Tres Zapotes area surely were more complex than a volcanic eruption alone, there can be little doubt that increased volcanic activity contributed to increased subsistence risk in the region. Furthermore, there is strong evidence for settlement shifts in the aftermath of volcanic eruption, both at the very local scale of the Arroyo Hueyapan floodplain within Tres Zapotes and at the sub-regional scale of the abandonment of the upper Río Catemaco valley. At Tres Zapotes, the timing of risk-diversification strategies in household contexts and diversification of formal architectural plans in the seats of ruling authority suggest that actual volcanic disaster and the perceived increase in the risk of volcanic hazard contributed to changes in the economic and political spheres. Among the apparent changes in the political sphere, as evidenced by developments at El Mesón, was a redrawing of political territories and a reduction in the control of Tres Zapotes over portions of its hinterland, including some of the likely sources of basalt used for manufacturing grinding tools as well as stone monuments.

Results of Ground Stone Analysis of Tres Zapotes

In the following pages, we seek to shed additional light on the mechanisms the epi-Olmec community of Tres Zapotes used to cope with the transformation of their physical and social landscape after a volcanic event through an analysis of the processes of production and consumption of basalt grinding stones that have been recovered by the Tres Zapotes Archaeological Project. The selected archaeological contexts provide insight into the diachronic

Table 7.1. Major Trends and Events in Volcanism, Demography, Political Organization, and Economy

Broad Categories of Change	*Specific Dynamic Processes*	*Early Formative*	*Middle Formative*	*Late Formative*	*Protoclassic*	*Classic*
Volcanism	Regional volcanic hazard (See table 7.2)	low	very low	low	high	Moderate
	Volcanic ashfall at Tres Zapotes				<u>X</u>	
Demographic trends	Regional Population	+	+	+	–	+
	Population at Tres Zapotes	+	+	+	+/–	–
	Floodplain abandoned				<u>X</u>	
Political Developments	Regional center at Tres Zapotes		X	X	X	X
	Factional competition at Tres Zapotes			<u>x</u>	<u>X</u>	X
	Factional alliance			<u>X</u>		
	Dissolution of alliance				<u>X</u>	
Economic trends	Mixed subsistence base	X	X	X	X	X
	Basic basalt technology	X	X	X	X	X
	Basalt recycling			<u>x</u>	<u>X</u>	?
	Bipolar obsidian technology	X	X	X	<u>x</u>	<u>X</u>
	Prismatic blade-core obsidian technology	x	<u>x</u>	<u>X</u>	X	X

Symbology:
+ increase
– decrease
+/– little or no change
x trend or characteristic is weakly present
X trend or characteristic is strongly present
Small-and-fast cycles and major transitions are underlined.

transformation of basalt stone tool production and consumption as well as technological change and resilience in the face of natural disaster.

Early and Middle Formative

Unit 8 (Operation 2B) tested a low platform of the Early Classic period in Group 2. Below this construction phase was a Terminal Formative platform of sandstone rubble and a series of floors and domestic refuse extending back to the Early/Middle Formative transition at a depth of 430 centimeters below datum.

The Early Formative (3200–2950 BP) and Middle Formative (2950–2350 BP) levels exhibited basalt working with secondary or fine flaking, as well as use of thermic shock to rough out pre-forms. The sizes and morphologies of products and byproducts of manufacture resemble artifacts that we have observed from contemporaneous Olmec contexts at sites such as El Manatí (state of Veracruz), and San Andrés (state of Tabasco). Flakes, macro-flakes, and natural quartz pebbles were present throughout the sequence in Unit 8 (table 7.2). Beginning in later Middle Formative levels, however, there is an increase in the discard of *metates* (stone grinding slabs), fragments of metates, and *manos de metate* (rolling pin–shaped grinders used with metates). Later, around the beginning of the Late Formative period, the morphology of metate fragments and debitage flakes suggests an increase in recycling.

Unit 12 (Operation 2C) tested stratified cultural deposits 15 meters north of Mound 9—a long, elite residential/administrative structure built in the Late Formative period—and provided one of the longest occupational sequences at Tres Zapotes. Flakes, natural quartz pebbles, and fragments of metates and manos all occurred in the lower levels that correspond to the Early and Middle Formative periods (table 7.2). Flakes and modified quartz pebbles continue in late Middle Formative levels (especially levels 33–27). In contrast to Unit 8, however, there is a decrease in the number of fragments of grinding tools that were discarded in these levels.

The knowledge of the basic repertoire of techniques represented in Units 8 and 12 was present in the long-term Olmec stone tool tradition. Since the Early Formative, we have evidence of the use of quartzite hammer stones for the manufacture of both utilitarian and elite artifacts in caches at El Manatí and San Andrés (Jaime-Riverón 2003). The Olmecs also used fire for specific steps in the manufacture of basalt implements (Casellas Cañellas 2004) and gabbro celts (Jaime-Riverón 2003). Recycling as a technique is also demonstrated in particular ritual and elite contexts of the Olmec area, such as occur at San Lorenzo (Cyphers Guillén 1997; Porter 1989).

Table 7.2. Relative Frequencies of Ground Stone Artifact Classes by Period in Selected Excavation Units at Tres Zapotes

		Early Formative		*Middle Formative*		*Late Formative*						*Protoclassic*	
Artifact Class	*Code*	*Unit 8 Levels 43–60*	*Unit 12 Levels 50–60*	*Unit 8 Levels 21–42*	*Unit 12 Levels 25–49*	*Unit 8 Levels 6–20*	*Unit 12 Levels 9–24*	*Unit 9*	*Unit 34*	*Unit 15 Plaza, Levels 1–10*	*Unit 15 Domestic, Levels 15–38*	*Unit 28 Levels 11–17*	*Unit 28 Levels 1–10*
metate			20.0%	6.7%	16.7%	3.8%	25.6%	0.6%	6.3%	1.5%	0.8%	35.7%	4.9%
mano			10.0%	2.2%	5.1%	7.7%		1.6%	6.3%	6.1%	2.4%		0.7%
stone vessel	4						2.3%		31.3%			42.9%	49.3%
polishing stone	5		10.0%	2.2%				0.2%					3.5%
pestle	7				1.3%								0.7%
bifacial ovate grinding stone	34							0.2%					
miscellaneous grinding stone	11							0.2%					
reamer-polisher	36			6.7%				0.4%					
stone ball	61			2.2%				0.6%					
disk	64										0.8%		
celt	10.1							0.2%					
unidentified worked stone	15							1.0%	18.8%				27.8%
metate preform	45		20.0%	13.3%	10.3%	15.4%	9.3%				1.6%		
natural metate preform	46							3.0%	6.3%				
flake	20	75.0%	40.0%	35.6%	32.1%	30.8%	7.0%	6.1%	31.3%	24.2%	21.4%	14.3%	10.4%
flake-possible	20.5							1.6%					
natural quartz pebble	21	25.0%		31.1%	32.1%	38.5%	37.2%	84.2%		66.7%	31.0%	7.1%	1.4%
unworked stone	19												1.4%
undefined raw material	25				2.6%	3.8%	18.6%			1.5%	42.1%		
	n	4	10	45	78	26	43	493		66	126	14	144

Late Formative

Unit 15 (Operation 2E) confirmed the Late Formative construction of the plaza of Group 2, which was preceded by earlier Late Formative domestic occupation. As was the case in Units 8 and 12, basalt flakes and modified quartzite cobbles occurred throughout this unit's stratigraphy (table 7.2). Strong continuity in technology is particularly notable in this location because it transcends the transition from domestic to ceremonial space. Discard of residues in the plaza may correspond to activities attached to the main buildings. Interestingly, few specimens from Unit 15 correspond to discarded grinding tools in either domestic or plaza contexts.

Unit 9 (Operation 2D) tested a low elevation physically attached to the northeastern corner of Mound 9. Previous surface collections in this area detected high densities of pottery, overfired sherds, and fragments of highly fired mud, which appeared to be fragments of ceramic kilns. The excavations confirmed ceramic production (e.g., vitrified kiln debris and overfired sherds). Metates, basalt flakes, and micro-debitage offered evidence of ground stone production as well.

During Late Formative occupation, the distribution of debitage and stone tool remains follow a production pattern of fine flaking of pre-forms (table 7.2). Some flakes come from shaping pre-forms for grinding tools, especially for curving the silhouette; ethnographically we have seen the production of this kind of debitage in Puebla and Guatemala. In later flakes we see an increase in recycling and continued use of fire to facilitate fracture. Moreover, some discarded fragments of metates show previous reutilizations, suggesting multiple incidents of reuse.

Operation 6 (Unit 34) tested Late Formative deposits around Mound 50, the long mound of the Nestepe Group. Unit 34 encountered three episodes of platform construction and Late Formative refuse deposits containing used stone grinding tools. This unit also shows the residues of production that were sealed by a floor that also covered a platform adjoining Mound 50. They are mostly discarded remains of grinding tools, some flakes from rejuvenating artifacts, and some cobbles (table 7.2). There is no evidence for the use of fire in this context.

Protoclassic

The Protoclassic context is important because it covers the transition of the immediate aftermath of the volcanic event and the transformation of the political economy of Tres Zapotes.

Operation 3B (Unit 28) consists of six contiguous units placed on a modified natural sandstone terrace to investigate non-elite ceramic production in a domestic context. Excavation yielded evidence of obsidian blade production and basalt working, including recycling of metates into stone vessels. Associated ceramics and a radiocarbon date indicate an early Protoclassic date for lower levels, while upper levels appear to post-date the Protoclassic eruption.

In the lowest levels of Pit 28 the most common ground stone tools are stone vessels and metates (table 7.2). In later levels stone vessels remain the most common artifact class. Metates however, nearly disappear as the archeological correlates of recycling increase along with various kinds of debris that correspond with multiple reuses of basalt as a raw material. Among some of the most striking qualitative differences in this aftermath of the volcanic hazard is the employment with basalt of techniques traditionally associated with obsidian, such as prismatic blades. In contemporary Puebla, where such resources are scarce, we were also able to document this process. When the system is faced with natural disasters and scarcity, memory from other ambits of social life may be applied to the specific problem of adapting to the changing landscape.

In table 7.3, we can see a summary of the main changes that occurred during the Formative period in the archaeological site of Tres Zapotes. We compare the lithic techniques used, the kinds of artifacts that were produced, the economic contexts, and the comparison in the implementation of strategies in a diachronic perspective.

Beginning in the Late Formative period, there was greater maximization of techniques that recycle and reuse raw materials during times of factionalism, which increased in the aftermath of a volcanic hazard (see also table 7.1). Implementation of a flexible political economy to mediate factional interests allowed Tres Zapotes to flourish after other Olmec centers collapsed. That flexibility also made Tres Zapotes more resilient to the changes originated by volcanic activity. Whereas a volcanic eruption forced the emigration of less complexly organized Protoclassic period communities from the central Tuxtlas, Tres Zapotes was able to persist in its same location. Contributing to the resilience of the Tres Zapotes polity were changes implemented in the productive sphere, including the production of basalt implements. Although all the basalt-working techniques were known by the Olmecs, their successors selected those strategies most useful for coping with the changing political economic landscape.

Table 7.3. Formative Period Changes in Production and Consumption, Tres Zapotes

Period	*Techniques of Production*	*Artifacts*	*Use, Recycling, and Discard*	*Economic and Political Strategies*
Early Formative	• secondary flaking • use of fire for fracturing nodules • polishing for shaping	• uniform flake distribution • rounded cobbles • discarded grinding tools	• rounded quartzite pebbles for fracturing basalt, polishing pottery, and funerary use	• large-and-slow cycle. Use of basic techniques for production of quotidian artifacts observed at earlier sites
Middle Formative	• more preparation for platforms of percussion in order to produce more squared volumes • flaking produced • Intensified use of fire for fracturing blocks. Thermal shock was more controlled.	• longer (straight and curved) macro-flakes for the manufacture of domestic grinding tools • squared and thinner cuts of rock • discarded grinding tools (metates and mortars)	• domestic units display evidence of implementation of flaking and polishing.	• large-and-slow cycle continues with use of techniques that are now part of the repertoire of the Olmec social memory. • techniques for producing thin blocks were implemented in both stelae and quotidian artifacts.
Late Formative	• lithic techniques emphasized recycling. • reduction of discarded artifacts	• thinner flakes • smaller nodules	• reduction of discarded artifacts • in the households, the remains of basalt smaller	• implementation of a small-and-fast cycle of recycling • dissolution of Olmec trade networks • factionalism and power-sharing within Tres Zapotes.
Terminal Formative	• maximization of raw materials • recycling • continuous decline of discarded artifacts	• smaller nodules and pre-forms. • reused and maximized artifacts	• few examples of discarded artifacts. • thinner and smaller artifacts	• small-and-fast cycle of recycling intensifies. • increased volcanic activity heightens subsistence risk locally and regionally in the Tuxtlas • factionalism intensifies, power-sharing declines.

Discussion and Conclusion

The volcanic eruptions in the Tuxtlas affected ancient communities in many ways. In the large-and-slow processes of the southern Gulf lowlands, the basalt technological subsystem accumulated a series of techniques beginning in the Early Formative. Later, the epi-Olmecs of Tres Zapotes implemented small-and-fast cycle changes in their stone tool technology when the political economy of Tres Zapotes transformed in the aftermath of a volcanic eruption. This is an example of a process of regional economic (craft) diversification and specialization to adapt to hazard-induced loss of production materials. In the Tuxtlas, there was a contested scenario not only for acquisition of rocks, but also in the disruption of social networks which unified communities integrated in the whole productive sequence.

This long-term archaeological perspective allows us to see that a volcanic event exacerbated differentiation, competition, and factionalism at the polity level. The cooperative power sharing among factions at Tres Zapotes in the Late Formative period was interrupted by a Protoclassic volcanic eruption, and the trend toward the wider distribution of stone resources for manufacture was accelerated, with diminished centralization of production. Tres Zapotes as a confederacy was able to implement technological changes in the economic sphere that previous polities were not able to accomplish. The small-and-fast change was not emigration outside of the affected area, as is seen in the central Tuxtlas. Instead, the factionalized groups reorganized the way in which they reproduced daily life.

Tres Zapotes is an interesting example of variation in response to natural disasters compared to Bezuapan and Matacapan in the Central Tuxtlas. The Protoclassic eruption of Mount Nixtamalapan there caused abandonment of Bezuapan and central Río Catemaco Valley as inhabitants migrated into uplands where volcanic ash more rapidly eroded away, exposing usable soils (Pool and Britt 2000; Santley et al. 2000). That historical event cleared the way for new occupation by a Teotihuacán enclave, which created a disjunction in the transmission of technical knowledge. This severing of the strong tie to long-slow cycles governing local technical knowledge opened up the opportunity for—and need for—changes in technological strategies for pottery production, particularly the expansion of kiln firing (Pool and Britt 2000).

In contrast, an eruption in the same period and—judging from thickness of ash deposits—of the same intensity did not result in the abandonment of Tres Zapotes. In this respect, the Tres Zapotes situation was more like Middle Classic Matacapan, which weathered a series of eruptions due to the more extensive political and economic network of which it was the center (Santley et al. 2000). Classic Period Matacapan, it should be noted, also had a more centralized political system than the confederacy of factions at Tres Zapotes. Consequently, the effects of the Protoclassic eruption on the Tres Zapotes polity also differed. There the eruption contributed to a process of intensified factional competition and the weakening of the Tres Zapotes ruling alliance.

The reconfigured political economic landscape also posed a challenge for the transformed institutions in their ability to cope with the availability of raw materials. Likewise, epi-Olmec groups of Tres Zapotes increasingly used a set of resource-maximizing strategies already present in the repertoire of the ancestral Olmec stone tool technological tradition—including the reuse and recycling of quotidian artifacts as well as of stone monuments like Stela C.

References

Braswell, Geoffrey E.

2003 Introduction: Reinterpreting Early Classic Interaction. *In* The Maya and Teotihuacán: Reinterpreting Early Classic Interaction. G. E. Braswell, ed. Pp. 1–43. Austin: University of Texas Press.

Casellas Cañellas, Elisabeth

2004 El Contexto Arqueológico de la Cabeza Colosal Numero 7 de San Lorenzo, Veracruz, México. [The Archaeological Context of Colossal Head Number 7 of San Lorenzo, Veracruz, Mexico.] PhD Dissertation, Universitat Autonoma de Barcelona, Spain, Barcelona.

Coe, Michael D.

1965 Archaeological Synthesis of Southern Veracruz and Tabasco. *In* Archaeology of Southern Mesoamerica, Part Two. G. R. Willey, ed. Pp. 679–715. Handbook of Middle American Indians, Vol. 3. Austin: University of Texas Press.

Córdova, Carlos, Ana Lilian Martín del Pozo, Javier López Camacho

1994 Paleolandforms and Volcanic Impact on the Environment of Prehistoric Cuicuilco, Southern Mexico City. Journal of Archaeological Science 21: 585–596

Cowgill, George L.

1988 Onward and Upward with Collapse. *In* The Collapse of Ancient States and Civilizations. Norman Yoffee and G. W. Conrad, eds. Pp. 87–114. Santa Fe, NM: School of American Research Press.

1997 State and Society at Teotihuacán, Mexico. Annual Review of Anthropology 26:129–161

Cyphers Guillén, Ann.

1997 El Contexto Social de Monumentos en San Lorenzo. *In* Población, Subsistencia y Medio Ambiente en San Lorenzo Tenochtitlán. [The Social Context of Monuments in San Lorenzo. *In* Population, Subsistence and Environment at San Lorenzo Tenochtitlán.] A. Cyphers Guillén, coord. Pp. 163–194. México City: IIA-UNAM.

Drucker, Philip

1943 Ceramic Sequences at Tres Zapotes, Veracruz, Mexico. (Bureau of American Ethnology Bulletin 140). Washington, DC: Smithsonian Institution.

Jaime-Riverón, Olaf

2003 El Hacha Olmeca: Biografía Y Paisaje. [The Olmec Celt: Biography and Landscape.] Unpublished Master's thesis, Universidad Nacional Autónoma de México.

Justeson, J. S., and Kaufman, T.

1993 A Decipherment of Epi-Olmec Hieroglyphic Writing. Science 259:1703–1711.

Knight, Charles, and Christopher A. Pool

2008 Formative to Classic Period Obsidian Consumption at Tres Zapotes, Veracruz, Mexico. Paper presented at the 73rd Annual Meeting of the Society for American Archaeology, Vancouer, Canada, 2008.

Loughlin, Michael L., and Christopher A. Pool

2007 The Late Preclassic and Protoclassic Periods in the Eastern Lower Papaloapan Basin, Veracruz, Mexico. Paper presented at the 72nd Annual Meeting of the Society for American Archaeology, April 25–29, Austin, Texas.

Nelson, S. A., and E. González-Caver.

1992 Geology and K-Ar Dating of the Tuxtla volcanic field, Veracruz, Mexico. Bulletin of Volcanology 55:85–96.

Plunket, Patricia, and Gabriela Uruñuela.

1998 Appeasing the Volcano Gods. Archaeology 51(4): Abstracts.

Pool, Christopher A.

1992 Strangers in a Strange Land: Ethnicity and Ideology at an Enclave Community in Middle Classic Mesoamerica. *In* Ancient Images, Ancient Thought: The Archeology of Ideology. A. S. Goldsmith, ed. Pp. 41–55. Calgary, Alberta: University of Calgary.

2003 Centers and Peripheries: Urbanization and Political Economy at Tres Zapotes. *In* Settlement Archaeology and Political Economy at Tres Zapotes, Veracruz, Mexico. C. A. Pool, ed. Pp. 90–98. Monograph 50. Los Angeles: Cotsen Institute of Archaeology, University of California.

2008 Architectural Plans, Factionalism, and the Protoclassic-Classic Transition at Tres Zapotes. *In* Classic Period Cultural Currents in Southern and Central Veracruz. P. J. Arnold, III and C. A. Pool, eds. Cambridge: Dumbarton Oaks and Harvard University Press.

nd Residential Pottery Production in Mesoamerica. *In* Archaeological Papers of the American Anthropological Association. K. G. Hirth, ed. Washington, DC.

Pool, Christopher A., and Georgia M. Britt

2000 A Ceramic Perspective on the Formative to Classic Transition in Southern Veracruz, Mexico. Latin American Antiquity 11:139–161.

Pool, Christopher A., and Michael Anthony Ohnersorgen

2003 Archaeological Survey and Settlement at Tres Zapotes. *In* Settlement Archaeology and Political Economy at Tres Zapotes, Veracruz, Mexico. C. A. Pool, ed. Pp. 7–31. Monograph 50. Los Angeles: Cotsen Institute of Archaeology, University of California.

Porter, James B.

1989 The Sculpture and Hieroglyphs of Tres Zapotes, Veracruz, Mexico. PhD dissertation, Deparment of Anthropology, University of California, Berkeley.

Redman, C. L., and A. P. Kinzig

2003 Resilience of Past Landscapes: Resilience Theory, Society and the Longue Durée. Conservation Ecology 7(1):14.

Reinhardt, Bently K.

1992 Volcanology of the Younger Volcanic Sequence and Volcanic Hazards Study of the Tuxtla Volcanic Field, Veracruz, Mexico. MS Thesis, Department of Geology, Tulane University, New Orleans.

Richardson, Gill, and Jerome Keating

2002 Volcanism and Mesoamerican Archaeology. Ancient Mesoamerica 13:125–140.

Sanders, William T., Jeffrey R. Parsons, and Robert S. Santley

1979 The Basin of Mexico: Ecological Processes in the Evolution of a Civilization. New York: Academic.

Santley, Robert S.

1994 The Economy of Ancient Matacapan. Ancient Mesoamerica 5(2):243–266.

2003 The Tuxtlas as Volcanic Hazard: Volcanism and its Effects on Site Founding and Abandonment in the Tuxtla Mountains, Southern Veracruz, Mexico. *In* The Archaeology of Settlement Abandonment in Middle America. Takeshi Inomata and Ronald R. Webb, eds. Pp. 163–180. Salt Lake City: University of Utah Press.

Santley, Robert S., Philip J. Arnold III, and Thomas P. Barrett
1997 Formative Period Settlement Patterns in the Tuxtla Mountains. *In* Olmec to Aztec: Settlement Patterns in the Ancient Gulf Lowlands. B. L. Stark and P. J. Arnold, III, eds. Pp. 174–205. Tucson, Arizona: The University of Arizona Press.

Santley, Robert S., Thomas P. Barrett, Michael D. Glascock, and Hector Neff
2001 Pre-Hispanic Obsidian Procurement in the Tuxtla Mountains, Southern Veracruz, Mexico. Ancient Mesoamerica 12:49–63.

Santley, Robert S., Barbara Hall, and Clare Yarborough
1987 Enclaves, Ethnicity, and the Archaeological Record at Matacapan. *In* Ethnicity and Culture. R. Auger, M. F. Glass, S. MacEachern, and P. H. McCartney, eds. Pp. 85–100. Calgary: University of Calgary Archaeological Association.

Santley, Robert S., and Philip J. Arnold, III
1996 Prehispanic Settlement Patterns in the Tuxtla Mountains, Southern Veracruz, Mexico. Journal of Field Archaeology 23:225-249.

Santley, Robert, S. Nelson, B. Reinhardt, C. Pool, and P. Arnold, III
2000 When Day Turned to Night: Volcanism and the Archaeological Record from the Tuxtla Mountains, Southern Veracruz, Mexico. *In* Environmental Disaster and the Archaeology of Human Response. Anthropological Papers, 7. G. Bawden and R. Reycraft, eds. Pp. 143–162. Albuquerque: Maxwell Museum of Anthropology, University of New Mexico.

Shepard, Anna O.
1952 Technological Analyses. *In* La Venta, Tabasco: A Study of Olmec Ceramics and Art. La Venta, Tabasco: A Study of Olmec Ceramics and Art. by Philip Drucker. With a chapter on structural investigations in 1943, by Waldo R. Wedel, and appendix on technological analyses by Anna O. Shepard. Smithsonian Institution. Bureau of American Ethnology. Bulletin 153 (Smithsonian Institution).

Sheets, Payson
1980 Archaeological Studies of Disaster: Their Range and Value. Boulder, CO: Institute of Behavioral Science, University of Colorado.
2002 Before the Volcano Erupted: The Ancient Cerén Village in Central America. Austin, TX: University of Texas Press.

Stoner, Wesley Durell
2008 Tepango Valley Archaeological Survey: Tuxtla Mountains, Southern Veracruz, México. Report submitted to the Foundation for the Advancement of Mesoamerican Studies, Inc., Coral Gables, FL. www.famsi.org/reports/07049/index.html.

VanDerwarker, Amber M.

2006 Farming, Hunting, and Fishing in the Olmec World (The Linda Schele Series in Maya and Pre-Columbian Studies). Austin, TX: University of Texas Press.

Walker, B., C.S. Holling, S. R. Carpenter, and A. Kinzing

2004 Resilience, Adaptability and Transformability in Social-ecological Systems. Ecology and Society 9(2):5.

Wendt, Carl John

1998 Intra-Community Settlement Organization at Tres Zapotes: The Perspectives from a Sub-Surface Testing Program. MA Thesis, Department of Anthropology, University of Kentucky.

2003 Buried Floodplain Deposits at Tres Zapotes: The Results from an Auger Testing Program. *In* Settlement Archaeology and Political Economy at Tres Zapotes, Veracruz, Mexico. Christopher A. Pool, ed. Pp. 32–46. Los Angeles: Cotsen Institute of Archaeology, University of California at Los Angeles.

Williams, Howell, and Robert Heizer

1965 Sources of Rocks Used in Olmec Monuments. University of California Archaeological Research Facility Contributions, 1. Pp. 1–39. Berkeley, CA: University of California, Berkeley.

Wolf, Eric R

1959 Sons of the Shaking Earth. Chicago: University of Chicago Press.

8

If the Pyroclastic Flow Doesn't Kill You, the Recovery Will: Cascading Impacts of Mt. Tungurahua's Eruptions in Rural Ecuador

LINDA M. WHITEFORD AND GRAHAM A. TOBIN

Some 55 volcanoes, at least 17 of them active, are strung for 190 miles along Ecuador's northern Andean spine, known as the "Avenue of the Volcanoes." More than a quarter of Ecuador's 12 million people live within 15 miles of an active volcano, making the northern and central Andes one of the most densely populated volcanic zones in the world. Particularly vulnerable is Ecuador's economic lifeblood, two pipelines, pumping about 535,000 barrels of crude oil a day, which parallel each other as they snake through the heart of volcano territory. Oil, with its $2.3 billion in annual foreign sales, accounts for more than half of the country's export income and 30 percent of the government's budget.

Equally significant to Ecuador's economy is the tourist industry. Ecuador has some of the greatest diversity in birds anywhere in the Ameri cas; tourists come for birding, they come for hiking the snow-capped volcanoes like Cotopaxi; they come to see the tropical life-forms of the Amazon; and they come to experience the thermal waters that bubble from the base of volcanoes like Tungurahua. And, they come to watch the volcanoes erupt.

In 1999, the town of Baños, a tourist destination situated at the base of Tungurahua, was forcibly evacuated because the volcano appeared about to erupt. The evacuation of the town caused significant and prolonged economic and emotional losses to the people living in and around Baños, resulting in their reluctance to evacuate again (Lane et al. 2004; Tobin et al. 2005; Whiteford et al. 2002). This reluctance and peoples' perception of the risk associated with the volcano has been one of the foci of our ongoing research. In this chapter, we discuss consequences of prolonged exposure to chronic disaster. We are particularly interested in the recovery efforts and the costs they hold for individuals and communities near the volcano.

Case Study: The Tungurahua Volcano

In August 2006, we stood close enough to the explosions to see several pyroclastic flows rush down the mountainside through villages where we had stood one day before. Pyroclastic flows—superheated material of rocks, gas, and debris—destroyed roads, blocked three rivers, and forced the shutdown of the hydroelectric power plant, leaving four Amazonian provinces without power for hours until energy could be rerouted to reach them (AP-CNN 2006). The flows destroyed settlements: houses were gone, crops were burned, and animals incinerated. What the pyroclastic activity did not destroy, the ash did. Following five hours of intense volcanic activity that included volcanic bombs, boulders thrown so far in the air that we could see them miles away, explosions that rattled windows, floors, and roofs, and the constant and unending roaring of the volcano, the visible blazes stopped. And then the ashfall began and continued heavily for seven hours, and then more lightly for days afterwards. We watched the explosion and listened to the roaring of the volcano, with windows shaking, for most of the night. In the morning, the noise was over, but the ash had begun to fall, as in James Russell Lowell's (2001) "First Snow Fall":

> The snow had begun in the gloaming,
> And busily all the night
> Had been heaping field and highway
> With a silence deep and white.

Except our "snow fall" was poisonous. "Tens of thousands of acres of pasture and crops were destroyed according to the Ecuadorian CD" (AP Newswire 2006).

The ash coated everything, things that moved and those that did not. It covered the tin roofs of the houses, later to corrode them. Pasture for animals was ruined; the volcanic ash destroys animals' stomachs and mucus membranes, and adversely affects eyes, throats, and skin. In addition, ash irritates lungs; people coughed for days, weeks, and months; throats became inflamed and asthma attacks more severe. The ash was dense and accumulated in thick deposits; it was difficult to remove, coating streets and clogging car engines. The area was closed to incoming and outgoing traffic, and commercial flights were stopped for fear of the ash getting into the engines and bringing the planes down. Furthermore, water systems were compromised and the supplies of gasoline were disrupted. Those going out covered their faces with scarves, but still the ash got into mouths, noses, and eyes. And the volcano was still not quiescent. The Pan American Health Organization (Reliefweb 2006) estimated that more than 676,000 people (out of a total population of 930,000) were affected during this time. Once the volcano stopped roaring like an oncoming train and throwing incandescent boulders and hot gases and the ash no longer looked like a gray snowstorm, the recovery began.

Mount Tungurahua is located just south of the Equator, with a summit at 5,023 meters that rises imposingly from the surrounding countryside. Two major rivers cut along the western and northern flanks of the volcano: the Pastaza and the Chambo. Smaller rivers that drain the volcano include the Ulba, immediately to the east of Baños, and the Vazcún, which runs along the westernmost portion of the town. Baños is the largest town in the area, but numerous small villages are located in close proximity to the rivers and along Tungurahua's western and northwestern flanks, which are intersected by a series of *quebradas* (gorges) that emanate in a radial pattern from the summit (figure 8.1).

The geologic processes that have been building, destroying, and rebuilding Tungurahua extend back more than 40,000 years (Yepes 2002). These eruptive and formative activities, and their associated hazards, have been studied extensively by scientists at Ecuador's *Escuela Politécnica Nacional* and by members of the French *Institut de Recherches Pour le Dévelopment* (Hall et al. 1999). The present volcano, Tungurahua III, has been growing for about 2,300 years and has rebuilt about half the cone of its predecessor (Tungurahua II). Nearly eight centuries before the arrival of the Spanish around 1500, Tungurahua began a cycle of erupting approximately every 100 years. The geologic record and eyewitness accounts confirm the volcano's destructive capacity, including pyroclastic flows, lava flows, lahars, and tephra falls. Heavy ashfalls were recorded during three historic events in 1773, 1886, and 1916–1918.

Monitored scientifically since 1989 (Instituto Geofisico nd), Tungurahua gave serious indications in September 1999 that it was awakening from an 80-year sleep; seismic activity and gas emissions increased. On September 8, the Instituto Geofisico issued a technical alert, following persistent tremors and the emission of a vapor column that extended to two kilometers above the crater. The alert level then was raised to yellow

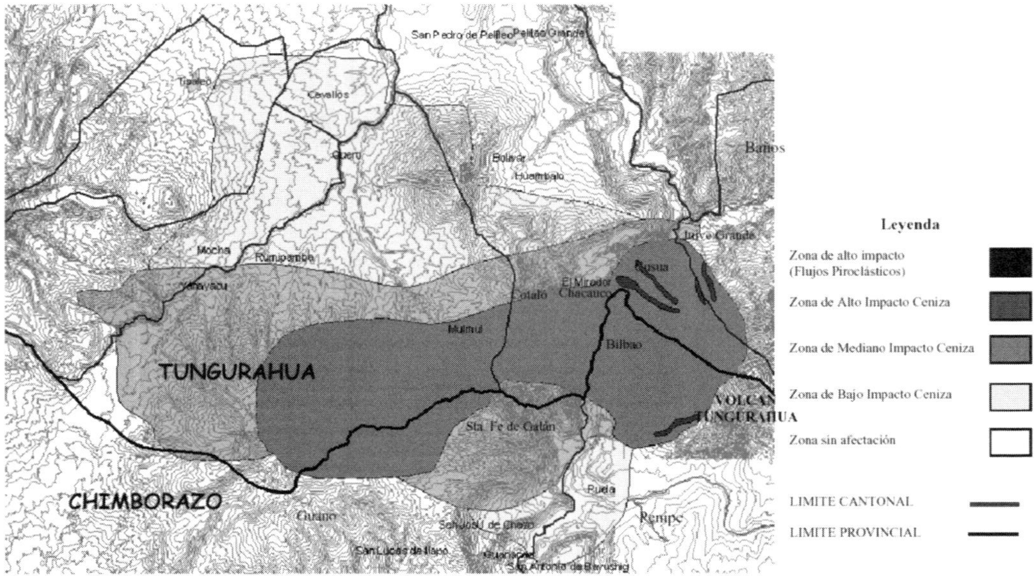

Figure 8.1. Ashfall from the July 2006 eruption (Source: Defensa Civil 2006).

signifying that a major eruption was possible within weeks. Three explosions on October 5 with increasing gas emissions and incandescence at the crater convinced authorities that an explosive eruption was imminent, and the alert level was raised to orange on October 16 (Global Volcanism Program 2008). This action resulted in a mandatory evacuation order for the town of Baños and several smaller rural communities in similar high-risk areas neighboring the volcano. Some inhabitants had voluntarily started to leave the area before the official order was issued (Tobin and Whiteford 2002a, 2002b).

While the major explosive eruption anticipated in mid-October 1999 has not occurred (as of May 2008), Tungurahua has been continuously, and sometimes intensely, active. Activity has included pyroclastic flows, the ejection of incandescent material to heights sometimes several hundred meters above the summit, numerous lahars, frequent ashfalls, tremors, and loud explosions, one of which was heard 75 kilometers away in Quito, the capital of Ecuador (Global Volcanism Program 2008). The 2003 Annual Summary Report from the Instituto Geofisico suggested that Mount Tungurahua is becoming more active (Instituto Geofisico 2003).

In 2003, episodes of intense volcanic activity included ashfall and increasing tectonic activity associated with the volcano (Instituto Geofisico 2003). June of 2003 began a particularly active cycle with tectonic events followed by swarms of tremors that lasted for long periods. Ashfall was heavy, resulting in then President Gutiérrez declaring the cantons of Peliléo, Quero, Cevallos, Mocha, Guano, and Penipe as a state of emergency (Instituto Geofisico 2003). In August and September of 2003 the activity intensified; ashfall continued, volcanic seismic events registered 4.5 and 4.7, and volcanic explosions were frequent (Instituto Geofisico 2003). This activity once again intensified in July and August 2006 and February and March 2007.

Methodology

Between 2000 and 2006, we conducted three research projects concerning risk, resilience, and vulnerability to chronic volcanic hazards in Ecuador, all centered on Mt. Tungurahua. A mixed methods approach was undertaken. An extensive literature review covered the following topics: acute respiratory infection, sociological and public health studies of social capital, women's health in high-risk environments, effects of stress on physical health related to disaster zones, infectious and communicable diseases, information on resilient and sustainable communities, rapid assessment procedures and other methodological instruments, and anthropological research design. These specific topics were identified in the literature from medical anthropology, public health, geography, and epidemiology. Background data on health characteristics were collected through retrospective reviews of records from 1995 to 2004 in the Health Centers in Baños and Quero looking for information specifically on upper respiratory infections and women's health. The team also collected information on infectious disease patterns in general for the same period.

In addition, we collected geophysical data on the frequency of eruptions, their relative intensities based on the Volcano Explosivity Index, and the spatial patterns of the ashfalls. In addition, samples of ash were collected and the chemical content correlated with the epidemiological patterns in the different study sites (Yepes 2002). We also conducted more than 850 interviews with members of nine communities, in shelters during the evacuation and in their homes when they returned. We held 6 focus group interviews totaling more than 80 people, interviewed more than 60 Civil Defense personnel and elected and appointed authorities, carried out more than 300 retrospective medical chart reviews, and applied geographic information system mapping to three communities. Focus groups were employed to elicit local terms, concepts, and concerns that were then integrated into the questionnaires. Interviews and questionnaires focused on disaster experiences, perceptions of risk, social conditions, health outcomes, social networks, and mitigation and recovery strategies. Interviews with authorities (elected, appointed, and Civil Defense) were designed to gather contextual information about tensions, actions/responses, relations between authorities and community members, and other information not available in secondary sources.

The research team also maintained a database of hazard documents, including volcanic activity reports from the Ecuadorian Geophysical Institute and the Global Volcanism Program at the Smithsonian Institution. The team also regularly reviewed Ecuador's major newspapers, including *La Hora*, *El Comercio*, *El Telegrafo*, and *Hoy*, to follow economic, political, and social issues related to the research project, in addition to tracking natural hazards both in the study area and nationally.

Cascade of Effects Model

We began to develop a Cascade of Effects Model based on our work in Ecuador, particularly on the effects of an initial disastrous event which is followed by continual exposure to ongoing disastrous conditions. We believe that such a set of chronic events has a cascading effect as impacts create additional problems that are exacerbated throughout society. For instance, Mt. Tungurahua has continued to erupt with varying degrees of intensity, creating problems for surrounding communities through periodic ashfalls, mudslides, pyroclastic flows, and lahars along with the ongoing threat of a total collapse of the volcano. The effects of such exposure are long-term in nature with significant structural devastation, yet aid and support from the government have been minimal. Our goal is to model how chronic exposure to disasters affects the health and overall well-being of people living in exposed communities.

Results from our third study are particularly relevant to this understanding as they set the stage for our understanding of the results of reconstruction—or the lack thereof—following the 2006 eruption. A brief overview is provided in table 8.1 of the results as they are structured by the Cascade of Effects model. These results are clustered in the following domains: *economic*, for example agricultural/rural recovery; *environmental*, for instance short-term soil contamination; *social capital*, including household reciprocity; and *health*, such as increased stress and compromised immune systems. The outcomes of each of these components are intricately linked with the others such that impacts on recovery are either exacerbated or diminished, depending on how the other variables play out. It is suggestive, for example, that strong social networks reduce the negative impacts of disaster and facilitate recovery. In contrast, chronic exposure to continuing ashfall leads to increased economic and health problems. These issues are highlighted in more detail.

Tungurahua Eruptions of 2006

Following the July 2006 eruptions, 940 people were living in shelters and more than 7,000 people had been evacuated to other places (Ecuadorian Red Cross September 4, 2007). While many people slept in the shelters, they returned daily to the hillsides to tend their crops and animals, often leaving their children in the shelters during the day. It has been estimated that when the August 2006 eruption began, as many as 1,600 people who initially left returned to rescue a few animals and belongings (Ecuadorian Red Cross September 4, 2006) in the ten hours of explosive rumblings that preceded the major eruption. After the eruption and the heavy ashfall, the conditions under which people lived and worked deteriorated, a situation complicated by rains and the limited

Table 8.1. Tabular Model of Cascade of Effects from Chronic Exposure to Volcanic Hazard

Specific Impact	*Cascade of Effects in Other Aspects of Life*
Economic	
Reduction in household economic reciprocity.	Social disarticulation.
Chronic ashfall damaged crops and reduced land productivity and stock animals.	Out-migration (and disruption of social support system); compromised nutrition and health.
Decreased ability to provide food (or having access to a reasonable level of resources on a day-to-day basis).	Lower levels of health and mental health; lowered perception of authorities and community leaders; increased perception of risk.
Women in both communities were less likely to have adequate food than men.	Negative social capital by impacting caregivers responsible for children and older adults.
Environmental	
In Quero, chronic environmental exposure (along with sustained economic loss, the loss of social capital, and negative health outcomes) has impeded recovery.	Exacerbated poverty and decreased standard of living.
Increased safety from short-term evacuation.	Disrupted networks and reduced economic output if evacuated for extended period of time.
Short-term soil contamination.	Abandonment of production (including possible forest regrowth), or alteration in type of crops planted.
Loss of crops, negative effect on next year's crop season, and loss of the seed crop.	Failing agricultural system; out-migration; exacerbated poverty.
Social Capital	
In Baños, strong social capital was essential to the recovery efforts following the evacuation.	Directly related to economic resources and level of health.
Cohesive family unit, especially non-absent spouses, although a traveling spouse also appeared to enhance social networks in Baños, potentially increasing resilience.	Better physical and mental health.
In Baños, personal social networks were the primary source of aid. Aid in Quero was primarily from government sources.	Redevelopment of economic reciprocity versus not, respectively.
Health	
Since the initial eruption in 1999, there have been increases in: total hospital consultations; upper respiratory tract infections, pneumonia, and diarrhea rates in children five years and younger; hypertension and gastric illnesses in older adults.	Reduced economic productivity.

Specific Impact	*Cascade of Effects in Other Aspects of Life*
Illnesses were significantly higher in Quero than in Baños.	Creation of greater dependence on external resources.
Increased anxiety and stress.	Reduced capability to fulfill social obligations.
Mitigation	
Negative experiences during and following evacuations.	Many inhabitants of both Baños and Quero will not evacuate again, although they recognize the risk of living near Mount Tungurahua.
Those who evacuated voluntarily had less difficulty finding food, possibly because of more resources and stronger social networks.	Government mandated evacuation, as opposed to voluntary evacuation, is associated with higher levels of stress and more negative perceptions of local governments.
In Quero, those who evacuated multiple times for short periods experienced better health conditions, more help, and less weight loss when compared with those who did not evacuate or only evacuated once.	However, confidence in the local government was lower than those who evacuated once or not at all.

number of shelters available. As the UN Office for the Coordination of Humanitarian Affairs (2006) noted:

- The main road between Baños and Riobamba was destroyed;
- There were 15,000 people in shelters, which lacked sufficient sanitary facilities, kitchen sets, and hygiene kits and water jerry cans;
- Shelters were set up in schools, delaying the start of the school year or forcing the relocation of the displaced people;
- Respiratory and skin problems were reported;
- Farmers lost cattle and crops (as well as chickens, sheep, goats, and pigs); and
- The Civil Defense declared that eight villages would have to be resettled in less than 60 days to safer areas (this was in August 2006).

Following the 2006 eruptions, two decisions significant for recovery were made by the authorities. One decision focused on the role of the Ecuadorian government in moving people out of the hazard zone, and the other decision reflected the role played by international humanitarian aid groups like UNESCO in recovery efforts.

According to one source (AP Newswire 2006), during the eruption in August 2006:

> Volcanic ash fell 140 miles west of Tungurahua; rivers of fire in the cold mt [mountain] night. Between 30,000 and 40,000 people inhabited the western slopes (faldas) of the volcano, the areas most heavily damaged and from which everyone was evacuated. At least 12 communities on the western slope were severely damaged with four totally destroyed. Juive Grande's 100 homes were destroyed by the molten flows, but the 600 residents escaped . . .
>
> Some, however, refused to leave their communities—3 higher up the slopes stayed. Fifty people from Penipe were treated for burns caused by lava flows and incandescent rocks that burned them as they tried to flee. They were also burned by vapor and the elevated heat in the zone. It was a scene of chaos, "a Dantesque situation" Dr. Hernan Aylala said.

In the interviews we conducted hours after the pyroclastic flows destroyed the villages, we asked people in shelters about their views on resettlement and their concerns about being in the shelters. Most of the people interviewed were from Cusua, one of the communities destroyed by the pyroclastic flows. At the time of the interviews, however, they did not yet know that much of their community was gone. When they were asked if they would be willing to move, they said:

> it would be very difficult to move because the soil of Cusua was so fertile that we fill the markets with our produce; there are no other soils on the other side [of the volcano, away from the flows].

Another man told us that the government only wanted to give them a little house, yet the people still needed their land; another woman added that they could sleep in the little government house, but would walk back every day to tend their land on the fertile side of the volcano. The displaced persons clearly recognized the power of the volcano, but generally distrusted the government and feared that they would be forced to leave their lands.

> We have had experience with the government and we no longer believe them. Last time when the roofs were destroyed, the government refused to give us new roofs for our homes because they said we lived in an extremely dangerous, high-risk zone and the roofs could be destroyed again anytime the volcano became very active.

Table 8.2. Losses Reported by Individuals in 2001 and 2004, by Community

Losses	*Penipe 2001 (percent)*	*Banos 2001 (percent)*	*Banos 2004 (percent)*	*Quero 2004 (percent)*
Crops	45	15	34	98
Animals	20	17	28	92
Equipment	5	—	10	20
Work	11	29	63	82
Business	6	17	—	—
Credit	7	1	33	47
House/Land	8	1	24	99
Total	N = 105	N = 106	N = 126	N = 104

Sources: Tobin et al. 2005; Whiteford et al. 2002

> These are our lands. They are the lands of our parents. If the volcano kills me and buries me, I will lie with my grandfather.

Cusua was one of the eight communities destroyed by the pyroclastic flows on August 16, 2006. Other nearby communities were damaged, but not destroyed, by the less powerful explosion on July 15, 2006. In the following tables, we highlight some of the losses in Cusua (in the canton or county of Pelileo) and the losses experienced in Penipe. Table 8.2 shows ongoing losses from the eruption and evacuation of 1999.

Losses in Penipe and in Quero were primarily due to continued ashfall and subsequent deleterious effects on crops and livestock. These losses help underscore how extremely vulnerable those communities were *prior* to the 2006 eruptions. Data in table 8.3 illustrate how Penipe, Baños, and Pelileo (canton where Cusua was located) were affected by the 2006 eruptions.

Table 8.3. Losses Reported after July and August 2006 Eruptions of Tungurahua

Canton	*Community*	*Houses Destroyed*	*Estimated Loss ($)*
Penipe	Bilbao Palictagua Yuibug Choclotus	185	2,910,290
Guano	Cahuaji El Cuzo Chazo La Palestina	158	2,862,862
Baños	Pondoa Juive Grande	170	3,108,523
Pelileo	Cusua Chacauco	200	4,008,732

Source: Defensa Civil 2006

Table 8.4. Estimated Losses by Canton

Canton	*Estimated Loss ($)*
Baños	**392,262**
Cevallos	399,528
Guano	572,727
Mocha	384,528
Pelileo	**917,471**
Penipe	**1,124,349**
Quero	419,528
Tisaleo	289,607
Total	4,500,000

Source: Defensa Civil 2006

Recovery strategy is complicated by the fact the there are three cantons (Baños, Pelileo, and Penipe) in the affected area (table 8.4). Previous research (Whiteford and Tobin 2004) revealed that Penipe had already suffered losses of crops, experienced extensive out-migration of the adult working-age population, and suffered ongoing health problems due to continued exposure to ash, and economic depression; the eruptions in 2006 only exacerbated a chronic situation. As table 8.4 shows, Penipe experienced further losses in July and August 2006 estimated at $1,124,349. The eruptions of July and August also had broader impacts that led to large-scale evacuations of several communities and official estimates indicate that at least 346 families were removed to shelters in Baños, Mocha, and Pelileo (table 8.5). In addition, the Ministry of Agriculture and Farming suggested that $500,000 monthly would be required to feed animals during the emergency (El Comercio 2007a).

Table 8.5. Population Evacuated in August 2006

Canton	*Community*	*Families*	*Population*
Pelileo	Cusua	107	535
	San Juan	26	130
	Pillate	28	140
	Various	6	30
	Choconco	32	160
Mocha	Juive Grande	12	60
Baños	Juive Grande	61	305
	Barrio las Ilusiones	37	185
	Pondoa	37	185

Source: Defensa Civil 2006

Reconstruction and the Cascade of Effects

The Cascade of Effects models shows how changes in economic conditions shape and are shaped by concomitant changes in environmental, health, and social conditions, and how they mutually influence one another. What the model demonstrates is that recovery is based primarily on reconstruction efforts and emergency relief associated with evacuation shelters or resettlement communities, in effect showing that health, the environment, and social capital issues are far beyond the scope of most recovery planners.

The reconstruction plan engaged by the Ecuadorian government is to relocate communities that have been destroyed to new plots of land, and in time, when the government raises enough money, to provide agricultural assistance to those relocated. However, people wish to remain on or near their traditional lands. Thus, an alternative to moving populations away from the risks and benefits of volcanic land might be to apply a "sustainable livelihoods approach" to the return or resettlement groups. A sustainable livelihoods approach:

> comprises the capabilities, assets (stores, recourses, claims and access) and activities required for a means of living: a livelihood is sustainable which can cope with and recover from stress and shocks, maintain or enhance its capabilities and assets, and provide sustainable livelihood opportunities for the next generation; and which contributes net benefits to other livelihoods at the local and global levels and in the short and long term (Chambers and Conway 1992:21).

This sustainable livelihoods approach builds on work by Cannon et al. (2003), United Nations International Strategy for Disaster Reduction (2005), and Wisner et al. (2004). Risk reduction, an idea utilized by the United Nations International Strategy for Disaster Reduction (2005), is associated with the concept of living with risk. Environmental hazards, then, are seen as a part of life and concomitantly productive livelihoods are translated into local disaster risk reduction strategies that include pre-disaster activities (i.e., preparedness and mitigation), and post-disaster response and recovery. Local-level discussion and design are considered paramount for success.

In the case of volcanic risk communities, such an approach would shift the focus from removal of risk to living with risks, and would put attention on the development of intergenerational means to productivity, for instance infrastructural know-how or rebuilding processes, as well as examining agricultural practices. For instance, could ash-resistant crops such as onions be substituted for commonly grown produce like potatoes? Rather than being further impoverished by the reconstruction, inhabitants would have their productive potentials enhanced. Nevertheless, such a substitution must be planned with local knowledge, because high reliance on one crop, like onions, almost inevitably creates higher vulnerability to market forces and can limit local food supplies. Also, political economic elites in such a case could also easily be criticized for leaving people in harm's way, even if economic activities become more sustainable.

Social Capital

Relocation, a policy favored by many in governmental authority, can have significant negative effects on social capital (Oliver-Smith 1996; Whiteford and Tobin 2004). Indeed, disruption of social relations and social networks that often provide support and access to resources can severely endanger community cohesion. Thus, during disasters, particularly when people are displaced or forced to relocate, social capital may be diminished, which threatens recovery efforts. When considered in terms of extended family and friends, strong social capital provides a critical lifeline for recovery, with family members sharing economic and social resources like childcare, food, and mutual aid. Many people like the Meneses family described below may never recover from the August 2006 eruption, but if they do it will be in part because of the benefits they received from family ties.

> On a once-lush landscape made wasteland overnight by the Tungurahua volcano, five members of the Meneses family raised picks and hoes in unison to strike at ash-encrusted earth. The land looks like concrete and is almost as hard. The Meneses returned home to replant their crops, only to find it takes backbreaking work just to crack a furrow in the hardened gray shell. "I am not giving in," said Antonio Meneses, sprinkling a neat row of corn kernels into one narrow crack. (AP Newswire 2006).

Meneses, 52, acknowledged he may not be able to pay back the $5,000 in loans he took out before the eruption to plant his crops. "We took a risk planting here," he said. "Now we will test our luck. Let us hope this harvest will not fail." Meneses' 26-year-old daughter, Ximena, added, "We have suffered the unspeakable, but we are not going to leave." (AP Newswire 2006).

Similar results were found in Baños, where strong social capital, especially cohesive family units and family networks were essential to recovery efforts. Aid, for example, was frequently related to social networks and promoted resilience in the face of the ongoing impacts of the disaster.

In addition, the incidence of disease was lower in Baños than in many other communities. These findings contrast with those in Quero canton, where communities were generally less wealthy and had limited access to most resources. The impacts of the volcano were more severe in Quero, where many rural communities had been evacuated on several occasions because of ashfalls. These chronic conditions have an almost daily impact on people's lives, and consequently many younger people left the area to find work elsewhere. Families came to rely heavily on government aid and generally exhibit higher levels of illness. The inevitable outcome has been a decline in community cohesion and a questioning of previous societal norms (Tobin et al. 2005).

Environment

While the biophysical environment has been modified due to the eruptions, the environment and people's relationship to it has also changed. Around the volcano, chronic exposure has severely impeded recovery as ongoing ashfalls and pyroclastic flows have disrupted agricultural practices, destroyed plants, and killed animals. In Penipe the mayor, Gonzalo Solano, identified the loss of crops, next year's agricultural season, and the seed crop as environmental losses (AP-CNN 2006). Lifestyles changed as farmers tried to adapt to prevailing disaster conditions. The inability to grow potatoes and the death of many small animals led to other costs related to health and economic outcomes (Tobin et al. 2005; Whiteford et al. 2005). Furthermore, types of crops might change from more intensive to less intensive, where people are not living on their land to provide the necessary care. Furthermore, contaminated water supplies precipitated more problems (Whiteford and Tobin 2007).

All these impacts have consequences on community sustainability and contribute to the cascade of effects. Environmental damage from the geophysical forces leads inevitably to economic pressures, in this instance farming, that in turn disrupt social networks and aggravate health problems. The chronic nature of the disaster further exacerbates the impacts.

Health

Health is a fourth factor in the cascade of effects. Health outcomes in areas of high ashfall have been well documented from around the globe (see for example, Allen et al. 2000; Forbes et al. 2003; Malilay et al. 1996; Searle et al. 2000) and from displaced person/refugee shelters (Cernea 1997, 2000).

Undoubtedly, the chronic conditions around Tungurahua have both directly, through ashfalls, and indirectly, though compromised lifestyles, increased health problems. Previous research has demonstrated higher incidence of various illnesses in those communities that are exposed regularly to ash and those residing in shelters (Tobin and Whiteford 2001, 2002b, 2004; Whiteford and Tobin 2002, 2004).

In the recent eruptions, the impacts were immediate. According to the appeal sent out by the website Reliefweb (2006), there were significant health problems, such as breathing difficulties, conjunctivitis, dermatitis, and acute diarrhea. The Ecuadorian Ministry of Food and Agriculture confirmed that 2,607,541 livestock and small animals had been affected, and that 1,297,994 hectares of land were covered in ash, destroying crops of corn, potatoes, beans, and barley.

Relocation and economic development incentives might provide improved sanitary conditions and closer access to health facilities, a positive, but the negative health consequences of greater sedentary lifestyles and easier access to processed foods might balance the benefits of relocation. Again, the cascade of effects must be considered in its local context.

Political Economy of Relocation

Ecuador has selected relocation and reconstruction activities rather than employing a sustainable livelihoods approach to those most affected by Tungurahua's ongoing explosions and chronic hazards. The plan has two stages. The first involves collaboration between the municipal government of Bilbao, in Chimborazo province, the Esquel Foundation, and the Ministry of Housing and Urban Development. The plan involves three housing developments with construction of 350 houses for those displaced. According to newspaper accounts, many residents were confused as to how to register for housing. Many evacuated people attempted to register in multiple assistance programs, because the government has only guaranteed homes for 553 families from Penipe and Guano, Chimborazo, and Pelileo and Baños, Tungurahua (El Comercio 2007c). In La Paz (Pelileo), the plan calls for the construction of 200 homes (El Comercio 2007c). While people wait, many have rented apartments in Baños or Riobamba, paying from $40 to $100 per month (El Comercio 2007c) while others remain in shelters. In February 2008, the first 65 houses of 185 houses in one relocation settlement were handed over to families by the government; other sites were still under construction (Hoy 2008). These houses are on 120 square meters of land and each house is to be 62 m^2 with two bedrooms, a kitchen, and a bathroom. One hundred houses were estimated to cost the government $1.5 million (El Comercio 2007c). In essence, these houses are miniature middle-class houses, but they are intended for rural agricultural families—perhaps not a best fit.

The second stage of the reconstruction plan—for which there is no timeline at the moment—is to provide for livelihood needs, although the government is still trying to raise money for this effort. This second stage involves delivery of productive technical assistance for agriculture activities in the new locations. According to Cornelio Marchán, president of the Esquel Foundation, the foundation had raised $350,000 in 2007, but was working to raise $500,000 to begin the second stage of the relocation project (El Comercio 2007d). In the meantime, the state of Tungurahua had given $1.3 million to aid with reconstruction; another $10 million had been released from Federal funds for distributing food for people and animals. Of this, $1.2 million had been assigned to the Geophysical Institute at the Polytechnic University to monitor the volcano, and $8 million promised to the Ministry of Housing (El Comercio 2007b).

Despite the promised aid, for those Ecuadorian communities most severely impacted, the challenges may be insurmountable. A sustainable livelihoods approach, for instance, would require comprehensive planning, the addition of significant resources, and considerable intergovernment cooperation within the country, not all of which can be achieved in a nation that is confronted by many other concerns. Indeed, political turmoil, economic crises, and social upheaval at the national level have been frequent for the past decade.

Conclusion

The data demonstrate that communities like Penipe—already weakened and vulnerable because of ongoing ashfalls in the years 1999–2008 and where the environment had already been negatively affected by the volcano—again suffered from economic and environmental depredations that cumulatively impacted one another. The synergistic effect has been to put pressure on people to move out of the area to non-ash-affected communities, thus impacting social capital both of those who left and those who stayed. Health, damaged by the ongoing exposure to ash, is further compromised by the loss of friends and kin as parents move to Riobamba to work, leaving children at home with elderly parents or grandparents (Whiteford and Tobin 2004). The data also suggest that there is a range of activities that these agriculturalists pursue, including: out-migration; different forms of land tenure; a change in crops grown; changes in the relative contributions of human labor, machines, animals, or chemical inputs; and household reformation.

Anthropologically, we can model disaster-related consequences of changed economic processes due to modified farming potential—including reduced household reciprocity, limiting celebration and event spending, reduction in agricultural activities, while perhaps increasing livestock husbandry. Although our study does not involve long-term or intergenerational adaptations to the eruptions over the past millennium, it is worth keeping in mind two things: the possibility that "built in" cultural responses do exist that allow for economic continuity, and that contemporary political economic strategies of elites place specific pressures on those cultural tendencies. As such, it would be useful to assess the costs to the Ecuadorian government of the volcanic eruption in terms of loans acquired, indebtedness increased, deepening impoverishment of rural communities, and the loss of land, homes, crops, and animals. In addition, from a political economic perspective, we need also to situate the Ecuadorian experience in the larger picture of global disaster recovery, which has seen an 800 percent increase in World Bank lending since 1980 (Klein 2005). Neoliberalism policies have moved international aid resources away from the public sector, accelerating the privatization of public sector services (Whiteford 2005; Whiteford and Bennett 2005). It is not just a question of distal and proximal causes or consequences of disasters, it is a question of what kind of outcome is desired and who should be in charge of designing it.

Nonetheless, most recovery interventions are too little too late. Questions, therefore, arise of the resiliency of communities under chronic disaster and to what extent sustainable conditions are even possible (Tobin 1999). As others have pointed out that reconstruction follows the same fault lines that exist prior to the disaster and hence serve to perpetuate problems rather than mitigate the real causes. Vulnerable populations then are left exposed to ongoing hazardous conditions.

In the final analysis, therefore, it is the context in which those people living in the communities around Mt. Tungurahua find themselves that presents the most significant challenges—challenges that can be found worldwide. The two main challenges are to: 1) undo the prevailing conditions that create such vulnerabilities and lead to the cascade of effects associated with chronic disaster, 2) revamp reconstruction, which, rather than being transformative and taking advantage of liminal moments to improve the lives of those affected by marginalization, prejudice, and poverty by reducing structural violence, tends toward the recovery of the same kind of inequality that existed before.

Acknowledgments

Funding provided by: the Global Center for Disaster Management and Humanitarian Action, Institute for the Study of Latin and America and the Caribbean, The College of Arts and Sciences at the University of South Florida. We would also like to acknowledge the help we received from A. J. Faas, doctoral student in anthropology at the University of South Florida.

References

Allen, A. G., P. J. Baxter, and C. J. Ottley

2000 Gas and Particle Emissions from Soufriere Hills Volcano, Montserrat, West Indies: Characterization and Health Hazard Assessment. Bulletin of Volcanology, 62: 8–19.

AP-CNN

2006 AP-CNN. August 18, 2006. Electronic document, www.cnn.com, accessed August 30, 2007.

AP Newswire

2006 August 21, 2006. Electronic document, www.lasvegassun.com, accessed August 30, 2007.

Cannon, T., J. Twigg, and J. Rowell

2003 Social Vulnerability, Sustainable Livelihoods and Disasters. London: Conflict and Humanitarian Assistance Department and Sustainable Livelihoods Support Office, Department for International Development.

Cernea, M. M.

1997 The Risks and Reconstruction Model for Resettling Displaced Populations. World Development, 25(10): 1569–1587.

2000 Risks, Safeguards, and Reconstruction: A Model for Population Displacement and Resettlement. *In* M. M. Cernea, C. McDowell (Eds.) Risks and Reconstruction: Experiences of Resettlers and Refugees. The World Bank: Washington, DC.

Chambers, R., and G. R. Conway

1992 Sustainable Rural Livelihoods: Practical Concepts for the 21st Century. Discussion Paper, 296. Brighton, UK: Institute of Development Studies, University of Sussex.

Defensa Civil

2006 Informe preliminar del desarrollo, efectos e incidencias de la actividad del proceso eruptivo del volcan Tungurahua Año 2006. [Preliminary Report on the development and affects of the eruptiveprocess of the Tungurahua volcano.] Ecuador Civil Defense document. 73 pages.

El Comercio
2007a 34 poblaciones de Chimborazo en líos por el volcán. [34 Communities in risk areas around the Chimborazo volcano.] March 13.
2007b Donde Esta el Dinero? Volcán: La espera por los recursos todavía sigue. [Where is the Money? The hope for resources continues.] March 23.
2007c 553 casas para los afectados del Tungurahua. [553 Houses affect by the Tungurahua volcano.] March 29.
2007d Apoyo del Ayuntamiento de Madrid para afectados por el volcán. [The Madrid Government to help those affected by the volcano.] March 30.

Ecuador Red Cross: International Federation of Red Cross and Red Crescent Societies, Ecuador
2007 January 9, Operations Update: Ecuador Volcanic Eruption. Appeal No. MDREC002, Update #3, http://www.ifrc.org/cgi/pdf_appeals.pl?06/MDREC00203.pdf.

Forbes, L., D. Jarvis, J. Potts and P.J. Baxter
2003 Volcanic Ash and Respiratory Symptoms in Children on the Island of Montserrat, British West Indies. *Occupational and Environmental Medicine* 60(3):207–211.

Global Volcanism Program
2005 www.volcano.si.edu/reports/usgs/. Accessed May 8, 2008.

Hall, M. L., C. Robin, B. Beate, P. Mothes and M. Monzier
1999 Tungurahua Volcano, Ecuador: Structure, Eruptive History and Hazards. Journal of Volcanology and Geothermal Research 91:1–21.

Hoy Online
2008 Chimborazo: Casas para 65 familias. [Chimborazo: Homes for 65 Families.] www.hoy.com.ec/NoticiaNue.asp?row_id=288682.

Instituto Geofisico
nd www.igepn.edu.ec/. Accessed May 8, 2008.
2003 Volcanes. Escuela Politécnia Nacional. [Volcanoes: National Polytechnic Institute.] Available at www.epn.edu.ec-igeo/sismologia/sismologia.html.

Klein, N.
2005 The Rise of Disaster Capitalism. The Nation. www.thenation.com.doc/20050502/klein. Accessed January 13, 2007.

Lane, L. R., G. A. Tobin, and L. M. Whiteford
2004 Volcanic Hazard or Economic Destitution: Hard Choices in Baños, Ecuador. Environmental Hazards: Human and Policy Dimensions, Global Environmental Change, Part B 5(1):23–34.

Lowell, J. R.
2001 The First Snow Fall. *In* An American Anthology 1787–1900. No. 351. Electronic document, www.bartleby.com, accessed April 7, 2007.

Malilay, J., M. G. Real, A. Ramirez Vanegas, E. Noji, and T. Sinks
1996 Public Health Surveillance After a Volcanic Eruption: Lessons from Cerro Negro, Nicaragua, 1992. Bulletin of the Pan American Health Organization. 30(3): 218–226.

Oliver-Smith, A.
1996 Anthropological Research in Hazards and Disaster. Annual Review of Anthropology 25:303–328.

Reliefweb
2006 Situation Report August 18, 2006. Electronic document, www.paho.org/Spanish/DD/PED/tungurahuaAug_06.htm, accessed August 30, 2007.

Searle, A., A. Nicholl, and P. J. Baxter
2002 Assessment of the Exposure of Islanders to Ash from the Soufriere Hills Volcano, Montserrat, British West Indies. Occupational and Environmental Medicine, 59: 523–531.

Tobin, G. A.
1999 Sustainability and Community Resilience: The Holy Grail of Hazards Planning? Environmental Hazards: Human and Policy Dimensions, Global Environmental Change, Part B 1(1):13–25.

Tobin, G. A., and L. M. Whiteford
2001 Children's Health Characteristics Under Different Evacuation Strategies: The Eruption of Mount Tungurahua, Ecuador. Papers and Proceedings of the Applied Geography Conferences 24: 183–191.
2002a Economic Ramifications of Disaster: Experiences of Displaced Persons on the Slopes of Mount Tungurahua, Ecuador. Papers and Proceedings of the Applied Geography Conferences 25: 316–324.
2002b Community Resilience and Volcano Hazard: The Eruption of Tungurahua and Evacuation of the Faldas in Ecuador. Disasters: The Journal of Disaster Studies, Policy and Management 26(1): 28–48.
2004 Chronic Hazards: Health Impacts Associated with On-Going Ash-Falls Around Mt. Tungurahua in Ecuador. Papers of the Applied Geography Conferences 27:84–93.

Tobin, G. A., L. M. Whiteford, L. M., and C. Laspina, with H. M. Bell, A. L. Freidus, and H. Yepes
2005 Perception, Social Support and Chronic Exposure to Hazards: Human Health and Community Well-Being. Technical Report. University of South Florida: The Global Center for Disaster Management and Humanitarian Action. 280 pages.

UN International Strategy for Disaster Reduction
2005 Hyogo Framework for Action 2005–2015: Building the Resilience of Nations and Communities to Disasters. Geneva: United Nations International Strategy for Disaster Reduction.

UN Office for the Coordination of Humanitarian Affairs
2006 OCHA Situation Report, 6. August 21.

Whiteford, L. M.
2005 Casualties in the Globalization of Water: A Moral Economy Perspective. *In* Globalization, Water and Health: Resources in Times of Scarcity. L. M. Whiteford and S. Whiteford, eds. Pp. 25–45. Santa Fe, NM: School of American Research Press.

Whiteford, L. M., and L. Bennett
2005 Applied Anthropology and Health and Medicine. *In* Applied Anthropology: Domains of Application. S. Kedia and J. van Willigen, eds. Pp. 119–149. Westport, CT: Greenwood Publishing.

Whiteford, L. M., and G. A. Tobin
2004 Saving Lives, Destroying Livelihoods: Emergency Evacuation and Resettlement Policies. *In* Unhealthy Health Policies: A Critical Anthropological Examination. A. Castro and M. Springer, eds. Pp. 189–202. Walnut Creek, CA: AltaMira Press.
2007 Potable Water as the Key to Survival and Community Recovery: Disaster Planning and Practice. *In* S. Wang, G. Tang, J. Zhang, W. Song, J. Ammann, and C. Kux (eds.) Strategy and Implementation of Integrated Risk Management. Proceedings of the International Disaster Reduction Conference, Harbin, China: Qunyan Press. pp. 447–453.

Whiteford, L. M., G. A. Tobin, C. Laspina, and H. Yepes
2002 In the Shadow of the Volcano: Human Health and Community Resilience Following Forced Evacuation. Technical Report. University of South Florida: Center for Disaster Management and Humanitarian Assistance. 548 pages.
2005 Research on People Affected by Chronic Exposure to the Eruptions of Mount Tungurahua 2000–2005. Ecuador: Building a Model for Hazard Mitigation. Summary Technical Report: Prepared for Ecuadorian Government Officials and Community Leaders. 23 pages.

Wisner, B., P. Blaikie, T. Cannon, and I. Davis
2004 At Risk: Natural Hazards, People's Vulnerability and Disasters, 2nd ed. London: Routledge.

Yepes, H.
2002 Personal Communication. Director, Instituto Geofisico, Escuela Politecnica Nacional, Quito, Ecuador.

PRODUCT DISTRIBUTION IN HAZARDOUS SETTINGS

IV

When the Lights Go Out: Understanding Natural Hazard and Merchant "Brownout" Behavior in the Provincial Philippines

9

TY MATEJOWSKY

THE INTERPLAY BETWEEN NATURAL HAZARD and undercapitalized infrastructure holds particular sway over Philippine society. Like populations in other less developed countries, Filipinos have grown accustomed to dealing with recurrent breakdowns in public services brought on by catastrophes and their secondary effects. Perhaps no group exhibits more resilience and self-reliance when confronting disaster-related adversities than local street store merchants. Shopkeepers' ability to recover and operate amid chronic power disruptions reveals a level of adaptability that is all but essential for negotiating the complexities of local commerce under the persistent threat of natural catastrophe. The underlying costs of disasters and intermittent grid failures on economic productivity and profitability are generally recognized by small shop owners as simply part of doing business in today's Philippines.

Surely, it seems almost paradoxical that something as crucial as basic infrastructural support has proven so unpredictable for local retailers. Years of mismanagement and under-investment in essential services like electricity, water, and transportation have left the Philippines with an infrastructural base that is in many ways fragmented and overburdened. In no circumstances are these structural deficiencies more starkly exposed than in the context of natural catastrophes. Underlying flaws in the country's infrastructural apparatus invariably emerge whenever disasters strike. Hindering recovery efforts in the short-term, these deficits also produce complications when comprehensive preparedness strategies are implemented in vulnerable communities over the long-term (Allen 2003).

Infrastructural failures take on an added dimension in the Philippines as it is widely regarded as a "culture of disaster" (Bankoff 2002). This label reflects the country's volatile position along the Pacific's "Ring of Fire" and "Typhoon Alley." Each year Filipinos are subject to a wide range of severe seismic and meteorological activity that frequently disrupts the rhythms of workaday life. The International Federation of Red Cross and Red Crescent Societies estimate that no other country experiences natural disasters of both types at a higher frequency than the Philippines (Philippine Daily Inquirer 2003). With 18 active volcanoes, at least 6 earthquakes per day, and some 20 tropical storms and typhoons sweeping through the islands annually, the immediate and lingering effects of extreme geophysical phenomena are realities that beset Filipinos on an almost daily basis (Wilson 2006).

With such high incidence and minimal safeguards, the human costs of recent natural disasters are nothing if not profound. Roughly 34,000 Filipinos were killed by natural catastrophes between 1971 and 2000, with around 35 million severely affected by disasters during the 1990s (Population Reference Bureau 2006:2). Indeed, the Philippines rank as the world's fourth most accident-prone country behind China, India, and Iran when counting both man-made and natural disasters (Philippine Daily Inquirer 2003). The Philippines likely would have placed much higher if not for China and India's massive populations.

Beyond human losses, natural disasters wreak havoc on the government's ability to provide essential services to affected areas. Such capacity falls woefully short given the often overwhelming and unpredictable nature of the typhoons, floods, mudslides, earthquakes, volcanic eruptions, and other natural hazards that recurrently impact the Philippines. Beginning in the early 1990s, when the Ramos administration began to provide financial guarantees mitigating many of the economic risks faced by investors (Abrenica and Llanto 2003:255), policymakers have worked actively to improve the country's infrastructural apparatus by privatizing and streamlining many of its regulatory agencies. Nonetheless, structural deficiencies continue to plague the Philippines' public works and utilities. A rapidly expanding population—85 million with an annual growth rate of 2.36 percent, one of the highest in Asia—and slow but significant economic growth threaten to exacerbate the situation by creating conditions where the state's capacity for meeting infrastructural needs is continually outstripped by public demand. The dramatic rise in global oil prices over recent years makes matters worse for the Philippine government, as energy costs become increasingly difficult to incur (Sun Star 2005).

Disasters and their residual effects present particular types of challenges for local residents when it comes to basic services like electrical utilities. Problems affecting power delivery in the Philippines are pervasive and long-standing (McBeth 1993; Tasker 1992). Grid failures are especially acute during the wet season (August–January) when torrential rains and typhoons seriously disrupt power flows. So prone to irregularities is electrical service that, rather than being viewed as anomalous, sporadic losses in voltage or "brownouts" have emerged as quasi-normative aspects of local life. Their frequency and persistent threat play a surprisingly influential role in shaping the political, economic, and cultural sensibilities of contemporary Filipinos (Bankoff 2002:1–2).

The effects of recurrent power failures touch nearly all levels of Philippine society. Inconveniences generated by brownouts are rarely experienced uniformly across local communities. The marked socioeconomic disparities underlying today's Philippines leave many Filipinos particularly vulnerable to these intermittent breakdowns. As is the case with natural disasters, low-income households tend to bear the brunt of brownouts' more disruptive effects. Ill-equipped to deal with the sometimes prolonged repercussions of power glitches, this group stands in contrast to more capitalized commercial ventures. For local merchants, cuts in electrical service typically introduce new wrinkles into their retail operations that vary in severity. Notably, these disruptions can occur without causing severe work stoppages. Grid failures have become so common for shopkeepers, in fact, that they are now sufficiently integrated into the fabric of everyday commercial life. Even in the aftermath of natural disasters when power delivery is spotty at best, store operators can almost always regain some semblance of a normal business routine.

In many ways this acceptance of unreliable infrastructure mirrors Bankoff's description of the roles that natural catastrophes and their looming threats play in shaping Filipinos' overall outlook toward life. He argues that characteristics like fatalism, ambivalence about the government, and strong religious faith are all aspects of a generalized Filipino mind-set that is significantly influenced by the persistent threat of earthquakes, typhoons, volcanic eruptions, and other natural disasters.

In many respects, erratic electrical service is something that merchants have learned to effectively manage rather than entirely overcome. Over the years, traders have developed an impressive array of coping mechanisms, stopgap business strategies, and risk aversion/management techniques to mitigate the disruptions caused by infrastructural failures and natural hazards (Robles 1991). Since local storeowners generally serve as the lifeblood of provincial communities in post-disaster recoveries, they are highly attuned to the risks and breakdowns stemming from brownouts and limited disaster response. Merchants are compelled to pursue various coping strategies, the specifics of which reveal an underlying economic rationality aimed at minimizing potential losses.

Objectives and Methodology

Broadly speaking, this study is a descriptive account of: 1) how weak infrastructural support exacerbated by disaster phenomena impacts local merchants; and 2) what strategies shopkeepers employ to alleviate the economic disruptions arising from such structural deficiencies. The underlying organization of merchant commerce in provincial Philippine cities is clearly influenced by previous encounters with disaster-related adversities. In significant ways, it determines how shopkeepers expend capital, utilize technology, extend credit, and employ labor.

An anthropological account of local responses to brownouts is long overdue. While there has been ample work on economic decision-making approaches in the post-disaster context, little has been written about how the residual effects of disasters like power outages are negotiated by affected communities. Over the past three decades, anthropologists have compiled several closely drawn profiles of individual, group, and institutional behavior at various stages of disaster phenomenon and the strategies employed to better cope with stress brought on by such extreme geophysical events. Notably, these studies focus on adjustments in religion and ritual, technology, economy, politics, and patterns of cooperation and conflict as they emerge both at disaster impact and over subsequent stages (Bode 1989; Button 1992; Gordon 1999; Guillette 1991; Rossi 1993).

Vulnerability to grid breakdowns elicits a range of responses at the community level that are articulated in both profound and subtle ways. Taken as a whole, these countermeasures reflect a level of grit that is essential for dealing with the unpredictable nature of urban commerce in the third world. Besides undercapitalized infrastructure, municipal retailing in less developed countries generally entails limited financial safeguards such as insurance, formal contracts, and government oversight, not to mention somewhat heightened vulnerability to street crime and minimal reliance on technological inputs.

Before taking a closer look at merchant brownout behavior, I will outline some problems confronting the country's infrastructural system, specifically addressing political and socioeconomic implications for both the Philippine state and local communities. This is followed by an overview of the country's power system, recently the focus of reorganization efforts by the government to improve the delivery of basic services and utilities throughout the archipelago. After developing that political economic context, I narrow focus to consider how electrical service glitches are experienced directly as Filipino merchants cope with the inconveniences born out of a disjointed power system. Finally, I situate one community's experience with brownouts into the larger context of the Philippines and other less developed countries. Matters of economic life such as capital outlays, technological inputs, credit extension, sales strategies, and store security all figure prominently in this section.

Ethnographic data collected in Dagupan City, Pangasinan, in 1994, 1997, 1998, 2005, and 2006 add empirical depth to a process affecting many local merchants. Dagupan's trade community provides an ideal context for such work as urban shopkeepers have repeatedly weathered brownouts stemming from natural disasters over the past two decades. Information culled from participant observation, informal interviews with local storeowners, and other field methods illuminate local brownout responses and the interplay between natural hazard and undercapitalized infrastructure.

Participant observation regarding brownouts occurred over several field sessions including June–July 1994, June–December 1997, January–May 1998, July–August 2005, and July 2006. The bulk of the informal interviews within Dagupan's local trade community took place in 1994 and 1997 with questions focusing mainly on post-disaster recoveries following a major 1990 earthquake. Interview participants were selected primarily through a cluster sampling technique whereby sellers specializing in different product lines were approached and secured for interviews. Interviews typically lasted 30 minutes to an hour with occasional follow-up conversations taking place over subsequent research periods. All told, some two dozen merchants were interviewed in this manner. I should note that one storeowner in particular has served as a viable source of information regarding local shopkeepers' dealings with brownouts. The knowledge that he shared with me in 2005 and 2006 on the subject of power outages and the coping mechanisms he and his competitors utilize in dealing with them proved indispensable to this study. Supplementary information was culled from local newspapers, especially Dagupan's *Sunday Punch*. Anthropologists interested in pursuing similar research in other cultural contexts would do well to balance ethnographic information collected via interviews and participant observation with local news coverage to better enhance the longitudinal character of their studies.

In order to better understand merchant brownout behavior in the post-disaster context, I follow Oliver-Smith's (1996:305) definition of disaster:

> a process/event involving a combination of a potentially destructive agent(s) from the natural and/or technological environment and a population in a socially and technologically produced condition of environmental vulnerability. The combination of these elements produces damages or loss to major social organizational elements and physical facilities of a community to the degree that the essential functions of the society are interrupted or destroyed. This results in individual and group stress and social disorganization of varying severity.

This definition underscores the far-reaching implications of natural catastrophes on impacted populations. When disasters strike they tend to be "totalizing" events that leave few, if any, aspects of community life unaffected. While secondary effects may vary in intensity, they can still serve as sources of frustration and inconvenience for locals in the weeks, months, or even years after disaster initially strikes.

Infrastructural Problems—State Level

At the national level, disruptions in electrical and water service, coupled with unreliable transportation and communication systems, stand as major obstacles for the state in attracting direct foreign investment and generating crucial economic development. Few international firms looking to expand operations in Southeast Asia will consider long-term investment in the Philippines without more infrastructural stability. In many ways, this situation represents something of a catch-22 for the Philippines. On the one hand, improved basic services and utilities rests largely on increased capital inputs from abroad. On the other hand, direct foreign investment will only intensify once the country's infrastructural base is more fundamentally sound. Notably, some of the only instances where foreign capital investment has flowed into the Philippines over the last 15 years are the former U.S. military installations of Clark Air Base (Angeles City, Pampanga) and Subic Naval Station (Olongapo, Zambales). While these demilitarized areas now mainly serve industrial and commercial purposes, their core infrastructure and operational systems remain a direct holdover from the American era.

The Philippine government has taken steps in recent years to rectify this situation of undercapitalized infrastructure by adopting a more neoliberal economic approach that favors open markets and the privatization of state-owned assets (Balisacan and Hill 2003). Moving from a government-controlled system to one that is for profit and privately run does not come without its share of risks. The government typically holds a disadvantageous position when it comes to negotiating with outside interests. In the past, the state has rushed into agreements with private companies only to find the long-term costs of these deals outweigh their short-term benefits (Bello et al. 2004:189–214). Even if recent efforts to privatize state resources prove successful, however, it will be years before their substantive benefits take effect. To date, few investors have expressed interest in taking on the state's more problematic public industries and utilities (Bello et al. 2004:210–12).

No less relevant, infrastructural deficiencies act as a drain on state resources and capital. Over the past 35 years the Philippines has accrued billions in foreign debt to bankroll a number of ill-conceived public works projects.[1] Funds that could be applied toward health care, education, and other vital matters are instead used to service interest on decades-old development loans. The Aquino administration's (1986–1992) emphasis on repayment of foreign debt has been roundly criticized as starving the Philippines of much-needed capital and contributing to economic stagnation (Bello et al. 2004:4). According to Supreme Court Associate Justice Reynato Puno, serving the country's debt now rates as the largest expenditure in the government's 2005 national budget of P907.56 billion (Cabreza 2005).

Finally, the Philippines' fragmented infrastructure works to undermine the state's ability to gain and keep the confidence of its citizens. In many ways, this flawed system serves as a vivid reminder of the government's inability to provide an adequate life for much of its population. The institutional integrity of the Philippine government becomes more tenuous every time one of its subsectors fails to deliver. Without a more robust infrastructure in place, many of the homegrown dissatisfactions that give rise to civil unrest, political instability, and other security concerns for the state will continue to germinate throughout Philippine society.

Infrastructural Problems—Local Level

Problems with basic services and infrastructure are no less significant at the local level. In fact, it is within local communities and households where the stark realities of a weak public infrastructure are most deeply felt. Erratic utilities exacerbated by natural catastrophes are just one of the structural adversities affecting living conditions in the Philippines. Such deficiencies superimpose themselves on nearly all aspects of local life; locals can almost always expect some kind of disruption in their daily routines due to such breakdowns. While this prevailing sense of uncertainty does not preoccupy residents to the extent that they suffer undue anxiety, it does exact an appreciable toll on them in terms of inconvenience, frustration, and lost productivity.

Philippines' Electrical Infrastructure

Few components of the Philippines' infrastructural base are more prone to inadequate service than the country's electricity industry. Problems affecting power generation and delivery are rooted in a number of wide-ranging factors including geography, weather, rising consumer demand, fiscal deficit, foreign oil dependence, and most especially, natural disasters. Their cumulative effect places the Philippines in a highly vulnerable position where there is no easy solution or quick fix.

Theoretically, the nation's capacity to produce energy should be much more reliable. The Philippines is endowed with abundant natural resources that if properly harnessed could augment the state's ability to generate electricity. Ample deposits of clean-burning natural gas remain untapped in parts of Mindanao and southern Luzon. Similarly, geothermal energy in volcanically active regions like the Visayas is woefully underutilized. Even though the Philippines are the second largest producer of this renewable resource,[2] efforts to capitalize on its geothermal potential have only been partially achieved. Environmentally friendly renewable energy sources like solar, wind, and micro-hydropower hold promise as future energy sources. Unfortunately, the benefits to be derived from these resources appear elusive as efforts to overhaul the Philippines' current energy system falter in the face of increasing consumer demand and entrenched political interests. Presently, the country meets most of its electricity needs through imported oil.

Some of the difficulties facing the Philippines' power sector are born out of its insular geography. The archipelago is comprised of over 7,100 islands strewn over a land area of some 300,000 square kilometers. From an operational standpoint, bringing power to remote areas, much less linking up the entire country under a comprehensive grid, represents a major logistical challenge. Compounding this is the extreme meteorological and seismic activity that impacts the Philippines with recurring frequency (Bankoff 2002). Typhoons, earthquakes, and other catastrophes wreak havoc on the operational capacity of the nation's utilities and basic services. Efforts to integrate the country's three major power grids (Luzon, Visayas, and Mindanao) into a more effective system are constrained by these natural events.

Notwithstanding disaster and geographical factors, problems affecting the delivery of electricity in the Philippines are also institutional in nature. The country's power supply network is basically a collection of loosely federated utilities that operate generation facilities—some successfully, some less so—at the regional level. At present, electrical service is dominated by the National Power Corporation (Napocor), which was created in 1936 as a non-stock public corporation that nationalized the hydroelectric industry. This state-owned utility and its smaller sister company, the Manila Electric Company (Meralco), which serves the Metro-Manila area, buy power directly from the private sector. The National Power Corporation and the Manila Electric Company hold contracts with a number of independent power producers that were negotiated under unfavorable circumstances for the state in the early 1990s.

Following the political and economic turmoil of the mid-1980s, the Philippine government was left with spiraling debt and insufficient investment capital to expand its energy reserves. With its generating capacity lagging behind consumer demand, the country began experiencing crippling power outages that lasted up to eight hours at a stretch in the late 1980s and early 1990s. At its worst, Filipinos suffered some 300 brownout days in 1993; up from 264 the previous year. This crisis was compounded by two major natural disasters: an earthquake measuring 7.8 on the Richter scale in July 1990, and the catastrophic Pinatubo eruption almost 11 months later. The most destructive and lasting effects of both disasters were concentrated in the towns and cities of central and northern Luzon.

The Aquino and Ramos administrations addressed the energy crisis by bringing in independent power producers to provide electricity. Initially, this move was hailed as a success since it significantly boosted the Philippines' production capacity, and power outages turned less severe. Over time, however, it became apparent that the opening of the state's electricity sector to private concerns had resulted in a number of unforeseen consequences for the state and utility consumers (Bello et al. 2004:189–214).

In order to attract independent power producers to the Philippines, the government was placed in the unfavorable position of having to offer serious concessions to private companies. The agreements that were eventually worked out between the state and independent power producers clearly favored the latter since they were shielded from substantial financial risk. Significantly, the government agreed to purchase all of the energy generated by independent power producers whether it was used or not. The National Power Corporation also pledged to sell fuel to private power firms at a set price whereby increases in petrol prices were assumed by the state. Similarly, payments to independent power producers from the state were negotiated in U.S. dollars at a fixed rate and not subject to any devaluation of the Philippine peso (Villanueva 2002). Thus, when the value of the peso began to slide against the dollar following the 1997 Asian Economic Crisis, the government was obligated to pay nearly twice as much to independent power producers for the same amount of power. Not surprisingly, many of these costs were passed to consumers in the form of increased utility rates. Filipinos presently pay some of the highest electrical rates in the world. To help streamline power delivery and reduce energy costs for Filipinos, a wholesale electricity spot market commenced operations in June 2006. In this scheme, prices are established through supply and demand and a free-market competition rather than by predetermined fixed rates (Yap 2006).

Brownouts

At the community level, flawed electrical service usually takes the form of brownouts. These intermittent power outages are known locally as *wala ng kuryinti*, which roughly translates from Tagalog as "no electrical current." Brownouts are caused by a temporary reduction or cutback in energy loads for a particular region. Unlike blackouts, which involve a complete grid failure over a large geographical area, brownouts tend to be more localized and usually less severe in terms of magnitude and duration. Filipinos almost always apply the term brownout to describe the recurring failures of electrical service that punctuate local life in the Philippines no matter how long they last. In some cases, electricity returns to normal levels within a matter of minutes or hours. In other instances, especially in a disaster's aftermath, days or even weeks can pass before full power capacity is restored.

Besides natural disasters, brownouts can be triggered by mechanical problems, power shortages, or consumer overuse. Sometimes electricity providers implement rolling brownouts over a coverage area in order to complete repairs or preventative maintenance on a power supply network. Consumers are usually notified through print or broadcast media about the expected timing and duration of such repair work. Frequently, utility companies overrun their initial time estimates or schedules by several hours or even days (Sunday Punch 2004). It is in the latter circumstances where the patience and coping skills of Filipinos are most severely tested.

Arguably, it is the erratic timing of brownouts and not so much their duration that works to disrupt the rhythms of workaday life. Not knowing if reliable electrical utilities will be available from one day to the next casts a shadow of unpredictability over living conditions in towns and cities across the Philippines. Power shutdowns usually occur without warning, leaving activities interrupted or unfinished. Even though a majority of Filipinos are still mired in poverty, few are so impoverished or geographically isolated that they live off-grid since it is common for people to steal electricity by siphoning off power from the electrical pylons that service their neighborhoods. The illegal wiring of homes to these external power sources is often done in ways that leave households vulnerable to fines, arrest, electrocution, or fire.

Dagupan City: Demographics, Disasters, and Power Disruptions

To better understand how brownouts affect local commerce in the post-disaster context, I now consider the recent experience of Dagupan City, Pangasinan. Dagupan is one of north-central Luzon's foremost commercial, educational, and financial centers. Situated along the Lingayen Gulf some 210 kilometers north of Metro-Manila, the coastal city has about 130,000 residents living in approximately 25,000 households spread across 31 precincts. Like other towns and cities in the Philippines, urban retailing is dominated by a small but influential community of Chinese-Filipinos.[3] While comprising less than 5 percent of Dagupeños, this tight-knit group has remained at the forefront of downtown's street store trade for generations, especially in subsectors like groceries, textiles, hardware, and appliances. Thus, it is among ethnic Chinese shopkeepers where merchant brownout behavior gains maximal expression.

Numerous power glitches have interrupted the smooth and effective flow of municipal commerce over the past two decades. By 2005, it was estimated that Dagupeños experienced around four or five brownouts each month. This figure appears lower than earlier years when electrical disruptions were much more common. While concrete numbers are hard to come by, it is clear that brownouts peaked locally sometime in the early to mid-1990s. This high incidence coincides with the country's post-BNPP energy crisis and Dagupan's recovery from the July 1990 earthquake.

Dagupan was one of the communities hardest hit by the disaster (Dannhaeuser 2004:234). Spared major losses of human life and injury, the earthquake and its aftershocks caused severe structural damage to downtown. Geological fissures generated by the temblor released massive deposits of subterranean water to the street surface. This abrupt seismic activity liquefied layers of loose alluvial soil on which the city rested, destabilizing foundations and compromising the structural integrity of many built features (Asiaweek 1990:34). About 90 percent of downtown buildings either tilted significantly or sank by a meter or more because of rapid soil liquefaction. All told, the city suffered some P6 billion in damages (Melecio-Prieto 2005:143–144).

In the earthquake's aftermath, hundreds of roads, bridges, and buildings lay in partial or complete ruin. Groundwater and black silt inundated city streets with a thick deluge of sandy muck. Wet conditions, quickly compounded by seasonal rains, proved especially dangerous as utility pylons and electrical wires lay exposed amongst the rubble. The psychological impact felt by most Dagupeños following the earthquake was one of intense desperation. Many wondered if the community could ever recover from the devastation. Numerous merchants considered abandoning their downtown shops for more secure locations beyond the urban periphery. Once government assurances about fast-track reconstruction were announced, however, most traders opted to stay (Matejowsky 2002). Ultimately, the economic dislocations brought on by the earthquake in terms of distribution and labor were only short in duration, several months to a year, as most merchants benefited from post-disaster redevelopments.

Unsurprisingly, electricity was unavailable for the rest of July and most of August. Power delivery remained patchy for Dagupeños throughout 1990 as neighborhoods gradually came back on grid. If electrical utilities were inconsistent before the earthquake, they were doubly erratic in the post-disaster context. Damaged facilities and large-scale repair projects became a persistent source of outages during the 1990s.

Prolonged brownouts were also triggered by heavy seasonal rains and the onslaught of several major typhoons, most notably September 1998's devastating Typhoon Gading. With registered winds over 100 mph, Gading carved a path of destruction across Dagupan that left hundreds homeless, streets flooded, and utilities knocked out for days. Looting of downed transmission lines in Gading's aftermath further hampered efforts at restoring power to affected areas. When considered all together, the sporadic brownouts punctuating Dagupan's commercial life from 1990 onward are certainly linked as corollary effects if not directly caused by natural disasters.

Brownouts and Merchant Coping Mechanisms

Against this backdrop of disaster impact and renewal, Dagupan's street store merchants have frequently grappled with the disruptive effects of brownouts. Indeed, it is now almost second nature for shopkeepers to weather the challenges of sporadic grid failures with something of a seasoned familiarity. Just how individual merchants ultimately deal with power outages is determined almost as much by the products (perishable versus nonperishable goods) as it is their level of capitalization. While highly capitalized ventures have more options when it comes to blunting the effects of brownouts, few stores are so well off that they can completely ignore the complications arising from electrical losses. Suffice it to say, variation exists in how traders choose to contend with brownouts.

Disaster-related coping mechanisms in the Philippines generally fall into two main categories: preventative and minimizing (Bankoff 2002:166–167). The former entails preemptive measures meant to avert the reoccurrence of past disruptive events, while the latter involves those efforts seeking to reduce potential losses and facilitate recovery. While local responses to electrical outages can vary significantly, those of downtown retailers tend to more readily correspond with impact-minimizing strategies. Notably, they reflect decision-making processes based on unambiguous motives of economic self-interest. Such countermeasures generally expose an underlying rationale that is less profit-driven and more about minimizing potential losses in the short term. Conceptually speaking, impact minimization efforts reveal just how much natural hazards and the residual effects have been assimilated into the framework of daily business operations.

Thus, downtown's merchant community is not thrown into turmoil nor does it grind to a halt whenever power fails. Rather, it merely shifts into a kind of stopgap mode whereby the pace of commerce slackens, attention to detail becomes more important, and responsibilities are renegotiated with more flexibility. As the next few sections make clear, brownouts precipitate slight to significant alterations in local merchants' sales tactics, reliance on technology, profit expectations, modes of payment, employee duties, and store security.

Light and Heat: Brownouts, Power Surges and Portable Generators

Power outages are normally greeted with audible groans and sighs. As lights flicker erratically and then fade, store operators begin disengaging from tasks already under way. Typically, the first order of business is unplugging all shop appliances and machines. This is done as a preventative measure to guard against the power surges that invariably occur when electrical service finally resumes. Voltage levels tend to fluctuate wildly as grid systems come back online. Merchants usually wait about five to ten minutes before reconnecting cables once power resumes. Since one energy spike can completely overwhelm the circuitry of store appliances, not to mention spark a fire, shopkeepers are generally quite cautious about possible over-voltage. With devices like surge-protected power strips not widely available until only recently, disconnection remains merchants' most effective and pervasive means of safeguarding equipment.

Essential supplies like candles, flashlights, lamps, radios, and portable generators are then brought out with a weary familiarity as workers transition into a new retailing mode. Just how much merchants utilize these items depends largely on the brownout's duration and time of day. Natural light plays a significant role in determining how to best deal with electrical losses. Most downtown street stores have open frontages that look out onto sidewalks instead of closed facades or glassed doors. Shops depend on available sunlight for illumination almost as much as they do overhead fluorescent lights. Even when standard light fixtures work properly, their pallid glow does little to alter the semidarkness at the back of most stores.

So, if power goes out during peak business hours on a sunny day, most retailers are not particularly inconvenienced. If it becomes unavailable at other times or in overcast weather, interior darkness makes workaday operations much more difficult. Practically speaking, the longer shops go without electricity the more important off-grid light sources become, especially if brownouts strike as nightfall approaches. Without some kind of illumination in the late afternoon or evening, many merchants will simply cut their losses and head home early.

The hum of diesel-powered generators can be heard reverberating throughout downtown Dagupan within minutes of a brownout. Most highly capitalized retail outlets employ generators as backup power sources to protect against slowdowns in productivity that regularly accompany grid failures. Groceries, pharmacies, and other ventures dealing in perishable and refrigerated goods are pretty much compelled to have generators at the ready. These shopkeepers clearly stand to suffer losses in both profits and merchandise if they cannot quickly tap into external energy sources. Likewise, modern eateries such as Dunkin' Donuts and Mister Donut that feature air-conditioning, glassed frontages, and other amenities typically employ generators to better maintain high levels of consumer comfort.

For less capitalized merchants, however, purchasing portable generators involves much more careful consideration. The viability of these machines must be weighed against their potential drawbacks. On the one hand, generators are easily activated, require minimal upkeep, and can generally restore operations to near or full capacity within minutes. If attainable, these devices can also give individual storeowners a competitive edge over rival shops. On the other hand, generators are typically priced beyond what most mom-and-pop operations can afford. The prohibitive costs of generators, not to mention steep fuel prices, are usually enough to keep small shop owners from investing in equipment that will only be used sporadically. While most diesel powered generators are imported duty-free from Japan or Taiwan, they still range in price locally from around P30,000 ($600) to upward of P65,000 ($1,300) with the exchange rate remaining $1 = P50 in recent years. Portable generators were notoriously difficult to find during the 1990 earthquake's aftermath. Shopkeepers fortunate enough to have a functioning unit still contended with scarce supplies of diesel fuel. Notably, beyond the purchase of portable generators, brownouts do not really affect stores' overhead expenses. The fixed costs of running a shop, such as advertising, labor, upkeep, and rent, are not greatly altered by the incidence of power failures. If anything, electrical losses mean less expensive power bills.

Beyond price concerns, portable generators do not really jibe well with downtown's cramped shop spaces. Machines normally run hot and loud, emitting powerful smells and smoke. Their unwieldy presence makes it difficult for employees to maneuver store space much less allow customers to browse. More significantly, they also raise the risk of accidental fires. Store blazes increase in likelihood during brownouts since generators often draw together highly flammable elements like candles and petrol cans. Faulty generators and careless usage have been at the root of several downtown fires over the last few decades. In this way, even those without generators become concerned about brownout fires as their shops are increasingly exposed to the potentially negligent actions of neighboring store personnel.

With most of Dagupan's merchants lacking generators, a number of less expensive lighting options come into play during brownouts. Again, how much shop owners rely on these alternatives depends on individual preference and sunlight availability. Battery-operated lanterns, flashlights, or emergency candles are frequently employed; however, most shopkeepers prefer makeshift hurricane lamps that burn kerosene and cost only around P20 each. While open flames clearly increase the possibility of accidental fires, these inexpensive lamps are easy to use and last for hours.

Brownouts, Sales Strategies, and Security Concerns

Shopkeepers begin focusing on sales matters once lighting issues have been sufficiently resolved. Grid failures elicit new strategies within Dagupan's street store community minimally aimed at averting profit losses. Retailers recognize that Dagupeños are not really inclined toward shopping when electrical service is unavailable. With locals more concerned about how brownouts affect them personally, store operators resort to more novel approaches of attracting customers.

The temporary inconveniences of power disruptions, thus, give rise to increased merchant flexibility on matters of sales; especially those related to price and credit. For traders, a willingness to extend favorable terms to customers is less about moving merchandise and more about implementing effective stop-loss responses during these periodic episodes of market dysfunction. In a very real sense, brownouts generate a kind of buyers' market whereby merchants are compelled to offer consumers deals that they would probably not do otherwise.

Credit extension is a common aspect of local commerce in towns and cities across the Philippines. Dagupan's merchant community engages in this practice mainly as a way of attracting and retaining clientele. The interest added to purchases through credit can be significant, if not sometimes severe, for customers under normal circumstances depending upon previous dealings. As retailers confront low consumer turnout amid brownouts and disaster recoveries, they are prone to knock off several interest percentages on credit payments to effectively jump-start sales. In some cases, they may even extend credit to individuals with whom they have not previously dealt. This scenario obviously introduces new elements of uncertainty that merchants would normally seek to avoid. Whether or not shopkeepers elect to modify terms of credit is contingent upon a number of factors directly and indirectly related to a brownout's intensity and duration: these include level of capitalization, capacity for recovery, time of year, and recent sales growth or lack thereof.

Similarly, Dagupan's shopkeepers are apt to lower prices when electrical service seems unlikely to resume promptly. Temporary sales reductions are easily accomplished as most retailers sell merchandise through a negotiable price system. The absence of fixed prices gives local street store merchants considerable latitude when it comes to cultivating customer relationships and closing deals. No less significant, it also affords them opportunities for making snap adjustments to store operations when brownouts strike.

To drum up sales during power outages many shop owners will hold impromptu sidewalk sales. In concrete terms, this means transferring merchandise outside from darkened store interiors to better catch the attention of passers-by. Storefronts can quickly take on a bazaar-like atmosphere if crowds gather and buyer-seller bargaining ensues. Typically, merchants dealing in ready-to-wear clothing, textiles, small electronics, watches, jewelry, household wares, and other high traffic items engage in this kind of selling. As a sales ploy, improvised sidewalk vending demonstrates considerable savvy on the part of affected traders. Pedestrian flows increase dramatically with downed electricity as locals prefer open city streets to being stuck indoors.

Grid disruptions are also believed to correlate with temporary rises in crime. While the veracity of this notion remains unsubstantiated, merchants do feel increasingly exposed to petty offenses like shoplifting and grab-and-run robberies. Such concerns are possibly linked to the lack of control that shop workers feel as attention is diverted away from selling toward brownout issues when electricity first goes out. The presence of armed security guards at various downtown locales helps allay these fears somewhat; however, most merchants feel compelled to step up their own policing efforts. Frequently, this means that store staff is expected to shadow customers inside shops or disallow personal baggage on sales floors.

Let There be Light: Electricity Returns

There is an almost palpable sense of relief amongst Dagupeños when power delivery finally resumes. Shop personnel ease into more normalized routines rather cautiously as electrical service sputters back to life. Workers usually take their time in reverting back to standard business operations as false starts in grid support are common. Once it becomes clear that utilities are operating up to speed, expressions of merchant brownout behavior become increasingly difficult to discern.

Conclusion

The economic success of the Philippines is in many ways constrained by the recurrent impact of natural catastrophes. Already saddled with considerable foreign debt and sluggish economic growth, the Philippines lacks sufficient capital to develop and maintain an infrastructural apparatus that can contend with such high disaster incidence.

The Philippines' long history of grappling with extreme geophysical phenomena and undercapitalized infrastructure has shaped many aspects of local life, including retail commerce. Street store merchants have distinguished themselves as a highly pragmatic and flexible segment of Philippine society, in part because of their dealings with disaster-related adversities. Their ability to rebound from major natural catastrophes, much less integrate phenomena like brownouts into their daily business schema, highlights a level of resilience that is indispensable given the scale and frequency of natural hazards in the Philippines.

As the case of Dagupan demonstrates, intermittent electrical failures provide a window into the behavioral and cognitive responses that Filipinos employ to better cope with disaster-related stresses. A vast storehouse of knowledge exists beneath the surface of everyday commercial operations that street store merchants call upon whenever electrical service fails. Far from esoteric, this knowledge has steadily accumulated over the years, pervading nearly all sectors of local retailing. Previous dealings with brownouts provide shopkeepers invaluable insights into how to address new challenges precipitated by infrastructural breakdowns. In fact, the economic decision-making processes of many urban traders are significantly influenced by, if not inextricably linked to, the looming threat of grid failures and natural disasters.

The experience of Dagupan's merchant community in dealing with the disruptive effects of disaster-related brownouts is not an aberration. If anything, it is representative of how local traders in other Philippine communities and throughout the developing world cope with the external pressures brought on by intersections of natural hazard and undercapitalized infrastructure.

Acknowledgments

I wish to thank Rubin Ang, Josie Gonzalez, and Mario Granada for their valued contributions to this work. Moreover, gratitude is extended to the University of Central Florida (UCF)'s Office of Commercialization and

Research, and the UCF Southern Region for their support during this research endeavor.

Notes

1. Most infamous is the mothballed Bataan Nuclear Power Plant located some 100 kilometers north of Metro-Manila. Built by America's Westinghouse and financed by lenders like the World Bank and U.S. Export-Import Bank, this uncommissioned light water reactor has come to be seen as a damning indictment of the Marcos administration's highly corrupted "edifice complex" (George 1988). The 2.3 billion dollar facility remains offline primarily because its construction near Mount Pinatubo (an active volcano) and along major geological fault lines renders it too dangerous for activation. Filipinos continue paying upward of $170,000 in daily interest for BNPP loans even though it has yet to generate a single watt of electricity.

2. Some 16 percent of the Philippines' generating capacity is derived from geothermal power.

3. Considering that ethnic Chinese have maintained a significant presence in Dagupan for almost five generations, the vast majority of Dagupeños with Chinese heritage today more closely identify with Filipino culture than did their immediate ancestors. Yet, divisions between the Chinese-Filipino community and more general Filipino populace still exist in the eyes of many area residents. For the purposes of this chapter, I intend the terms "ethnic Chinese" and "Chinese-Filipino" to be understood in the same way locals use them when differentiating ethnic Chinese from native Filipinos.

References

Abrenica, Ma. Joy V., and Gilberto M. Llanto

2003 Services. *In* The Philippine Economy: Development, Policies, and Challenges. A. M. Balisacan and H. Hill, eds. Pp. 254–280. Oxford: Oxford University Press.

Allen, Katrina

2003 Vulnerability Reduction and the Community-Based Approach: A Philippines Study. *In* Natural Disasters and Development in a Globalizing World. Mark Pelling, ed. Pp. 170–184. New York: Routledge.

Asiaweek

1990 Coping as Town Sinks. Asiaweek 16(31):34.

Balisacan, Arsenio M., and Hal Hill

2003 The Philippine Economy: Development, Policies, and Challenges. Oxford: Oxford University Press.

Bankoff, Greg
2002 Cultures of Disaster: Society and Natural Hazard in the Philippines. New York: Routledge.

Bello, Walden, Herbert Docena, Marissa de Guzman, and Mary Lou Malig
2004 The Anti-Development State. The Political Economy of Permanent Crisis in the Philippines. Quezon City, Philippines: University of the Philippines.

Bode, Barbara
1989 No Bells to Toll. Destruction and Creation in the Andes. New York: Scribners.

Button, Gregory V.
1992 Social Conflict and Emergent Groups in a Technological Disaster. The Homer Area Community and the Exxon-Valdez Oil Spill. PhD Dissertation, Department of Anthropology, Brandeis University.

Cabreza, Vincent
2005 Stop Paying the Country's Nuke Plant Debt, SC Justice Urges Gov't. Philippine Daily Inquirer. April 21.

Dannhaeuser, Norbert
2004 Chinese Traders in a Philippine Town. From Daily Competition to Urban Transformation. Quezon City, Philippines: Ateneo de Manila University Press.

George, Susan
1988 A Fate Worse than Debt: The World's Financial Crisis and the Poor. New York: Grove Press.

Gordon, Norma
1999 Children and Disasters. New York: Routledge.

Guillette, Elizabeth
1991 The Impact of Recurrent Disaster on the Aged of Botswana. Paper presented at the 50th Annual Meeting of Social Applied Anthropology. Charleston, SC.

Matejowsky, Ty
2002 Globalization and Retail Development in the Post-Disaster Context: A Comparison of Two Philippine Communities. *In* Research in Economic Anthropology, vol. 21: Social Dimensions in the Economic Process. Norbert Dannhaeuser and Cynthia Werner, eds. Pp. 311–341. Amsterdam: Elsevier-JAI Press.

McBeth, John
1993 Stripped of Power. Philippines Electricity Crisis Hits Small Firms the Hardest. Far Eastern Economic Review 156(25):60.

Melecio-Prieto, Carmen
2005 Nandaragupan. The Story of a Coastal City and Dagupan Bangus. Dagupan City, Philippines: Dagupan City Heritage Commission.

Oliver-Smith, A.
1996 Anthropological Research on Hazards and Disasters. Annual Review of Anthropology 25:303–328.

Philippine Daily Inquirer
2003 Philippines Tops Disasters: Red Cross. Philippine Daily Inquirer November 16.

Population Reference Bureau
2006 Making the Link in the Philippines: Population, Health, and the Environment. Washington DC, Population Reference Bureau.

Robles, Alan
1991 National Trauma. Prone to Calamity. Filipinos May Have Developed Their Own Coping Mechanisms. Manila Chronicle, August 3, 1(18):1,6.

Rossi, Ino
1993 Community Reconstruction after an Earthquake. Westport, CT: Praeger.

Sunday Punch
2004 Extended Brownout Denounced. Sunday Punch: Dagupan City, Philippines, July 18.

Sun Star Cebu
2005 "Energy Auditors" to Inspect Offices, Schools. Sun-Star Cebu, Cebu City, Philippines. August 26.

Tasker, Rodney
1992 More Heat than Light: Power Shortage Poses Big Test for Ramos. Far Eastern Economic Review 155(48):13.

Villanueva, Ruffy
2002 Public Awaiting Efficient Service. Business World Anniversary Report 2002: Manila.

Wilson, Karl
2006 Disaster-prone Philippines Suffering More and More: Analysts. Manila Times Bulletin, December 3.

Yap, Cecille
2006 Wholesale Electricity Market to See Annual Trade of P100B. Philippine Daily Inquirer, Quezon City, June 23.

Where Others Fear to *Trade*: Modeling Adaptive Resilience in Ethnic Trading Networks to Famines, Maritime Warfare, and Imperial Stability in the Growing Indian Ocean Economy, ca. 1500–1700 CE 10

RAHUL OKA, CHAPURUKHA M. KUSIMBA, AND VISHWAS D. GOGTE

AMONG THE MOST RESILIENT OF HUMAN ACTIVITIES, trader networks operate sustainable, profitable, and growing commercial ventures in both adverse and beneficial conditions and, throughout history, show continuous adaptation to natural and intentional disasters such as famines and/or warfare. In this chapter, we reevaluate the changing political economy of the Indian Ocean coastal polities during the decline period ca. CE 1500–1700. During this period, Portuguese and other European groups aggressively entered the Indian Ocean trading world (an intentional hazard for some, and an intentional disaster for others), and global climate changes brought by the Little Ice Age caused frequent droughts and famines (natural hazard/disaster). This chapter argues that this decline cannot be attributed to any single cause or variable, and introduces a third factor that proved disastrous over the long-term for the established Indian Ocean Trading Complex. This was the economic boom of the Imperial Détente in the sixteenth and seventeenth centuries that witnessed the growth of commerce-centered strategies such as import-substitution, centralization of entrepreneur-based production and distribution activities, and the rise of politically connected portfolio-capitalists and financiers. During this boom, most of the trade and economic infrastructures that had characterized the pre–sixteenth century economy were abandoned or were out-competed, leading to a large-scale decline of the Indian Ocean trading complex after the sixteenth century. As we show in this chapter, this process was not a simple restructuring of macro-economies but was in fact a disastrous transition, involving demand-driven competitive consumption of manufactured commodities, overexploitation of natural and human resources, large-scale loss of traditional livelihoods and subsistence activities, and the growth and institutionalization of disaster-era commerce, encouraged by the extant states.

Methodology

The data for this chapter comes from three sources: 1) semistructured interviews with traders, 2) archival data on the ivory trade, and 3) archaeological data on trade ceramics. First we interviewed people from ethnic trading groups of Africa and Asia to understand how traders respond to disasters and/or changes in the political economic environments. The data were collected from retailers, peddlers, wholesalers, brokers, money-lender/financiers, portfolio capitalists, industrialists, and entrepreneurs in South Asia, East Africa, Europe, and North America. The informants (n = 104) were from the Hindu, Muslim, Jewish, and Christian religious orders, and from the Gujarati Bania, Rajasthani Marwari, Sindhi, Tamil Vanniar and Chettiar, Mumbai-based Parsi, Ismaili, Bohri, Khoja, Armenian, Jewish, Lebanese, Yemeni and Omani Arabs, Somali, and Swahili trading communities. The religious and ethnic variation among the informants mirrors the ethnic makeup of the trading communities active in the Indian Ocean trade for most of the second millennium CE. The information sought included occupation and specialization, the number of years in trade, origin narratives, migration details (push versus pull factors), and the building and maintaining of relationships with other business groups, business decision-making strategies for moving, relocating, specialization, diversification, expansion, capitalization, credit and risk calculation, ideas on profit/loss, and ideas about business growth/sustainability.

Second, archival data CE 1500–1700 were drawn from the collections of the late Peter Thorbahn and C. M. Kusimba on the volume of ivory exports from East Africa calculated from the archives of the India House collections in London and the Casa da India in Lisbon, Portugal. These archival records were broken down by port (both export and import), date of exchange, volumes traded, and prices, which allowed us to compute trends in ivory exports from East Africa to South Asia and other parts of the world from the early sixteenth to the late nineteenth centuries as a general indicator of the changes in Indian Ocean exchange.

Third, we draw upon our archaeological analyses of excavated Chinese and Middle Eastern trade ceramics from the port sites of Chaul (Gogte

2003) on the Northwest Coast of India, and Mtwapa (Kusimba 1993, 1999a) on the southern Kenyan coast, focusing specifically on White and Blue-on-White porcelains from east and southeast Asia, regarded as good indicators of changes in trading patterns for Indian Ocean trade (Chittick 1974; Horton 1996; Kusimba 1999a). Figure 10.1 shows the location of the two ports, Chaul in India and Mtwapa in Kenya, within the Indian Ocean trading world.

Combining the ethnographically derived trading systems model with the ivory trade data, this chapter will show that the rise of politically stable and trade-friendly empires in Afrasia led to changes in global commercial production and distribution that privileged commercial production and trading specialists and ultimately led to a macro-scale disaster for the Indian Ocean trading complex, including the decline of more than 70 percent of the trading ports that had been active for at least half a millennium prior to the sixteenth century. The newly enriched and empowered trade specialists utilized their middleman positions to benefit from the European-Afrasian elite maritime hostilities as well as the commercial opportunities created by frequent droughts and famines (disaster-era commerce). Using archaeological data, we show how the emergent commercial economy of the seventeenth century consisted of highly centralized mass production and high-volume distribution of finished commodities from the East and South Asian core to the peripheral ports of East Africa, and of raw materials from East Africa to South Asia. We argue that this transition spelled disaster for the peripheral areas in the system by initiating major changes in elite consumption patterns and sponsorship of local production.

The Indian Ocean Trading Economy Prior to the Sixteenth Century

For over five thousand years the Indian Ocean has been the forum for exchange of peoples, goods, and ideas between Africa, Asia, and Europe (Pearson 2003). Trade specialists and their networks have crossed oceans, deserts, and mountains, linked coasts to their hinterlands, and facilitated transoceanic diplomacy in good times and bad (Curtin 1984; Stein 1999). By CE 1500, the world of Islam stretched from West Africa to Southeast Asia and Central Asia. Mecca and the Indian Ocean were the center of a truly global exchange in which elites across Afroeurasia actively sought Chinese porcelain, Southwest Asian glazed ceramics, south and east Asian silks, brocades, and cottons, as well as African ivory and precious gems (Chaudhuri 1990).

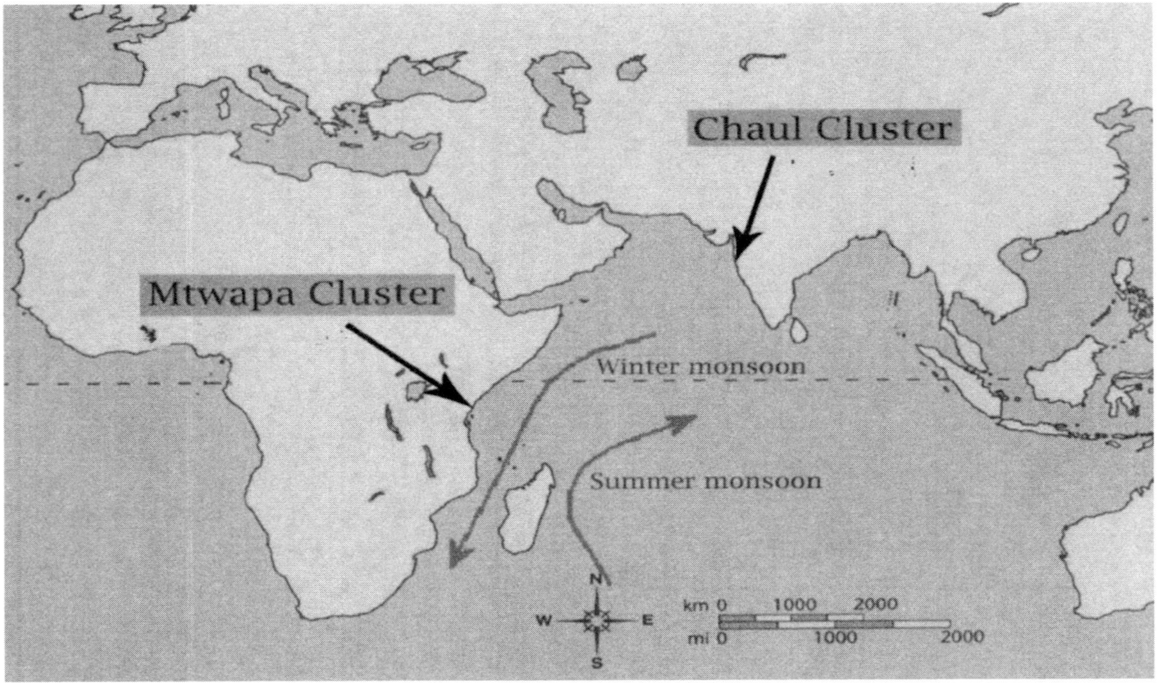

Figure 10.1. The Indian Ocean Trading World Showing Mtwapa (Kenya) and Chaul (India) Port Clusters (The arrows represent the direction of the Southwest (summer) and Northeast (winter) Monsoon winds).

The Indian Ocean littoral was dotted with port cities that facilitated exchange between their hinterlands and overseas demands (Pearson 2003). Along with providing services for a distribution economy, many of these ports also developed entrenched local craft production sectors for regional or overseas consumption. The ports hence depended on revenues generated by both the distribution sectors and value-added commodity production.

Prior to CE 1500, most of these Indian Ocean trading ports had developed basic similarities and characteristics due to overlapping functions (Gogte 2003; Kusimba 1999a). Regardless of the political economic nature of their hinterlands, the ports enjoyed a certain measure of autonomy (Fukuzawa 1991; Middleton 1992). The administration of the ports was generally in the hands of various elite families and could be described as a corporate oligarchy even though the visible political infrastructure resembled a hierarchy as that found in the larger territorial states in Africa, Asia, and Europe (Kusimba 1999b). As a corollary, the ports also had little, if any, control over their hinterlands, whether they were occupied by dispersed agro-pastoral or foraging groups as in East Africa or Southeast Asia, or large territorial states as in Southwest, South, and East Asia. The relationship between port elites and their hinterlands was one of negotiated alliances based on marital ties, inherited friendships, blood-brotherhoods, and fictive kinships, and were managed through calculated gift exchange (Herlehy 1984).

The realities of distance forced a pragmatic acceptance of the futility of long-term domination of any one area by another and created parity between most participants in this vast distributed network (Stein 1999). Within this network, the role of trade specialists and their networks was one of maintaining a sustained flow of goods between areas and facilitating political relationships between distal groups that benefited trade and commercial activities (Cohen 1971). However, the political and economic actors within this exchange frequently clashed in their differing ideas on the role of trade specialists, trade, and commerce in society (Oka 2008). As they do now, fully cognizant of their vulnerability as the wealthy "other" (Moore 2000) in times of fluctuating political economic stability, trade specialists kept to their networks and emphasized conservative and self-limiting strategies to protect themselves against political elites (Landa 1994).

A gradually increasing demand and the larger economic infrastructure of this pre-1500 Indian Ocean trading world faced limitations to growth because of the instability of the constant cycling of dynasties that resulted from the following: constant invasions by central Asian Turkic and Mongol groups into east, south, and southwest Asia and Europe, the vast demographic impact of the Black Plague in the fourteenth century, and the transitions from the warm and wet Little Climatic Optimum (CE 900–1350) to the colder and dryer conditions of the Little Ice Age (after CE 1350). However, this period—especially the fourteenth and fifteenth centuries—witnessed many attempts to implement and develop economic and commercial innovations that would later become institutionalized across Eurasia in the politically stable Early Modern era.

The Disastrous Early Modern Era

Into this Indian Ocean trading world, the sixteenth century brought three disasters. The first disaster was man-made, and consisted of the arrival of the Portuguese in the Indian Ocean in CE 1498 (Pearson 2003). The second disaster was climatic and consisted of sustained droughts and famines brought about by the Little Ice Age to the Indian Ocean (Kusimba 1999a). The third disaster was the economic boom precipitated by the rise of the Imperial Détente[1]: the stability resulting from the rise of three powerful Islamic Empires: Ottoman (Southwest Asia and North Africa), Safavid (Persia, Central Asia), and Mughal (South Asia; Oka 2008).

First, the arrival of the Portuguese heralded a militarization of maritime interactions in the Indian Ocean as port after port was sacked and plundered, their inhabitants massacred and enslaved, and their hitherto elite families brought down into subjugation (Mbuia-Joao 1990). The Portuguese introduced and institutionalized state-sponsored piracy and protection racketeering and preyed on the defenseless ships of the Afrasian potentates and trade specialists (Pearson 1987). Second, the Little Ice Age effects on the Indian Ocean littoral and the hinterlands between the fifteenth and eighteenth centuries were characterized by increasing incidence of droughts resulting in frequent famines (McAlpin 1983; Nicholson 1979). According to historian Irfan Habib (1999:110), these famines led to migration of peasants and abandonment of rural settlements. Third, between CE 1500 and 1600, the stability caused by the Imperial Détente and liberal trade policies of Islamic elites of the three empires changed the relationship between trade specialists and political elites, giving the traders larger roles in the production economy, as well as growing political clout and overt influence over state economic policies. In this booming economy, the Little Ice Age famines and the activities of the Portuguese can be seen as *multipliers* or "focusing devices" that exacerbated and added to the growing strength of trade specialists.

Ethnography of Trade Specialists

The empirical data for the proposed modeling approach was drawn from Oka's (2008) ethnographic research in India, Kenya, the United States, and Mexico. Between 2000 and 2004, oral histories from 104 trade specialists (peddlers [n = 17], retailers [n = 41], wholesalers [n = 24], bankers, money-lenders, and brokers [n = 13], entrepreneurial industrialists [n = 9]) and 7 political specialists were conducted. The method followed was to ask respondents specific questions of how, when, and why trade specialists respond to given political economic circumstances, how they react and adapt to good times and bad times, and for them to give us specific examples when such strategies were used. For the peddlers, retailers, and wholesalers, the numbers of informants were large enough that confirmation of strategies by 90 percent of the informants was deemed sufficient. However, we made sure that all the strategies used by bankers, moneylenders, brokers, entrepreneurs, and industrialists were repeated and confirmed by all the individuals interviewed. Along with checking the responses across individual informants, we also confirmed that the strategies used overlapped across ethnic, religious, and national boundaries, and that the strategies were not products of local histories and conditions.

Resilient[2] and Adaptable Trader Networks

Ethnic trading communities lend themselves particularly well to cross-cultural comparative research (Cohen 1971; Curtin 1984; Falzon 2004; Markovits 2001; Oka and Kusimba 2008; Stein 1999). Studies on trader network culture (Bellina 2003:287) suggest that trading specialists operate under similar social and political conditions irrespective of the culture or time: traders usually are regarded as wealthy and the "other;" they tend to be regulated by the suspicious political elite and populace during peacetime; and they are victimized during difficult periods (Alexander and Alexander 1991; Seidenberg 1996).

During data analysis, it became clear that trader networks showed great similarities to latent sub-systemic or semi-groups identified by Mary Douglas (1986) and subsequently in contemporary studies on network-centric organizations (Allenby and Fink 2005). Regardless of our informants' ethnicity, religious background, and professional or commercial specialization, we isolated seven characteristics of trader networks that emphasized the importance of adaptive resilience as a desired feature of socioeconomic interactions. Our analysis also suggested that as emergent properties of hundreds of years of trading activities, these seven factors represent the convergent evolution of conservative strategies, constantly kept in play because of their repeated successes in the face of social, political, and environmental changes. Since the focus of this chapter is to show how trader networks respond to disaster, we have listed the factors as generalizations. The ethnographic justification for these generalizations, in the informants' own words, has been placed in the endnotes for each factor. A fuller explanation of these factors can be found in Oka (2008) and Oka and Kusimba (2008a).

First, the distribution of information, agents, and thus risks and resources across space grants the whole trader network an enhanced capacity to react to shocks and/or changes in any one part of the network (Cohen 1971). This dispersal is the result of both resource distribution and continual tension between the political elites and trade specialists.[3] The decentralized transregional ties permit resource mobility while avoiding the attention of political elites and scapegoating (Girard 2001; Moore 2000). The distributed network is also important for recouping losses or rebuilding during crises, or for exploiting crisis-based opportunities—disaster-commerce (Keen 1994).

Second, trader networks have fewer hierarchies; they exist as semi-independent agents linked to other agents through embedded social, commercial, and political relationships including consanguinal, marital, and fictive kin alliances, friendships, as well as professional relationships maintained through gift-exchange and other social rituals[4,5] (see Curtain 1984). In case of shock or change, new or restructured relations can be sought in other areas and/or with new agents.

The third characteristic is that trader networks self-organize and self-regulate. Within each node, agents react to local and regional information, perform adaptive reconfigurations and/or repairs in case of shock or change to the network, and feed resources and information back to the larger network. The resources shared across networks include information, capital, credit, favors, commodities, and goods, as well as protection and legal redress.[6]

Fourth, trader networks are less concerned with monopolies and monopsonies than with generating sustainable production and distribution of resources. The size of individual operations in these networks is self-limited unless sanctioned by the majority of the agents in a specific network.[7] Even in the case of sanctioned growth in any one zone, the concentration and dispersal of resources are balanced to guard against shock to the zone. The self-restraint in trader diasporas is managed through informal leveling mechanisms based on trust and consensus rather than hierarchical decision-making (Landa 1994).

Fifth, stable conditions mean expansion and diversification of activities (within certain limits),[8] while political instability means exercising utmost restraint and prudence. Under the intensification of both favorable and unfavorable conditions, there is a positive cascade of information and resources flowing through the various nodes. When the flow reaches critical mass, agents respond to local and regional events, leading to network reorganization to maintain information, resources, and goods flow.

Sixth, a trader network is the emergent property of interacting individuals who share a general concern with maintaining production/distribution activities and do not draw stark boundaries between their social, political, and ideological concerns.[9] Nobody goes out to establish a trader network controlled by well-framed policies. The strategies that govern the interactions are chosen for their efficacy in sustaining the network.

Seventh, agents are concerned with maintaining core competence of each node or particular business specialization. The core business is the predominant focus despite autonomy within each trading family for individuals or groups to diversify, expand, or relocate. The core competence is the fallback position involving the least amount of risk that will see the trading group through a crisis. The core business also keeps generating capital and both generalists and skilled personnel who can then help expand, relocate, or diversify operations.[10]

Trading Systems and Trading Complexes

Based on our ethnographic research and Complex Adaptive Systems thought, we regard the trader network as an emergent property of interactions between economic agents using simple strategies (Albin 1998; Lansing 2003; Padgett 2001). As confirmed by the range of specializations of our informants, these behaviors apply across the scale of trading specialists (peddlers to entrepreneurs and industrialists) and activities (from distribution-centered activities such as peddling, retail, and wholesale to production-based activities such as crafts-manufacture and commercial agricultural production, as well as service activities such as banking and brokerage). Each agent uses various strategy sets to interact with others and bases the sustainability of the business on the continuity of these relationships, as seen in figures 10.2 and 10.3.

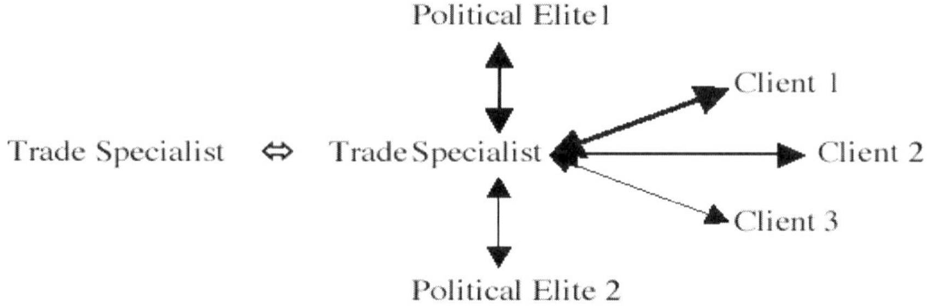

Figure 10.2. Schematic Representation of Personalized Networks of Influence (Trading System 1).

The strategies used to build and maintain such networks have been termed Trading Systems (Oka 2008; Oka and Kusimba 2008). A trading system is a:

> particular set of strategies developed in response to certain political economic conditions, and is used by individuals to mobilize resources and minimize risk as well as build alliances and social relations.

The consensus opinion from our data analyses produced a few basic axioms for setting up trading networks:

a) All traders look to controlled growth of the network and not individual business growth. Involvement of family is seen as a good thing,
b) Controlled growth is self-limiting as trading communities are vulnerable to political and social forces. Sustainability is generated by lack of exposure,
c) Trading communities see their occupation as their living culture, not just livelihood
d) Sustainability and growth of trade are only as good as the network's stability,
e) The changes in trading behavior are dependent upon the presence of extra-network factors that determine stability and business conditions: **social sedentism**, **political economic stability**, and **trade-friendliness**.

From ethnographic research, Oka (2008) has identified four separate "trading systems" used by ethnic trading groups to manage their business affairs. We have used numbers "1, 2, 3, 4, . . . n" to denote trading systems to suggest the additive nature of trading systems in that 4 contains 3, 2 and 1; 3 contains 2 and 1; and 2 contains 1.

Trading System 1 (figure 10.2) is the reciprocity related strategy set guiding basic gift, favor, and service exchange that structures most human relationships and forms the underlying (but not only) basis for maintaining trade relationships. *Trading System 2* is the set of strategies used by mobile traders (peddlers, tinkers, pastoralists, gypsies) who operate in small quantities of commodities at high risk and use Trading System 1 to structure their commercial relationships with members of settled societies. *Trading System 3* dominates the activities of ethnic traders or trading families (merchants and brokers) who deal in larger volumes and invest time, effort, and resources to maintain and expand relational networks to keep goods and capital moving within their communities. *Trading System 4* is the strategy set used by established trading groups (entrepreneurs, large merchants, and portfolio capitalists) in conditions of growing trade stability and involves dealing with large volumes of commodities and capital, close nexus with and sponsorship of political elites, and heavy influence in economic affairs.

Our interviews confirm that trade specialists switch from one trading system to another in response to changes in sedentism, sociopolitical stability, and trade-friendliness. To clarify, political stability indicates that political elites have effective mechanisms for extracting wealth from commoners that does not result in rebellion or loss of legitimacy. Trade friendliness indicates the nature of relationships between political elites and trading specialists. As Stability Factor 1 (Sedentism) increases, good "traders" become the nodes through which long-distance commerce is channeled (Trading System 1 → Trading System 2). However, the emergent network is self-limiting due to political instability (Factor 2) that curtails the accumulation of resources at any particular node. As Factor 2 (Political Stability) is increased, trading agents can increase both the intensity of commercial activities and accumulate (still subject to restrain and control) wealth within each node (Trading System 3 → Trading System 3). As Factor 3 (Trade Friendliness) increases, trade specialists can minimize self-restraint and can visibly and heavily invest in entrepreneurial production and distribution activities (Trading System 3 → Trading System 4).

The intersection of any two, three, or all four Trading Systems results in an evolving trading network, or a Trading Complex. The Trading Complex is self-organizing and is subject to both positive (shift in Trading Systems) and negative feedback (maintenance of Trading System). Agents in a trading system are subjected to internal and external changes and react by adopting or switching strategies. When the emerging complexity cascades to the system critical mass due to positive feedback from changes and altered strategies, the behaviors that were minimized or insignificant in the previous state start dominating and solidifying.

Subsequently, continuation or cessation of Stability Factors 1–3 will determine whether the emergent behaviors are institutionalized or if agents slip back to a previous strategy set. However, even if the dominant trading system ceases to be significantly operational, the solidification of previous strategies means that they are carried over into the new system. Figure 10.3 shows the relationships between trading systems, external Stability Factors 1–3, and the emergent Trading Complexes with corresponding trade specialists.

In figure 10.3, trading specialists shift strategies as a function of political economic stability (continuity of political infrastructures and alliances) and trade-friendliness. In this model, generalized and large-scale disasters and hazards can contribute to declining conditions of stability and downshifting in Trading Systems (N → N-1). Alternatively, more localized disasters/hazards provide opportunities for trading specialists, allowing them to institutionalize and/or up-shift the use of Trading Systems (N → N+1). We have followed Oka's earlier strategy in also referring to the emergent Trade Complexes by numbers for the reason of mathematical simplicity:

Trade Complex N = Trading Systems (N, N-1, . . . 1)

Specifically,

Trading System (N-1, N-2, N-3, . . . 1) +
Stability → Trading System (N, N-1, . . . 1)

And,

Trading System (N, N-1, . . . 1) +
Instability → Trading System (N-1, N-2, . . . 1)

Although one may hypothetically generate infinite Trading Systems and Trade Complexes, we have discerned only four distinct Trading Systems used by our informants as well as within the historic Indian Ocean trading complex. The four trading systems and the three stability factors (sedentism, political stability, trade friendliness) are sufficient for modeling the various Trade Complexes of the Indian Ocean from the earliest periods

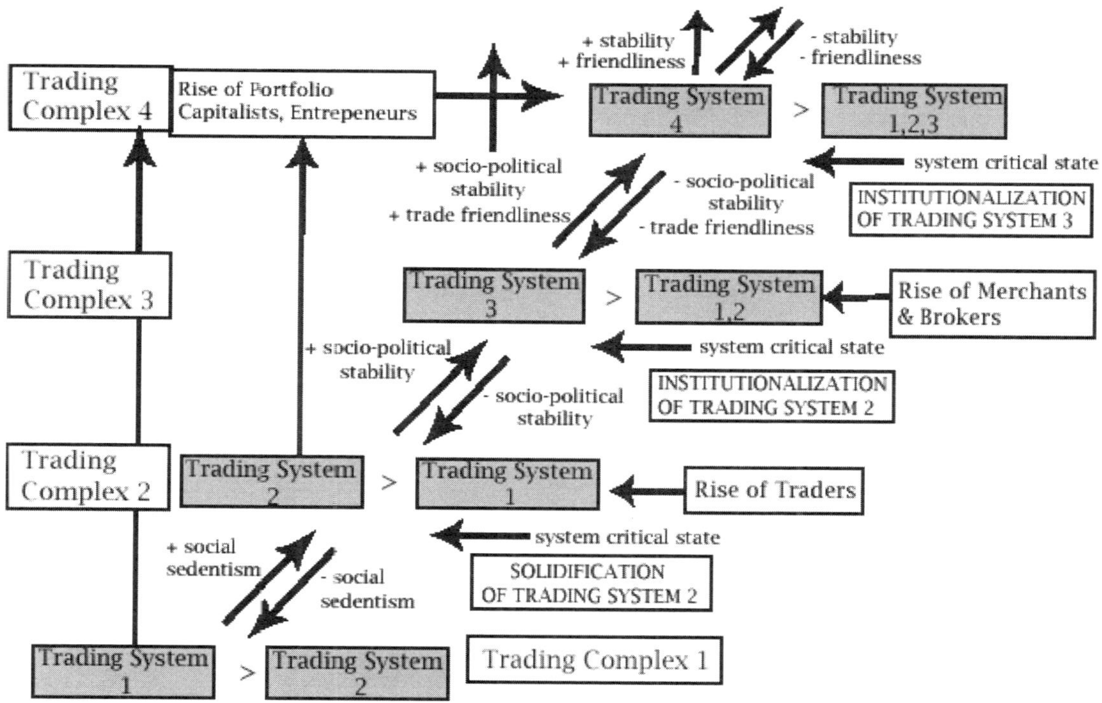

Figure 10.3. The Transition of Trad ng Systems and the Emergence of Trade Specialists in Response to Stability Factors/Events.

to the eighteenth century. In the next sections, we will apply this model to the Indian Ocean economy of the sixteenth and seventeenth centuries to show how the economic boom of the Imperial Détente, along with European Maritime Activity and the Little Ice Age, led to the decline of the Indian Ocean ports.

Imperial Détente: Political Stability and Trade-Friendliness

At the dawn of the sixteenth century, the larger Afroeurasian economy was adapting to the consolidation of the Ottoman Empire across Asia Minor, Europe, and the Arabian Peninsula followed by the rise of the Safavid and Mughal empires between CE 1501–1556. The resulting détente extended this stability to core production areas across Afrasia including Fustat, Alexandria, Mocha, Aleppo, Iznik, Izmir, Kirman, Gumroon, Hormuz, Isfahan, Kabul, Qandahar, Gujarat, Benares, and Bengal (Chaudhuri 1990; Das Gupta and Das Gupta 2001; Das Gupta and Pearson 1987; Pearson 1998). Combining factors of distance with military and economic might, the three empires achieved a dynamic parity in which no single political entity could maintain lasting ascendancy due to the complex segmentary and shifting alliances (Finkel 2006; Melville 1996; Richards 1993).

This stability allowed major changes in production infrastructures including mass production, specialization, and technological innovation. A boom in hinterland production was paralleled by a growth in maritime activities as all participants focused on profit and sustainability of trade rather than controlling the trade itself. Mughal, Safavid, and Ottoman nobility and royalty sponsored commercial activities and even invested in such activities themselves. They offered tax breaks and other incentives to trade specialists to draw in capital and investment in production activities.

The sixteenth and seventeenth century trade-friendliness was extremely beneficial for trading communities. Trading System 4 strategies dominated and led to intense entrepreneurial investment in production activities, diversification of ventures including investments in other groups' businesses, and increasing relationships with political elites through banking and fiscal services. These activities gave trade specialists greater say in commercial policies of state and indeed control over large amounts of capital that could be used to influence political decisions for their benefit (Chaudhuri 1996; Pearson 1988).

Many "great firms" moved away from earlier self-limiting strategies that focused on restraint and reinvesting in networks (Leonard 1998). Members of trading families became counselors, treasurers, and helped manage fiscal administration (Leonard 1998:408–410). The result of this close involvement was a rise in the social status of trading specialists and increasingly visible display of commercial activity as well as personal wealth: the emergence of portfolio capitalists and entrepreneurs (Morrison 1997). This rise would be exacerbated by the multiplier effects of Portuguese activities and the Little Ice Age.

Multiplier 1: The Portuguese and the Commerce-Centered Maritime Economy

The arrival of the Europeans in the sixteenth (Portuguese) and seventeenth (English and Dutch) centuries altered the hierarchy of naval strength and gave the Asian polities new allies and enemies useful for challenging each other and for building commercial relationships (Kling and Pearson 1979). The Safavids developed strong alliances with the English and the Dutch to help them against both the Portuguese and the Ottomans (Jabbari 2004). The Mughals allied with the Portuguese against the Ottomans and subsequently the Dutch, the Siddi and the Omani against the Portuguese (Farooqi 1989).

As Asian and African peoples adapted to the Portuguese in the sixteenth century, the maritime political landscape started to resemble that of the hinterland in that no one group, whether European, African, or Asian, could gain and maintain ascendancy for long. This led to a shift in the primary objective of the Portuguese and indeed most European traders in the sixteenth and seventeenth century. Rather than focus on the Europe-Indian Ocean run and on capturing Indian Ocean trade, the naval specialists chose to actively participate in the ongoing and lucrative trade within the Indian Ocean and South China Sea (Pearson 1987, 1988). However, the Europeans were heavily dependent upon their local African and Asian partners for their access to hinterland products as well as capital and/or credit mobilization (Gokhale 1979).

The Portuguese increased their influence and profitability by controlling maritime commerce through a licensing system called the *cartaz*, soon recognized as a standardized protection document across the Indian Ocean (Das Gupta 1979; Pearson 1987). Another common tactic was blockading ports for ransom, used whenever the Portuguese felt the need to renegotiate agreements or increase revenues (Farooqi 1989; Pearson 1988). Within these conflicts, trade specialists often acted as intermediaries (Das Gupta and Das Gupta 2001:271).

Both the maritime and the hinterland powers became increasingly dependent upon the revenues from the growing commercial production as well as customs, taxes, and *cartazes*. Other benefits of this growing commerce included banking and brokerage services. The interdependency of these two factors created a multiplier effect that enhanced the importance of middleman trade specialists. This rising influence along with the growing wealth from long-distance trading ventures fed back into the already booming production and distribution economy. This resulted in a radical alteration of the positions of the trade specialists and made them increasingly impervious to political control, from both the maritime and the hinterland groups. They gained power over state fiscal policies. The rise in wealth and status was also aided by the global effects of the Little Ice Age, ca. CE 1500–1800, through the generation and institutionalization of disaster-commerce in which trade specialists enriched themselves from crises-era distribution activities.

Multiplier 2: The Little Ice Age and Profits from Disaster

The fifteenth and sixteenth centuries marked the beginning of the global climate processes of the Little Ice Age (Fagan 2002). The warm and wet conditions of the Medieval Warm period ended in the late fourteenth to mid-fifteenth centuries, giving way to increasingly cold temperatures in the temperate zones and low precipitation regimes in the subtropics and the tropics (Dhavalikar 1996, 2002).

From the late fifteenth century, strong to moderate El Niño/Southern Oscillation (ENSO) events, low Nile floods, and deficient Southwest Monsoon precipitation in South Asia/East Africa indicate the rise of Little Ice Age conditions across Afrasia (Anderson et al. 1992; Cole et al. 1992; Dhavalikar 1999; Fagan 2002; Quinn 1992). Many areas in Africa and Asia underwent periodic droughts and drought-related famines (Alin and Cohen 2003; Allen 1993; Cohen et al. 1997; Nicholson 1979). Oral historiographies of various groups in East Africa as well as Portuguese and Swahili records in the sixteenth and seventeenth centuries mention numerous (and attested) famine events (1543–1544, 1560–1562, 1567, 1587–1589, 1621–1622, 1724, and 1787) in which wells dried out, crops failed, and hinterland pastoral groups such as the Oromo, Somali, and Masai migrated en masse in search of pastures, frequently clashing with each other (Allen 1993; Kirkman 1964; Middleton 1992; Schove 1977:39–41; Wood 1977:72). There was a drought or drought-related famine in East Africa and South Asia approximately once every four years between CE 1500–1700 (Habib 1999; McAlpin 1983:197; Quinn 1992). Contemporary chroniclers observed parents selling children into slavery, cannibalism, agricultural fields turned into wastelands, towns and cities abandoned, and the populace decimated or driven into exile (Allen 1993; Habib 1999). However, while these were devastating for many people and settlements, it is also true that no matter the disaster, there is always someone who benefits.

The collapse of political infrastructure is ultimately caused by elite inability to maintain managerial legitimacy during crises. In response to the changing climatic conditions in the sixteenth and seventeenth centuries, Afrasian rulers accordingly changed their redistribution and land settlement policies to bring more land under cultivation and to bring famine relief to their growing populations through redistribution (Grove 1997; Hardiman 1996; Will 1990).

However, such redistribution also strains the resources of the state from attrition due to the interdependent problems of corruption, hoarding, smuggling, inefficient transportation, and collapse of local and regional redistributive infrastructures. Governments are frequently unable to provide famine relief at all times. This is where trade specialists simultaneously play a significant role in food distribution and profit from these ventures.

Resilient trader networks allow trade specialists to operate in areas where political infrastructure and traditional social networks have collapsed (Dunstan 2006). Since information usually travels faster through trader networks than bureaucratic channels, trade specialists usually are the first to buy up the surplus food in unaffected areas and send it to the crisis zones where it is sold for profit (Keen 1994). These activities are carried out by portfolio capitalists, brokers, and entrepreneurs who speculate on grain prices and futures; and merchants (wholesalers to peddlers), who actually carry the goods into the targeted areas (Dunstan 2006; Watts 1991:46).

Bankers and moneylenders also thrive in times of severe crises as their expertise in marshalling capital and offering credit becomes increasingly important to elites and commoners alike (Bhatia 1967). Regulation of this "disaster commerce" is usually beyond the ability of political elites (Lobo and Correa-Afonso 1990:116). Faced with regulation and/or censure, trade specialists shift strategies and revert to informal and underground economies and start hoarding, price gouging, and smuggling commodities while making the state bureaucracy complicit in this process through bribery and corruption (Dunstan 2006).

The food shortages and famines of the sixteenth and seventeenth centuries offered ideal conditions for making quick high profits in an otherwise growing production and distribution economy (Hardiman 1996). These activities created great wealth for the trade specialists but brought little succor to the destitute, and many people migrated to the cities and served as cheap labor for the burgeoning urban economy based on mass production and distribution of commodities fueling further production and distribution (Habib 1999).

Modeling the Effects of the Boom, European Activities, and the Little Ice Age

The combination of the three interdependent extreme economic events created a positive cascade in which the network-centric and distribution-based infrastructure of the trader networks reached a critical point and went through stages suggested in figure 10.3, from institutionalized Trading System 3 to dominance of Trading System 4 strategies. The result was collapse and reorganization of the old trader networks (Trading Complex 3) into a dual-hierarchical economic infrastructure (Trading Complex 4). In the new economic complex, some individuals became "Great Firms" or portfolio capitalists controlling the lion's share of commercial production/distribution activities. These were linked to—yet also distinct in political power and status from—the surviving traditional trade specialists' networks that remained within the network-centric organization of the self-restraining and self-limiting trader network. Figure 10.4 models this shift to a dual-layered network through a schematic diagram.

From the late sixteenth century, the combined power of the trader network and the growing role of trade allowed Asian trade specialists to benefit as middlemen in the conflicting Portuguese-Islamic relationships and Little Ice Age disasters (Das Gupta and Das Gupta 2001:272; Pearson 1988). With greater political clout, and divorced from the former rules of self-restraint and network-centrism, trade specialists could act with impunity—and they did (Gokhale 1979). Such decisions included typical import substitution tactics by centralizing production activities in areas with cheap labor, thus providing settlement incentives to crafts-specialists, or by outsourcing. They also manipulated currency markets, speculated in grain futures, and dumped cheap and/or imitated commodities to force the decline of small-scale peripheral production economies (Clarence-Smith 2006; Leonard 1998; Newitt 1987; Palat and Wallerstein 1999; Veinstein 1999).

This led to the collapse of many local and restricted production industries as craftspeople were forced to relocate, work for lower wages, or turn to other activities. The settlements that depended upon the revenue from

Trading Complex 3 characterizing the Political Instability between 1200-1500 CE

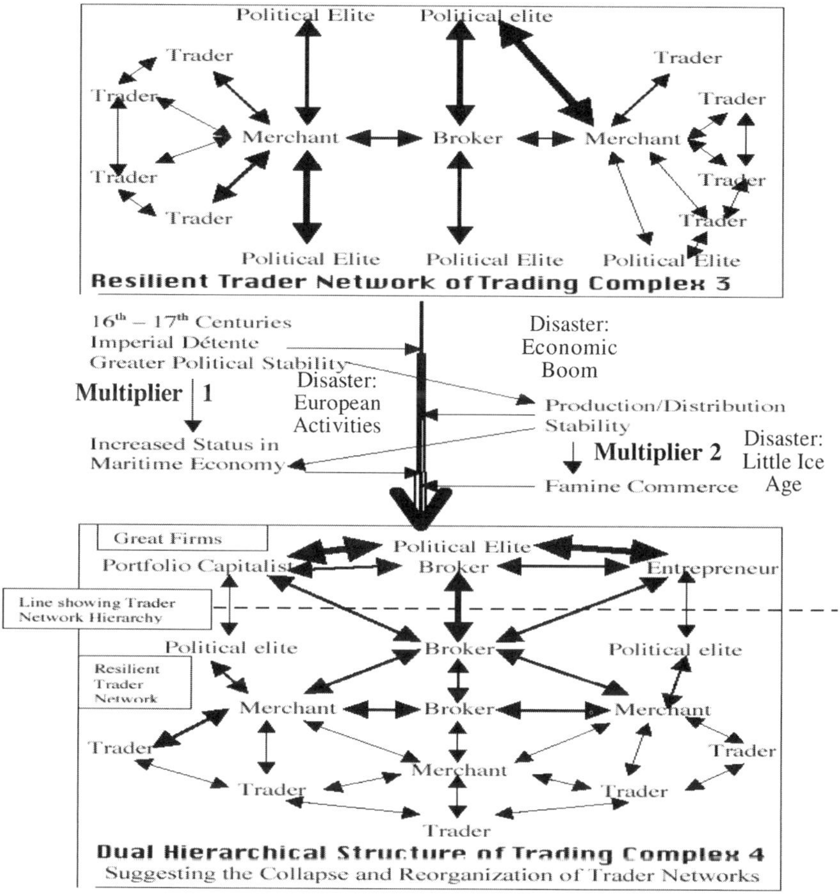

Figure 10.4. The Adaptive Cycle from Trading Complex 3 to Trading Complex 4 during the sixteenth and seventeenth centuries CE (The bidirectional arrows represent relationships and the thickness of these arrows represents the intensity of these relationships. Unidirectional arrows represent interdependent feedback from the four factors into the evolving complex. The path-dependent trajectory of the growing production and distribution economy is shown by the composite arrow linking Trading Complexes 3 and 4).

local production saw their civic infrastructure dwindle within decades, and the combined effects of declining revenues and out-flow of labor and capital resulted in a general decline of these settlements in the seventeenth century (Palat and Wallerstein 1999).

Furthermore, as the distribution area of the emerging production economy increased and trade specialists sought markets for their products, the vast amount of commodities dumped across the network ensured that the decline was felt across the peripheries. Trade specialists also took advantage of the significant resource differentials resulting from droughts/drought-related famines (Lobo and Afonso-Correia 1990). Many peasants who were affected by the famines migrated to production areas to work as cheap unskilled laborers, further boosting the mass-production economy of the urban areas (Habib 1999).

Fueled by the increasing capital investment and the constant injection of cheap labor into centralized areas, production of value added commodities approached proto-industrial levels (Clarence-Smith 2006; Palat and Wallerstein 1999). Trade specialists aggressively sought new markets for these products. South Asian trade specialists lobbied the Russian nobility to start mass distribution of Indian cloth into the Russian and Baltic states (Dale 1994). Chinese and Indian merchants dumped various commodities in Southeast Asian markets (Bronson 1990) and East Africa (Newitt 1987; Pouwels 1987). The increased demand for raw materials led to the direct purchase of raw materials as opposed to processed products (Newitt 1987). There were conscious efforts to undermine the value-addition into peripheral products.

Trade specialists who still remained within the trader networks shifted their roles accordingly, either as procurers of raw materials to send to the central production areas or suppliers of mass-produced commodities to the peripheral markets. Portfolio capitalists, entrepreneurs, and large merchants and brokers grew wealthier and more wealth was concentrated in the hands of fewer people. The most significant indication of this shift is the difference between the distribution of elite wealth in the pre–sixteenth century economy, and in the economy of the sixteenth and seventeenth centuries.

Given these indicators, how would this transition look in the archaeological record? Based on the aforementioned changes in behavior, we expect to find the following four correlates in the archaeological record at peripheral ports and trading centers of the Indian Ocean between CE 1500–1800, specifically at the ports of Chaul, India, and Mtwapa, Kenya:

1. There would be a corresponding rise in the exports of raw material from peripheral areas (e.g., ivory exports).
2. There would be a significant increase in the total number of late sixteenth and seventeenth century prestige artifacts given as gifts by trade specialists to political and economic elites (e.g., Chinese Blue-and-White porcelains).

3. Late sixteenth and especially seventeenth century prestige ceramics would be increasingly concentrated in fewer areas than pre–sixteenth century, suggesting vastly increased social distance between elites and non-elites, as well as transition from corporate oligarchy to a classic hierarchy.
4. There would be a drastic decline in local crafts-production and in the lives of local artisans and crafts-specialists (e.g., decreasing access to prestige goods).

The Growing Demand for Ivory from CE 1600–1800

The main demand for ivory in the global economy of the sixteenth to eighteenth centuries CE came from South Asia and East Asia. Trade specialists participating in the India-China trade realized that the almost limitless supplies of East African ivory could be procured very cheaply in exchange for mass-produced South Asian cotton cloth. In the drive to get more ivory to offset the silver flow from Europe to South Asia and thence into China, the trade specialists resorted to various strategies to corner both the ivory supply routes and the local cloth markets (Oka et al. 2007). This included dumping of cheap cloth and actively out-competing local cloth industries in East Africa (Newitt 1987). The success of this process can be seen in the rising volume of ivory exports from East Africa to South Asia between the sixteenth and eighteenth centuries as shown in figure 10.5.

As seen in figure 10.5, there was a tremendous increase in the amount of ivory exported from East Africa to Asia. Conservative estimates suggest that actual trade volumes ranged from 5 to 20 times the amounts reported by the European companies in the sixteenth, seventeenth, and eighteenth centuries (Chaudhuri 1996). As trade specialists invested in coastal plantation economies, they also sponsored slave-raiding ventures that piggy-backed on the ivory trade. The landscapes of the East African hinterland were steadily decimated by the coastal slave-raiders who by the late eighteenth century had reached the Great Lakes region in Central Africa. This trade reached its apogee in the nineteenth century after which supplies of both ivory and slaves fluctuated (Kusimba 2004; Kusimba et al. 2005). In figure 10.6, we show the results of our estimates of the total ivory and slave trade figures as conducted by Europeans, Africans, and Asians between CE 1507 and 1908.

When looking at figure 10.6, for every two tusks of ivory reported, four people were dragged from their homes, and one elephant slaughtered. In return, cheap South Asian cloth flooded the East African markets and

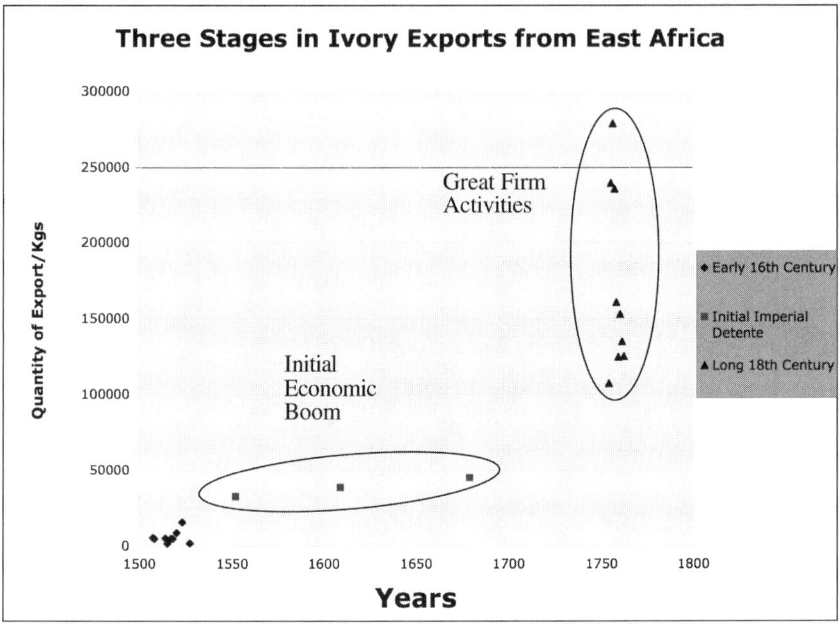

Figure 10.5. Ivory Exports from East Africa (Portuguese and British Records Collected by Peter Thorbahn).

rendered indigenous industries defunct. How did this process affect the East African ports? As trading polities, they should profit from such trade. In the next section, we show through analysis of prestige ceramics from Chaul and Mtwapa that trade did intensify at both ports. But at Mtwapa, the manufacturing base declined and Mtwapa changed from a traditional port offering both production and distribution services ruled by a corporate oligarchy into a specialized raw material supplying port with the elites engaged in stiff competition for alliances with the emerging Asian portfolio capitalists.

Prestige Ceramics and the Fates of Indian Ocean Ports

Prestige goods networks are good indicators of intra-societal wealth and power distributions (Schneider 1991). Apart from conveying the basic spatial differences between elites and non-elite groups within cities, the type and volume of prestige goods can also provide information on differential tastes, preferences, and networks within elite groups themselves and the changing fortunes of the different elite localities within a trading center

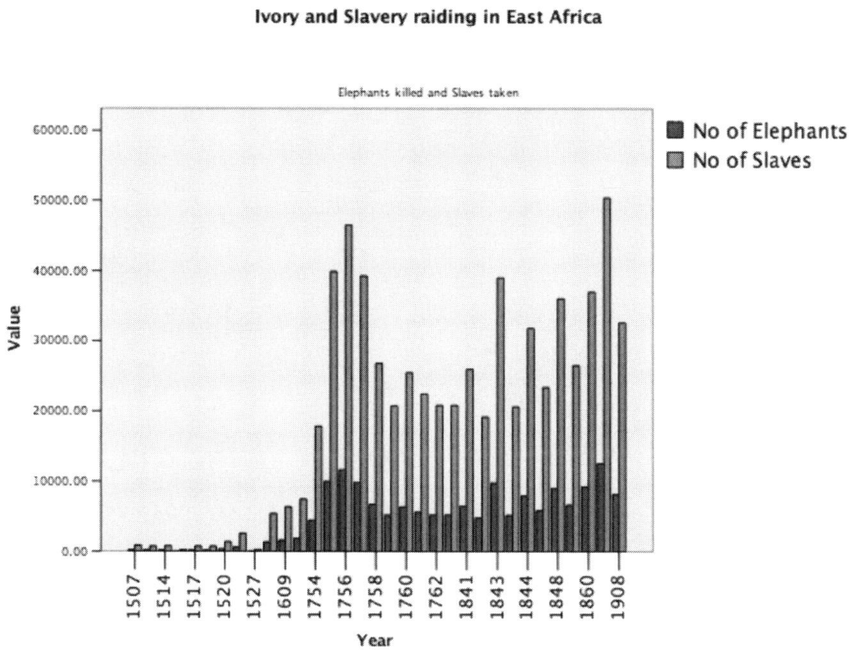

Figure 10.6. Ivory Trade Boom Corrected for Estimates of Total Trade.

(Junker 1999; Kusimba 1999b). The Islamic Era in the Indian Ocean (CE 700–1750) was characterized by a prestige goods network that linked elites from East Africa, Southwest, South, and Southeast Asia, leading to overlapping tastes and preferences for luxury consumables, commodities, and gifts (Carswell 2000; Watson 2004).

Since the early fifteenth century, Chinese Blue-and-White porcelains were the most desired prestige ceramics in the Western Indian Ocean, were also imitated across the Indian Ocean, especially in Southeast Asia (Vietnam and Thailand) and Southwest Asia (Persia, Syria, and Turkey; Carswell 2000; Middleton 2003). Using high-resolution laser-ablation inductively coupled mass-spectrometry, we first determined the chemical characterization of the white and blue glazed areas of the excavated Blue-and-White ceramics from Chaul and Mtwapa (Oka 2008; Oka et al. 2007; Swerts et al. 1994; Yu and Miao 1996, 1997), and then dated the selected Chinese sherds using manganese/cobalt and manganese/iron ratios and compared these results to studies already conducted on Chinese Blue-and-White porcelains (Cheng et al. 2002, 2005; Lin et al. 1992; Oka 2008; Wen et al. 2007; Yu and Miao 1999). The results of these tests are shown in tables 10.1 and 10.2.

Table 10.1. Distribution of Fifteenth, Sixteenth, and Seventeenth Century Chinese Blue-and-White Porcelains at Chaul and Mtwapa

Dynasty	*Chaul*	*Mtwapa*
Yongle/Xuande (1403–1430)	7	8
Mn/Co > 4.5		
Mn/Fe < 0.35		
Chenghua (1465–1488) & Jiajing (1521–1567)	3	0
Mn/Co = 0.2–2.5		
Late Ming- Qing (1600–1800)	31	43
Mn/Co > 4.5		
Mn/Fe > 0.35		

The results show with statistical significance, apropos the postulated correlate, that the majority of the Chinese Blue-and-Whites can be dated to the late sixteenth and seventeenth century and not before, indicating that both ports participated in the seventeenth century Trading Complex 4. Through archaeological analysis, we can explore the dark side of Trading Complex 4, such as how peripheral areas of the Indian Ocean such as Mtwapa lost out to the production and distribution economies of the core located in mainland Asia. Figures 10.7a–d show the changes in distribution of prestige ceramics across the 10 localities across the rise and decline of Mtwapa from ca. CE 1000–1700.

In figures 10.7a–d, prior to CE 1600, Areas 4, 5, and 6 show similarities in the distribution of all the prestige wares, especially glazed wares. Second, this trend is completely reversed in the post-1600 CE period. A majority of the prestige wares at Mtwapa are relegated to one area, Area 4 or the main Friday Mosque Complex. This was the main gathering zone where city elites would gather to display their status through displays of philanthropy, such as giving alms to the poor and public feasting. We can also see a decline in the former elite localities of Area 5 and Area 6 that were the centers of iron/steel working and cloth manufacture, respectively.

Table 10.2. Differences between pre-1600 and post-1600 Maritime Economy, Indicated by Chinese Blue-and-White Porcelains

Period	*Chaul (n = 41)*	*Mtwapa (n = 51)*
2: 1400–1600	10	8
1: 1600–1800	31	43
χ^2 (p = 0.5)	10.756	24.019
Significance Level	0.002 (1 df)	0.001 (1 df)

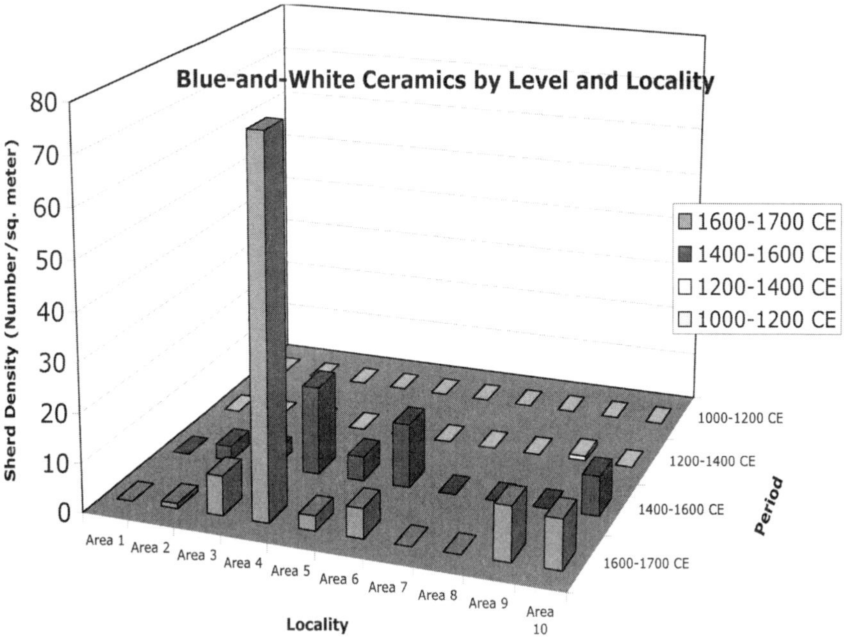

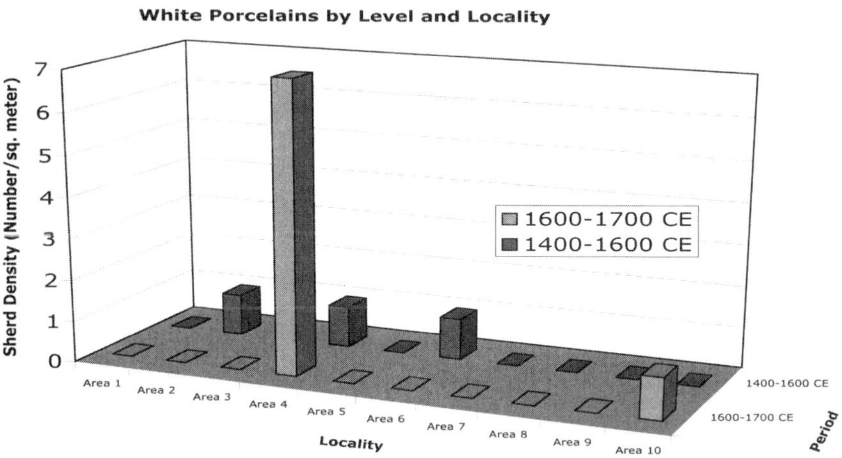

Figure 10.7. Spatial and Vertical Distribution of Sherd Density (Number per Cubic Meter) in Prestige Wares across Mtwapa

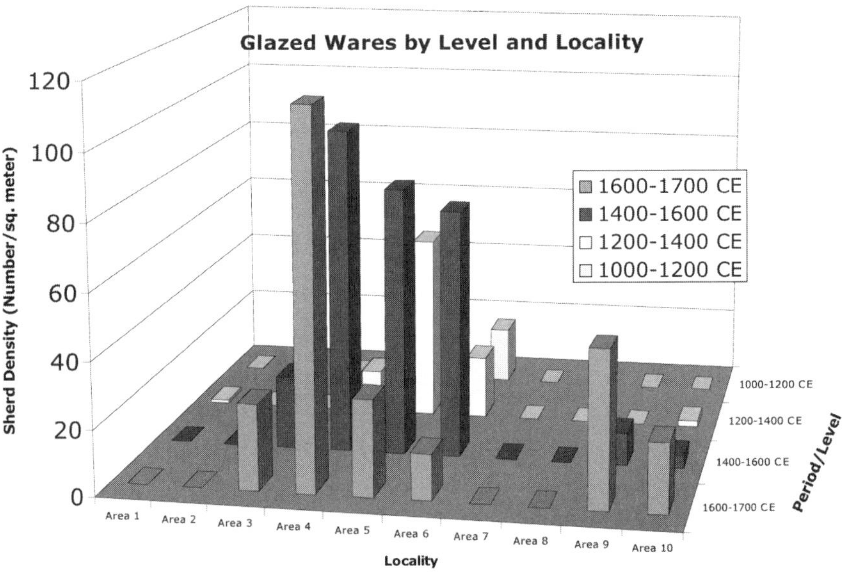

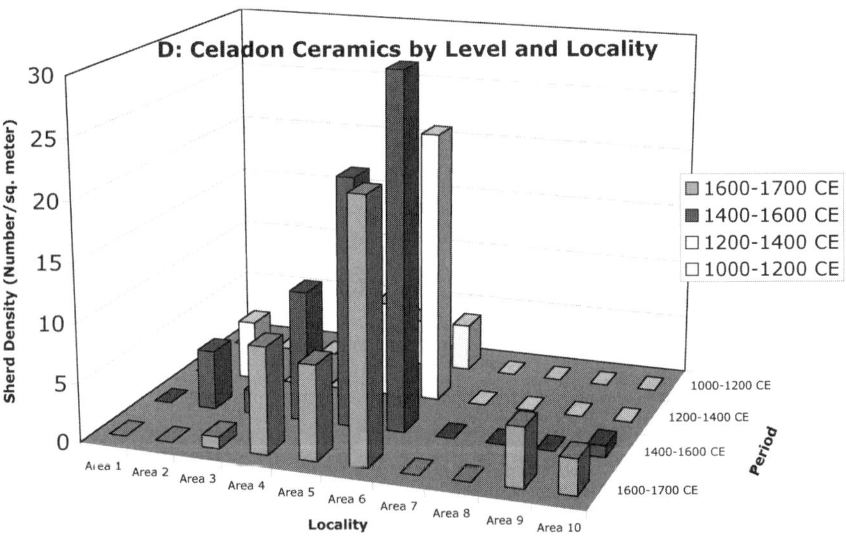

Figure 10.7. (*continued*)

The prestige goods received by elites through trade were not kept for display in their homes as before but were increasingly used for public feasting and gift-giving at the Friday Mosque. The resultant attrition of wealth was compounded by the decline of local crafts-production (e.g., iron, steel, cloth). However, even as the elites of Mtwapa competed amongst themselves for access to the raw material–based economy encouraged by their Asian partners, the same Asian trade specialists started shifting their activities to larger and more centralized ports, such as Mombasa, that could handle larger volumes and provide additional service infrastructure for the emergent Trading Complex 4 economy (e.g., warehousing, wholesaling, and large-scale banking and credit facilities).

Conclusion

One of the largest differences between Trading Systems 1, 2, and 3 and Trading System 4 is the role played by self-restraint and self-limitation in the generation of sustainable networks. However, under the politically stable and trade-friendly conditions that allow the emergence of Trading System 4, the need for these limiting strategies is increasingly nullified as trade specialists reinvent themselves as nobility, and start controlling and influencing state finance. Trading Complex 4 over time starts self-replicating, and actively generates the trajectory toward a weak political infrastructure beholden and indebted to the trade lobbies. Using their economic and political capital, the newly empowered portfolio capitalists constantly undermine any attempt at regulation by elites (Chaudhuri 1996). In the absence of regulation, the continued domination of Trading System 4 can create very wealthy predatory economic elites who control the economic aspects of the polity (Das Gupta 1998; Das Gupta and Das Gupta 2001).

The flip side of trade specialists' influence over and sponsorship of political elites means that if they are displeased or see better opportunities anywhere else, trade specialists can switch alliances and/or relocate toward better opportunities. This harkens to established theories of urban decline in which the productivity of the primary sector (and its tax revenues) determines the level of efficiency and quality of the administrative sector (Allen et al. 2000). Political elites of declining ports usually try and maximize revenue by increasing taxes and/or trying to draw capital back into the city. Increased taxes never draw trade specialists, thus attempts to attract capital usually focus on sociocultural aspects such as monumental architecture or sponsoring festivals (Yoffee 2004). However, without regenerating the primary sector by reinvestment in old or by adopting new economic activities, the decline of the city or port as a center for capital investment and business ventures is usually a foregone conclusion.

The peoples of the East African coast and South Asia might have been able to adapt to the Portuguese aggression in the sixteenth century and there is evidence to suggest that they started doing so using various strategies including resistance, alliances with other naval powers, and by migration (Pouwels 1987). They could also have dealt with the vagaries of the Little Ice Age as they had reacted to changes in climatic conditions for the previous three millennia (McCann 1999; Nicholson 1979). The effects of both extreme events together were catastrophic but perhaps nothing that peoples of Afrasia could not handle over the period of a century. As various scholars have pointed out, adversity becomes a normative condition when sustained over a long period of time, and people adapt in one way or another (Nicholson 1979). However, the drastic change was brought to these recovering and adapting ports in the form of Trading Complex 4 and its effect on local production economies. Within this emergent Trading Complex 4 global economy, the activities of the Portuguese and the Little Ice Age droughts/famines became multiplier or focusing devices that exacerbated the Imperial trade boom and boosted the role of trade specialists to one of dominance in both the hinterland and maritime economies, with the smaller trading centers—highly specialized for the pre–Modern Era trade—brutally out-competed or bypassed in the drive toward centralization and mass-production and distribution.

It is perhaps sobering to realize the enduring nature of the oft-cursed "disaster-based" commerce that inevitably follows contemporary disasters also, whether natural or anthropogenic. In this chapter, we have argued that such behaviors, while seen repeatedly in human history, became normative in the rising global commerce of the sixteenth and seventeenth centuries. As multipliers, they allowed those who have a lot to accumulate even more and profit off others' suffering. Practices such as profiteering, vulture fund portfolios (buying up weak countries' debts from donors and then extracting the full amount), hoarding, and market speculation, are not new and in fact may be a necessary part of our global economy. As we gather data on how people respond to disaster, we also need to pay heed to debilitating factors such as disaster commerce built into the very economy into which the afflicted peoples are meant to be reabsorbed. Adapting to natural and anthropogenic disasters hence would need close attention to such activities and their far-reaching and usually disastrous effects.

Notes

1. Because of the specific religious, geographic, military, and economic powers of these empires, we argue that the relationship between these empires could be characterized as that of détente, in which no one polity dominated another and pragmatic decision-making drastically decreased the number and intensity of outright warfare between and within these three empires (D'Souza 2001; Oka 2008).

2. Resilience is defined as "the capacity of a particular system (in this case the network) to absorb disturbance, undergo change and still retain essentially the same function, structure, identity, and feedback loops," and adaptive resilience in social networks includes the "capacity of actors in the network to manage resilience, either by moving the system toward or away from a threshold that would fundamentally alter the properties of the system, or by altering the underlying features of the stability landscape" (Holling and Gunderson 2002:28–29).

3. Econ S suggested: "each family operates for its own benefit but we also know that we are stronger as a group of linked families. If one of us is targeted, we can rely on our relationships to deal with the problem."

4. When Econ U got married, he said, "the match was between the families, and the strengthened relationships between [his] bride's families and [his own resulted] in greater business opportunities, greater collective trading strength and the chance for the crossing over of good smart people within the two businesses." Also, "since my other family [wants the] business to thrive, now [I have] 400 years of collective experience backing him; [his] worth has doubled." Econ R said: "the religion of the person does not matter, just how good are his business skills, does he learn fast and does he learn from his mistakes. The bottom line is, if I send my son to my friend and say 'teach him,' not only am I doing my son a favor as he is learning in a unprivileged environment [as an apprentice and not a son], but I am essentially giving great honor to my friend as I am saying, 'the future of my business and family [my son] is being taught by you, if you give him half knowledge or don't train him well, we will say: 'Is that the way you do business?' Word might get around."

5. Econ F asserted the importance of the distributed network, especially in crisis situations: "if [one venture] fails, we will be paid back because the main business will still be there, so they won't have lost everything. Sometimes, even if you have lost everything, you can still get investment if you were well-known, someone can vouch for you, if the business fails because of political trouble or violence, etc." He mentioned a well-known family that had lost their business during the Partition of India in 1947; that they had nothing when they came, "but they [found] other people who knew their past successes and were able to start from nothing and build a huge business."

6. Econ L stated: "Business strategies are decided within our family, taking into account things such as politics and also what other families may think. This gives us greater power over our own life than say, a public servant or a farmer. We

do not have to ask for permission for taking business decisions but we will work within a consensus." Econ T suggested, "our success is dependent on the fact that we do not have one head, one chairman, or one president; we are linked families who cooperate with each other to keep businesses going. It is like Hinduism, if we had one Pope, one holy book, we would have become Muslims or Christians long back."

7. Econ L said, "Interactions with other families are based on a code of morality. Hence, any person not behaving within acceptable norms will be shunned in social gatherings and will find it difficult to get business deals within the community. When we take apprentices from other families, we teach them the full range of business, because he is a future collaborator, not a competitor." Econ D placed the concept of profit within a context akin to the equilibrium theory of games formulated by economist John Nash: "there is profit in everything, but you need to be sure that everyone is happy, [. . .], satisfied, so some people do not have everything, and others do not have nothing. If we do not blindly go for maximum phaida (profit), but moderate it in favor of making sure that business can stay, then everyone involved gets something in return." Econ G stated that the concept of open competition and monopolistic ventures was harmful to the stability of the commercial landscape; he preferred to have arbitrary or artificial specialization of trading spheres or healthy relations between direct competitors. In his words, "if I deliberately stock what [competitors] have, they will do the same and we both lose out, if we keep to each other and help each other, we both win. There is an unspoken accord but we don't know if we can make them do anything by force." Furthermore, uncontrolled growth and price wars just affect the business, and usually negatively, as in a "growing market, there is enough for everybody."

8. Econ L said stability is a "prerequisite for growth" but that "real business lies in expansion." Expanding the business meant increasing the volume of trade, opening up new shops in different cities, and/or diversification. Econ I described this growth thus: "[traders] come in, first with the [generalized groceries], then with the [specialized shops such as sweets sellers], then with the appliances and hardware and then with small eating places." Echoing the majority of my informants, Econ C stated: "the customer is my god and the business is my religion. Those who [cheat] come to a bad end and we take steps to cure [cheats] even if the government does not. And think of this, when times are bad, who will the mob come after, the evil government? No, [they will come after] the *kanjus makkhichus* (miserly fly-sucking) trader. Why would we do ourselves harm? We are here to do business not harm. Good business means long-term business and we are willing to forgo immediate profit for staying in place, you see. Why do you think that we stay on even when everybody else leaves?" Econ N spoke about the constant danger of instability: "things are always changing in politics and our friends today might not be able to protect us tomorrow. Look at what happened to the Indians in Uganda and is going on in Fiji right now. Banking structures can fail or be destabilized. But we operate on friendships and reputations built over years and make sure that untrustworthy

people cannot do business, [. . .] so our relationships are the only way we can be sure of keeping our business safe, [. . .] in good times we can work with the politicians but in bad times, we can generate opportunities, credit and capitalization through *olakh* [personalized networks of influence]."

9. Econ P said: "business is in my blood, not just a livelihood, and there is no separation between business and family. For us, commercial activities are our destiny; and [our duty] is ensuring that not only our children but also our great-great grandchildren see and participate in our business. A well run business should run at five generations if not more." Reaffirming the entangled nature of business, commerce, and trader culture, Econ F said "what do you do business for if not a happy home and why do you create a family if not to continue bringing Lakshmi (wealth) into your life."

10. Econ L, P and R all stated that any expansion into a new area or increase in scale of business had to be done in the core business. This is followed even when the family or group might have diversified their interests. Econ L said, "we have the most knowledge and experience with our core business, and that gives us the highest chances for success and hence the reputation for competence. And our reputation is what allows us to maintain profitable relationships with other businessmen." As Econ G stated, core competence made "people come [. . .] based on our reputation not on the look of the place. They come here to see the merchandise. They look around and see me, the man who will sell to them. They know me personally and by reputation. They know my father and grandfather. They know, that we will give them the best price, and the best quality, and also that we care about their business, that our business [in fact, relationship] is not over with the transaction. They see a no-nonsense place that will do business with them. You see the other stores around you? They are all hardware stores. How do we survive and do well in all the competition? Because [we] have been here since [this commercial area] was started more than two hundred years ago."

Acknowledgments

Funding for our research came from the National Science Foundation (Grant nos. BCS 016664, BCS 9615291, BNS 9024683, and BNS 0320903), Provost's Grant for dissertation research from the University of Illinois at Chicago, the Deccan College Post-Graduate and Research Institute, Pune, India and the National Institute of Ocean Technology of the Government of India. We would acknowledge our debt to Christopher and Pauline Smith (Mexico), Uma Oka (India) and Pierrette Barut, CO, USA for their support in fieldwork in India and Kenya. We would like to thank Laure Dussubieux, Mark Golitko, and Ryan Williams from the Anthropological Science Laboratory at the Field Museum for their help and assistance with the chemical analysis. We would like to thank our colleagues at the Field Museum, the University of Illinois at Chicago, the University of Notre Dame, the Deccan College Post-Graduate and Research Institute, and Fort Jesus Museum, Mombasa, Kenya. In particular we thank Sibel Barut, Vania Smith-Oka, Bennett Bronson, Gary Feinman, John Terrell, Laura Junker, Slyvia Vatuk, Anna Roosevelt, Larry Keeley, Carolyn Nordstrom, Ian Kuijt, Agustin Fuentes, M.D. Kajale, Abhijit Dandekar, Rukhashana Nanji, Srikant Pradhan, Sachin Joshi, and Mohammed Mchula. We especially thank the late Peter F. Thorbahn for collating the archival data on ivory trade, and Barbara Thorbahn for her generosity in sharing this work with us.

References

Albin, P. S.
1998 Barriers and Bounds to Rationality: Essays on Economic Complexity and Dynamics in Interactive Systems. Princeton, NJ: Princeton University Press.

Alexander, J., and R. P. Alexander
1991 Protecting Peasants from Capitalism: The Subordination of Javanese Traders by the Colonial State. Comparative Studies in Society and History 33:370–394.

Alin, S. R., and A. S. Cohen
2003 Lake-level History of Lake Tanganyika, East Africa, for the Past 2500 Years Based on Ostracode-inferred Water-depth Reconstruction. Palaeogeography, Palaeoclimatology, Palaeoecology 199:31–49.

Allen, James de V.
1993 Swahili Origins: Swahili Culture and the Shungwaya Phenomenon. London: James Currey.

Allen, J. R. L., and K. Pye, eds.
2000 Coastal and Estuarine Environments: Sedimentology, Geomorphology, and Geoarchaeology. Special Publication, 175. London: Geological Society.

Allenby, B., and J. Fink
2005 Towards Inherently Secure and Resilient Societies. Science 309:1034–1036.

Anderson, R. Y., A. Soutar, and T. C. Johnson
1992 Long-term Changes in El Niño/Southern Oscillation: Evidence from Marine and Lacustrine Sediments. *In* El Niño: Historical and Paleoclimatic Aspects of the Southern Oscillation. H. F. Diaz and V. Markgarf, eds. Pp. 419–434. Cambridge: Cambridge University Press.

Bellina, Berenice
2003 Beads, social change and interaction between India and South-east Asia. Antiquity 77:285–297

Bhatia, B. M.
1967 Famines in India: A Study in Some Aspects of the Economic History of India (1860–1965). London: Asia Publishing House.

Bronson, B.
1990 Export Porcelain in Economic Perspective: The Asian Ceramic Trade in the Seventeenth Century. *In* Ancient Ceramic Kiln Technology in Asia. C. Ho, ed. Pp. 126–151. Hong Kong: Centre of Asian Studies, University of Hong Kong Press.

Carswell, J.
2000 Blue and White: Chinese Porcelain around the World. London: British Museum Press.

Chaudhuri, K. N.
1990 Asia before Europe: Economy and Civilization of the Indian Ocean from the Rise of Islam to 1750. Cambridge: Cambridge University Press.
1996 The Structure of the Indian Textile Industry in the Seventeenth and Eighteenth Centuries. *In* Technology and European Overseas Enterprise: Diffusion, Adaptation and Adoption. M. Adas, ed. Pp. 343–398. Aldershot, UK: Varorium.

Cheng, H. S., Z. Q. Zhang, H. N. Xia, J. C. Jiang, and F. J. Yang
2002 Non-Destructive Analysis and Appraisal of Ancient Chinese Porcelain by PIXE. Nuclear Instruments and Methods in Physics Research B 190:488–491.

Cheng, H. S., B. Zhang, D. Zhu, F. J. Yang, X. M. Sun, and M. S. Guo
2005 Some New Results of PIXE Study on Chinese Ancient Porcelain. Nuclear Instruments and Methods in Physics Research B 240:527–531.

Chittick, H. N.
1974 Kilwa: An Islamic Trading City on the East African Coast. No. 5. Nairobi: British Institute in Eastern Africa.

Clarence-Smith, W. G.
2006 Cotton Textiles on the Indian Ocean Periphery, c1500-c1850. Electronic document, www.lse.ac.uk/collections/economicHistory/seminars/Clarence-Smith-Textiles4.pdf, accessed August 30, 2007.

Cohen, A.
1971 Cultural Strategies in the Organization of Trading Diasporas. *In* The Development of Indigenous Trade and Markets in West Africa. C. Meillassoux, ed. Pp. 266–281. London: Oxford University Press.

Cohen, A. S., K. E. Lezzar, J. J. Tiercelin, and M. Soreghan
1997 New Paleogeographic and Lake-level Reconstruction of Lake Tanganyika: Implications for Tectonic, Climatic and Biological Evolution in a Rift-Lake. Basin Research 9:107–132.

Cole, J. E., G. T. Shen, R. G. Fairbanks, and M. Moore
1992 Coral Monitors of El Niño/Southern Oscillations Dynamics Across the Equatorial Pacific. *In* El Niño: Historical and Paleoclimatic Aspects of the Southern Oscillation. H. F. Diaz and V. Markgarf, eds. Pp. 347–375. Cambridge: Cambridge University Press.

Curtin, P. D.
1984 Cross-Cultural Trade in World History. Cambridge: Cambridge University Press.

Dale, S.
1994 Indian Merchants and Eurasian Trade. Cambridge: Cambridge University Press.

Das Gupta, A.
1979 Indian Merchants and the Decline of Surat: c. 1700–1750. 1. Steiner, Wiesbaden.
1998 Trade and Politics in Eighteenth Century India. *In* The Mughal State, 1526–1750. M. Alam and S. Subrahmanyam, eds. Pp. 361–397. Oxford: Oxford University Press.

Das Gupta, A., and U. Das Gupta
2001 The World of the Indian Ocean Merchant, 1500–1800, Collected Essays of Ashin Das Gupta. New Delhi: Oxford University Press.

Das Gupta, Ashin, and Michael N. Pearson, eds.
1987 India and the Indian Ocean. Oxford: Oxford University Press.

D'Souza, Rohan
2001 Crisis before the Fall: Some Speculations on the Decline of the Ottomans, Safavids and Mughals. Social Scientist 30(9/10): 3–30.

Dhavalikar, M. K.
1996 Environment: Its Influence on History and Culture in Western India. Indica 33(2):81–118.
1999 The Golden Age and After: Perspectives in Historical Archaeology. Presidential Address to the Indian Historical Congress Platinum Jubilee (60th) session, Calicut, India.
2002 Environment and Culture: A Historical Perspective. Pune, India: Bhandarkar Oriental Research Institute.

Douglas, M.
1986 How Institutions Think. Syracuse, NY: Syracuse University Press.

Dunstan, H.
2006 State or Merchant: Political Economy and Political Process in 1740s China. Cambridge, MA: Harvard University Press.

Fagan, B.
2002 The Little Ice Age: How Climate Made History, 1300–1850. New York: Basic Books.

Falzon, M. A.
2004 Cosmopolitan Connections: The Sindhi Diaspora, 1860–2000. Leiden: Brill Publications.

Farooqi, N. M.
1989 Mughal-Ottoman Relations: A Study of Political and Diplomatic Relations between Mughal India and the Ottoman Empire, 1556–1748. Delhi: Idarah-I-Adabiyat.

Finkel, C.
2006 Osman's Dream: The Story of the Ottoman Empire, 1300–1923. New York: Basic Books.

Fukuzawa, H.
1991 The Medieval Deccan: Peasants, Social Systems and States, Sixteenth to Eighteenth Centuries. Delhi: Oxford University Press.

Girard, R.
2001 I See Satan Fall Like Lightning. Translated by J. G. Williams. New York: Orbis Books.

Gogte, V.
2003 The Archaeology of Maritime Trade at Chaul, Western Coast of India. Man and Environment 28(1):67–74.

Gokhale, B. G.
1979 Surat in the Seventeenth Century: A Study of Urban History of Pre-Modern India. London: Curzon Press.

Grove, R. H.
1997 Ecology, Climate and Empire: Colonialism and Global Environmental History, 1400–1940. Cambridge, UK: White Horse Press.

Habib, Irfan
1999 The Agrarian System of Mughal India, 1556–1707. Oxford: Oxford University Press.

Hardiman, D.
1996 Usury, Dearth and Famine in Western India. Past and Present 152: 113–156.

Herlehy, T.
1984 Ties That Bind: Palm Wine and Blood-Brotherhood at the Kenya Coast During the 19th Century. International Journal of African Historical Studies 17(2):44–57

Holling, C. S. and L. Gunderson
2002 Resilience and Adaptive Cycles. *In* Panarchy: Understanding Transformations in Human and Natural Systems. Lance Gunderson and C. S. Holling, eds. Pp: 25–62. Washington, D.C.: Island Press.

Horton, M.
1996 Shanga: The Archaeology of a Muslim Trading Community on the Coast of East Africa. London: The British Institute in Eastern Africa.

Jabbari, H.
2004 Trade and Commerce between Iran and India During the Safavid Period (1555–1707). New Delhi: Indian Bibliographies Bureau.

Junker, L. L.
1999 Trading, Raiding and Feasting: The Political Economy of Philippine Chiefdoms. Honolulu: University of Hawaii Press.

Keen, D.
1994 The Benefits of Famine: A Political Economy of Famine and Relief in Southwestern Sudan, 1983–1989. Princeton, NJ: Princeton University Press.

Kirkman, J. S.
1964 Man and Monuments on the East African Coasts. London: Lutterworth Press.

Kling, B. B., and M. N. Pearson, eds.
1979 The Age of Partnership: Europeans in Asia before Dominion. Honolulu: University Press of Hawaii.

Kusimba, C. M.
1993 The Archaeology and Ethnography of Iron Metallurgy of the Kenya Coast. PhD Dissertation, Department of Anthropology, Bryn Mawr College.
1999a The Rise and Fall of Swahili States. Walnut Creek, CA: AltaMira Press.
1999b Material Symbols Among the Precolonial Swahili of the East African Coast. *In* Material Symbols, Culture and Economy in Prehistory. J. Robb, ed. Pp. 318–341. Carbondale, IL: Center For Archaeological Investigations, Southern Illinois University.
2004 Archaeology of Slavery in East Africa. African Archaeological Review 21(2):59–88.

Kusimba, C. M., S. B. Kusimba, and D. K. Wright
2005 The development and Collapse of Precolonial Cthnic Mosaics in Tsavo, Kenya. Journal of African Archaeology 3:345–365.

Landa, J. T. (ed)
1994 Trust, Ethnicity, and Identity: Beyond the New Institutional Economics of Ethnic Trading Networks, Contract Law, and Gift-Exchange. Ann Arbor, MI: University of Michigan Press.

Lansing, J. S.
2003 Complex Adaptive Systems. Annual Review of Anthropology 32:183–204.

Leonard, K.
1998 The "Great Firm" Theory of the Decline of the Mughal Empire. The Mughal State, 1526–1750. M. Alam and S. Subrahmanyam, eds. Pp. 398–419. Oxford: Oxford University Press.

Lin, E. K., C. W. Wang, P. K. Teng, Y. M. Huang, and C. Y. Chen
1992 Elemental Analysis with External Beam PIXE. Nuclear Instruments and Methods in Physics Research B 68:281–284.

Lobo, V., and J. Correia-Afonso, translators.
1990 Intrepid Itinerant: Manuel Godinho and His Journey from India to Portugal in 1663. J. Correia-Afonso, ed. Bombay: Oxford University Press.

Markovits, C.
2001 The Global World of the South Asian Merchant. Cambridge: Cambridge University Press.

Mbuia-Joao, T.
1990 The Revolt of Dom Jeronimo Chingulia of Mombasa, 1590–1637: An African Episode in the Portuguese Century of Decline. PhD Dissertation, Department of History, Catholic University of America.

McAlpin, M. B.
1983 Subject to Famine: Food Crises and Economic Change in Western India, 1860–1920. Princeton, NJ: Princeton University Press.

McCann, J. C.
1999 Climate and Causation in African History. International Journal of African Historical Studies 32(2-3):261–279.

Melville, C., ed.
1996 Safavid Persia: The History and Politics of an Islamic Society. New York: St. Martin's Press.

Middleton, J.
1992 The World of the Swahili, An African Mercantile Civilization. New Haven: Yale University Press.
2003 Merchants: An Essay in Historical Ethnography. Journal of the Royal Anthropological Institute 9:509–526.

Moore Jr., B.
2000 Moral Purity and Persecution in History. Princeton, NJ: Princeton University Press.

Morrison, K. D.
1997 Commerce and Culture in South Asia: Perspectives from Archaeology and History. Annual Review of Anthropology 26:87–108.

Newitt, M. D. D.
1987 East Africa and the Indian Ocean Trade. *In* India and the Indian Ocean. A. Das Gupta and M. N. Pearson, eds. Pp. 201–223. Oxford: Oxford University Press.

Nicholson, S. E.
1979 The Methodology of Historical Climate Reconstruction and its Application to Africa. Journal of African History 20(1):31–49.

Oka, R. C.
2008 Resilience and Adaptation in Trade Networks in East African and South Asian Port Polities, 1500–1800 CE. Unpublished PhD Dissertation, Department of Anthropology, University of Illinois–Chicago/Field Museum.

Oka, R. C., and C. M. Kusimba
2008 Archaeology of Trading Systems 1: Towards a New Trade Synthesis. Journal of Archaeological Research DOI: 10.1007/s10814-008-9023-5

Oka, Rahul C., Laure Dussubieux, Chapurukha Kusimba, and Vishwas Gogte
2007 The Impact of Imitation Celadon and Porcelain Industries on Chinese Ceramic Exports in the Indian Ocean Maritime Exchange, ca. 1200–1700 C.E. Lecture given at the Fourth Forbes Symposium on Scientific Research in the Field of Asian Art, Smithsonian Institution, Washington D.C., September 27–29, 2007.

Padgett, J. F.
2001 Organizational Genesis, Identity and Control: The Transformation of Banking in Renaissance Florence. *In* Networks and Markets. J. E. Rauch and A. Casella, eds. Pp. 211–257. New York: Russell Sage.

Palat, R. A. and I. Wallerstein
1999 Of What World-System was Pre-1500 "India" a Part? *In* Merchants, Companies and Trade: Europe and Asia in the Early Modern Era. S. Chaudhury and M. Morineau, eds. Pp. 36–37. Cambridge: Cambridge University Press.

Pearson, M. N.
1987 India and the Indian Ocean in the Sixteenth Century. *In* India and the Indian Ocean. A. Das Gupta and M. N. Pearson, eds. Pp. 71–93. Oxford: Oxford University Press.
1988 Brokers in Western Indian Port Cities: Their Role in Servicing Foreign Merchants. Modern Asian Studies 22(3):455–472.

1998 Port Cities and Intruders: The Swahili Coast, India, and Portugal in the Early Modern Era. Baltimore, MD: Johns Hopkins University Press.

2003 The Indian Ocean. New York: Routledge.

Pouwels, R.

1987 The Horn and the Crescent: Cultural Change and Traditional Islam on the East African Coast, 800–1900. Cambridge: Cambridge University Press.

Quinn, W. H.

1992 Study of Southern Oscillation-Related Climatic Activity for A.D. 622–1990, Incorporating the Nile River Data. *In* El Niño: Historical and Paleoclimatic Aspects of the Southern Oscillation. H. F. Diaz and V. Markgarf, eds. Pp. 119–150. Cambridge: Cambridge University Press.

Richards, J. F.

1993 The Mughal Empire. Cambridge: Cambridge University Press.

Schove, D. J.

1977 African Droughts and the Spectrum of Time. *In* Drought in Africa 2. D. Dalby, R. J. H. Church, and F. Bezzaz, eds. Pp. 38–53. London: International African Institute.

Schneider, J.

1991 Was There a Precapitalist World-System? *In* Core/Periphery Relations in Precapitalist Worlds. C. Chase-Dunn and T. D. Hall, eds. Pp. 45–66. Boulder: Westview Press.

Seidenberg, D.

1996 Mercantilist Adventurers. New Delhi: Vikas Publications.

Stein, G. J.

1999 Rethinking World-Systems, Diasporas, Colonies, and Interaction in Uruk Mesopotamia. Tucson, AZ: University of Arizona Press.

Swerts, J., with A. Aerts, N. de Biscop, F. Adams, and P. Van Espen

1994 Age Determination of Chinese Porcelains by X-ray Fluorescence and Multivariate Analysis. Chemometrics and Intelligent Laboratory Systems 22:97–105.

Thorbahn, P. F.

1975 The Precolonial Ivory Trade of East Africa: Reconstruction of a Human-Elephant ecosystem. Unpublished Ph.D. Dissertation, Department of Anthropology, University of Massachusetts at Amherst.

Veinstein, G.

1999 Commercial Relations between India and the Ottoman Empire (Late Fifteenth to Late Eighteenth Centuries): A Few Notes and Hypotheses. *In* Merchants, Companies and Trade: Europe and Asia in the Early Modern Era. S. Chaudhury and M. Morineau, eds. Pp. 95–115. Cambridge: Cambridge University Press.

Watson, O.
2004 Ceramics from Islamic Lands. London: Thames and Hudson.

Watts, M.
1991 Heart of Darkness: Reflections on Famine and Starvation in Africa. *In* The Political Economy of African Famine. R. E. Down, D. O. Kerner, and R. P. Reyna, eds. Pp. 23–68. Philadelphia: Gordon and Breach.

Wen, R., C. S. Wang, Z. W. Mao, Y. Y. Huang, and A. M. Pollard
2007 The Chemical Composition of Blue Pigment on Chinese Blue-and-White Porcelain of the Yuan and Ming Dynasties (AD 1271–1644). Archaeometry 49:101–115.

Will, P. E.
1990 Bureaucracy and Famine in Eighteenth Century China. Translated by E. Forster. Palo Alto, CA: Stanford University Press.

Wood, C. A.
1977 A preliminary chronology of Ethiopian droughts. *In* Drought in Africa 2. D. Dalby, R.J. H. Church and F. Bezzaz, eds. Pp. 68–73. London, International African Institute.

Yoffee, N.
2004 Myths of the Archaic State: Evolution of the Earliest Cities, States, and Civilizations. Cambridge: Cambridge University Press.

Yu, K. N., and J. M. Miao
1996 Non-destructive and Quantitative Investigation of Jingdezhen Blue-and-White Porcelains Using EDXRF. Journal of X-Ray Science and Technology 6:167–171.
1997 Locating the Origins of Blue-and-White Porcelains Using EDXRF. Applied Radiation and Isotopes 48:959–963.
1999 Characterization of Blue and White Porcelains Using the Mn/Fe Ratio from EDXRF, with Particular Reference to Porcelains of the Xuande Period (1426–1435 A.D.). Applied Radiation and Isotopes 51:279–283.

Madagascar's Cyclone Vulnerability and the Global Vanilla Economy

11

MARGARET L. BROWN

THIS STUDY IS AN EXPLORATION of the political economic foundations of cyclone vulnerability among Madagascar's vanilla farmers. An island nation of 19 million people off the southeast coast of Africa, Madagascar experiences more tropical hurricanes than any other African country (ProVention Consortium 2007). However, as the hazards and disasters literature well documents, human vulnerability is not merely a matter of living in a physical location that is exposed to a natural event (Cutter and Emrich 2006; Oliver-Smith 1996). Social vulnerability involves a complex of factors including poverty, social supports, technology, sources of livelihood, and the nature of the built environment (Cutter et al. 2003). The vulnerability of vanilla farmers in Madagascar is illustrative of this complexity.

Examining vanilla production within the framework of concern for social vulnerability encourages closer attention to the dichotomy between the concerns of most vanilla farmers and the concerns of many exporters and importers. The latter are concerned more with quality and price, while vanilla farmers express the most concern about security of income and property. Although it is true that impoverished farmers the world over express similar concerns, the frequent but unpredictable occurrence of cyclones in Madagascar complicates the assessment of vanilla farmers' needs. In thinking about what these farmers need to thrive, we need to consider what material and non-material resources would enable them to recover from a cyclone in addition to the factors typically considered by economic development professionals addressing rural poverty, such as increasing production yields and quality and providing cheaper inputs and greater technical support.

Vanilla is produced in one of the most cyclone-prone regions of this island nation that suffers from a profound level of poverty. Vanilla comprises approximately 10–15 percent of Madagascar's total exports, and Madagascar still provides between 60 and 70 percent of world demand. The physical infrastructure throughout Madagascar is in dilapidated condition, and the vanilla-growing region is no exception. These factors alone would add up to considerable vulnerability, but they are compounded by the particular position of vanilla both in Madagascar's own political and economic history and in the world economy. For a number of reasons explored in this chapter, Madagascar has, since 1960, made decisions regarding vanilla production and sale that have had the overall effect of destabilizing the economic lives of vanilla farmers. More importantly in terms of cyclone vulnerability, these decisions have attended to Madagascar's position in the world vanilla economy through a focus on price and quality. Cyclones, when at all considered by authorities, have largely been addressed as unpredictable events that affect short-term quality and supply, leading to a response that primarily focuses on farming techniques and international marketing intended to maintain a worldwide desire for Malagasy vanilla. At the same time, other countries have viewed cyclones as a price driver, and have acted accordingly in the past decade to try to take advantage of Madagascar's natural disasters. The resulting situation is one in which the focus upon yields and quality—without substantial attention to economic diversification or creation of social and economic support systems among vanilla farmers in Madagascar—has prevented them from being able to develop adequate resources to withstand cyclones.

Recent Cyclone Trends in Madagascar

The headline read, "Madagascar cyclones may be boon to vanilla market" (Mongabay.com 2007). The date on the byline was March 27, 2007, 12 days after Cyclone Indlala struck Madagascar's northeast coast. It was the fifth major cyclone to hit Madagascar in a three-month period, but it was the first of these five cyclones to have a severe impact on Madagascar's central vanilla growing region. By early April, the government was reporting that tens of thousands of hectares of rice and other food crops had been destroyed, 100,000 people had been displaced, and 80 percent of the country's vanilla crop had been lost. However, a few days later, Claude

Andreas, the president of Madagascar's vanilla exporting group GES, was declaring that the damage to vanilla had not, overall, been as bad as first reported. It was true that Antalaha, one of four major vanilla producing towns in northeastern Madagascar, had lost 80 percent of its vanilla crop; 2007 production estimates for Antalaha had fallen from 400 tons at the end of January to 80 tons after Indlala. But, the three other regions had suffered only minimal damage. One of the other towns, Sambava, had lost only an estimated 5 percent of its crop, dropping from a predicted 640 tons of production to 610 tons. Losses were similar for the other two regions, ultimately leading Andreas to predict that the major vanilla-producing and exporting SAVA region of the northeast (Sambava, Antalaha, Vohemar, and Andapa) would suffer losses of only 25 percent of its predicted vanilla production for the year, down from 1600 tons to 1200 tons (CIRAD 2007). While this would be a difficult year for the farmers in and around Antalaha, farmers in the other four towns had escaped complete devastation.

A few days after this report was made, Madagascar was slammed by Cyclone Jaya on April 3. When it made landfall, Jaya passed through the same general region as Indlala, but with a slightly more northward trajectory. One report quoted a local official saying, "Half Sambava town was devastated" (Reuters 2007). Although no official report was issued by government officials or vanilla exporters, it is almost certain that the second storm did significant damage to Sambava's crop, further reducing Madagascar's expected exports.[1]

The succession of violent storms battering Madagascar in 2007 were sadly familiar. In 2000, Madagascar was first hit by Cyclone Eline on February 17, followed by Cyclone Gloria two weeks later. Then, on April 2, Cyclone Hudah hit. Hudah was a tremendous storm, nearly 250 miles across with average wind speeds of over 140 miles per hour (Kriner 2000; Rakotoarivelo 2000), reaching as high as 186 miles per hour by some reports (CNN 2000). Both Gloria and Hudah followed paths directly through the center of Madagascar's vanilla-producing northeastern provinces. Four years later—the amount of time it takes for freshly planted vanilla vines to mature and produce fruit—an economic report on Uganda bore the headline, "Uganda's vanilla industry blossoms after Cyclone Hudah" (Japan Economic Newswire 2004). The report contained the following statement about vanilla farmers in Uganda: "The estimated 80,000 small-lot farmers are currently reaping the benefits from the misfortunes caused by Cyclone Hudah, which is now a common name to the Ugandan vanilla farmer."

Around the time most Ugandan vanilla farmers were preparing for their first significant vanilla harvest, Madagascar was hit again, this time by Cyclone Elita, on January 28, 2004. Then, on March 7, 2004, Cyclone Gafilo, a category 5 storm, made landfall, ripping across northern Madagascar, turning southwest, and unexpectedly rotating back east to tear across the southern part of the country before finally dissipating off the southeast coast. Again, economics news reports linked the cyclones to vanilla prices. Dairyreporter.com began its story on March 17 with the following sentence: "The key ingredient in the most popular ice cream in the world faces a rise in price, already at a 10-year high, as the worst cyclone in two decades hits Madagascar, the leading producer of vanilla" (Dairyreporter.com 2004). It is somewhat jarring to read news accounts, just days after a devastating storm that killed or displaced over 100,000 people, highlighting the effects of the storm on ice cream prices. At the same time, it is impossible to deny that cyclones in Madagascar have effects far around the globe.

Unfortunately for Madagascar's farmers, recent studies indicate that the increase in the severity and frequency of cyclones in Madagascar in the past two decades is likely to be a harbinger of the future. According to a study conducted by the ProVention Consortium (2007)—a consortium of international organizations, government agencies, and academic institutions aiming to improve disaster risk management and to reduce the impacts of disasters—the number and severity of cyclones hitting Madagascar in the past twenty years is showing a strong upward trend. Additionally, the vanilla-producing region of Madagascar is among the most frequently struck by cyclones. Predictions about global climate trends suggest that cyclones in this region are likely to continue to increase in frequency and severity over the next several decades (ProVention Consortium 2007).

Examining linkages between tropical cyclones, rural poverty in Madagascar, and the global economy, the remainder of this chapter flows as follows. After a brief description of the post-cyclone fieldwork I conducted in the northeastern vanilla-growing regions, I examine the history of vanilla production in Madagascar and briefly explore how the presence or absence of disasters in Madagascar affects vanilla farmers in other countries. In the end, I discuss how a variety of national and international trends in economic development initiatives have led to a failure to attend to the vulnerability of vanilla farmers to natural disasters such as cyclones. Ultimately, the argument I make is that decision-making regarding vanilla is left disproportionately in the control of political and economic elites with an eye on the export market. Failure to draw effectively on farmers' interests and experiences has led to an overemphasis on quality and efficiency in vanilla production at the expense of attention to income security and risk management.

Post-Cyclone Fieldwork in Madagascar

The data for this research are drawn from two periods of research in Madagascar, participation in an international vanilla conference, and reading of various newspapers and online sources related to the vanilla industry. From February 1995 to November 1996, I conducted ethnographic research in two villages in northeastern Madagascar, and traveled extensively throughout the region. During this period, I interviewed and observed the activities of farmers in a variety of settings and participated in village life. I also interviewed local and national government officials and had extensive discussions with conservation and development professionals working throughout Madagascar. During this first long period of fieldwork, the vanilla sector was only one part of a larger study of rural Malagasy social and economic life. Within that context, I gathered vanilla production and sales data from nearly 150 farmers in household economic surveys. I also learned about the operations of the export market in Antalaha through informal conversations with town residents. In June and July of 2001, I returned to Madagascar to conduct research specifically on vanilla production. During that time, I conducted semistructured interviews with twelve vanilla farmers in two villages in different locations of the vanilla belt. I also interviewed six traveling buyers of vanilla, three of the main vanilla exporters, and interviewed and traveled with a professional agricultural researcher. I also traveled with six farmers to survey their land and with one farmer during a vanilla harvest. The combination of formal interviews and informal conversations covered a broad range of topics related to the vanilla market. Formal interviews began with questions about the person's position in the market, length of time in that position, and an open-ended question about their concerns about vanilla. General themes emerged from these discussions. Middlemen and women buyers were most notably concerned about the prices they were being paid for their services, their mistrust of the exporters, and the threat of theft on the roads during buying trips. Farmers were most vocal about prices, security of their crops, and the ability to expand their production. Exporters expressed concern about having trustworthy buyers and farmers, the availability of high quality vanilla, and the need to keep prices down (and an accompanying suspicion that some of their fellow exporters were speculating in the vanilla market). I heard the exporters' views reinforced in public presentations by international export and import professionals at an international vanilla conference in 2004.

During my 2001 research, because of the occurrence of Cyclone Hudah in April of 2000, I had the opportunity to see how the villages of my previous research were coping post-cyclone. During that earlier research in the community 40 kilometers south of Antalaha, I had conducted a number of oral histories with village elders. All told how the early founding of the core village in the community began with a cyclone that led to flooding in their ancestors' former home village. That was in 1950. Some of them could remember the storm and how the residents decided to create a new village on higher ground, leading to the founding of Andranobe.[2] Andranobe was relocated one time between 1950 and 2000 due to an epidemic. But there were no other big storms besides the 1950 one that stood out in villagers' minds.

In 2001 when I returned to Andranobe and its sister village, Befotaka, many things were not as I remembered them. On the walk to Befotaka from the coast, I noted that there were barely any remaining signs of one village that had existed across the river from Befotaka and Andranobe. Its residents had scattered to other, less-devastated villages or moved to a new inland settlement. In Befotaka, I noted many changes, but also much was the same. Most residents of Befotaka had decided to rebuild in the same location after Hudah. Houses in these villages are almost uniformly built of reeds and tree limbs, with just a few that had had cement foundations from long ago. Rebuilding, residents said, had taken about three months. People had to travel nearly 4 kilometers into the forest searching for building materials, and getting into and through the forest was difficult because of the storm. Still, a new visitor to the village probably would not have been able to tell that nearly every house in the village had been knocked down in a cyclone just 15 months prior. However, careful attention, particularly inside homes, revealed some remnants of disaster and recovery efforts.

Many homes had plastic water storage containers, distributed by USAID. They also had now-empty containers of a water-purification substance that had been distributed by aid agencies after the storm. Throughout the region, there were gunnysacks stamped with the USAID name. These had once been filled with rice for a near-starving population that had lost most of its year's supply of rice in the storm. Now empty, they were used to cover gaps in walls and holes in thatched roofs. Although prone to develop the occasional hole under average conditions, roofs turned out to be very resilient in the storm. Walls were knocked down by the wind, but many of the roofs had remained intact and sat on the ground like tents. In fact, while rebuilding their houses, villagers slept under those roofs like they were tents, reinforced by plastic and canvas tarps provided by aid agencies. Some of those roofs were still on the ground, covered in tarps, during my visit.

Despite their proximity to one another, Andranobe had responded quite differently than Befotaka to the storm, and most of its residents had decided to move to higher ground. Now that it was farther from the river, the village also had two wells for drawing water; neither village possessed a single well in 1996. In interviews and informal discussion with several residents of Befotaka and Andranobe after the storm, I learned that even with the terrible destruction of property in both villages, not one person died in the storm. Some of the older women (in their 60s) who had headed their own households during my earlier research were, in 2001, living in households headed by their sons or daughters. For a few, this had meant moving from Befotaka to Andranobe. A few villagers had sold land to others as lots for building new homes. In terms of community resources, except for the wells, little infrastructural development occurred. There had been no school in 1996, and little progress had been made toward building one in 2001.

When I talked to villagers about their lives after the cyclone, their stories were a mixture of positive and negative. Some focused on how life had always been hard, and they were grateful now that they had some goods (such as the plastic jugs and the tarps) that would endure for a while. In my own observations, I was surprised by how many material goods had survived the storm. The home of my former assistant in the village was an example. Posted on her wall were photographs of her family I had given to her in 1996. She was wearing clothes from that same period, and her front door was covered with a cotton curtain I had given her just before I left.

Others focused on the immediate hardship. Food was scarce and not very varied as residents waited for new harvests and for fruits and vegetables to be restored. Papaya had been in abundance after the storm, as had bananas, but other staples were slower to return. Rice prices were high for those who still needed to buy. Everywhere, people talked about the pervasive stench in the village in the weeks after the storm from saturated mattresses and rotting food. Regardless of their relative optimism or pessimism, however, everyone gave the same answer to one question. They would not have any vanilla to harvest and sell in 2001.

A couple of weeks later, I was visiting villages on the road between Sambava and Antalaha. The differences between those villages and the area around Befotaka and Andranobe were stark. These more accessible villages were filled with vanilla-market activity. I interviewed planters who had thousands of vanilla vines that were bearing fruit, and thousands more they had just recently planted. Many had discovered that although some of their vines had been knocked down in Hudah, they had not been destroyed. Some of these vines were able to produce as soon as 2001, while others had provided cuttings for transplanting. Some farmers had also benefited from STABEX interventions (see below), although not to a great extent for the upcoming harvest.

Everyone was expecting vanilla prices in 2001 to be high. While villagers in Befotaka and Andranobe viewed restoring their food crops as their highest priority, villagers on the road to Sambava were concerned with protecting their lucrative vanilla crop. I talked with farmers whose entire families slept at their vanilla plantations at night in an effort to prevent theft, and I observed their temporary shelters set up at the fields. Men had rifles in their homes, also as protection against thieves. Some acknowledged to me that they had begun to harvest earlier than allowed by law, either to sell to speculators on the black market, or to prevent theft. Every time I went for a walk with one particular farmer, he planted a few more vanilla vines on his property. Everywhere I went, activity and talk were centered on the vanilla market.

History of Vanilla Production in Madagascar

Indigenous to Mexico, *Vanilla planifolia* was introduced to the Indian Ocean in the late eighteenth century. However, it arrived without the necessary pollinator, and it was several decades before a plantation owner on the island of Réunion discovered, almost by chance, how to pollinate vanilla by hand (Ecott 2004:123). After this discovery around 1840, France aggressively moved to establish vanilla plantations throughout its Indian Ocean island colonies. Ultimately, they were most successful in Madagascar. Madagascar's vanilla is now and has historically been regarded as the best quality in the world due to its high levels of vanillin (Ecott 2004).

During colonial times, the French instituted policies encouraging Malagasy people to work on French plantations or to establish their own in order to be able to pay taxes (Brown 2004; Cole 2001; Feeley-Harnik 1991; Sodikoff 2004). In the northeastern region of Madagascar, where the climate is especially well suited for vanilla production, vanilla took hold as one of the main crops produced for sale and export. Vanilla is not well suited to being grown on large plantations, so it was possible for Malagasy farmers to grow it on their own small farms (Laney 2004). Those farmers' active involvement in vanilla cultivation helped develop a tradition of vanilla knowledge among Malagasy farmers, and catapulted Madagascar forward in global vanilla production. In 1924, Madagascar was the primary producer in the world, a position it retained into the 1980s (MAEP 2005). However, by the early 1990s, the country's market share had plummeted to between 30 and 40 percent (Andreas 2006; Cadot, et al. 2006).

Madagascar's loss of control over the vanilla market was largely due, analysts say, to its pricing and marketing policies (Blarel and Dolinsky 1995; Cadot et al. 2006; Metzel et al. 1999). These strategies included creating a government-run program that centralized the export of vanilla, setting both producer and export prices and levying an export tariff. This program kept producer prices low, but export prices and government revenues remained high, taking advantage of Madagascar's near monopoly in the market. In a move that further served to maintain high prices, Madagascar joined with the islands of the Comoros and Réunion to form a cartel in 1962. The cartel managed prices and promoted the region's vanilla to the worlds' buyers.

These policies led to vanilla prices that depressed worldwide demand. Furthermore, the low prices being offered to farmers reduced their incentives to put in the necessary time and labor to produce a high quality crop, and the quality of Madagascar's vanilla, which had always been its strongest advantage, became less reliable. Importers of vanilla, mostly located in the United States, Great Britain, and France, began to search elsewhere for substitutes for Malagasy vanilla. Meanwhile, farmers in other parts of the world had began to enter the market to provide vanilla at a price more in line with what buyers were willing to pay (Blarel and Dolinsky 1995; Cadot et al. 2006; Metzel et al. 1999).

Starting in the 1970s, Indonesia began to increase its vanilla production. Because of its lower vanillin content, Indonesia's vanilla is generally judged to be of a lower grade compared to the vanilla produced in Madagascar. However, for industrial users of vanilla for manufacturing such things as soft drinks, perfumes, baked goods, and non-gourmet ice creams and candies, lower grade vanilla can meet many of their needs.[3] Thus, Indonesia's vanilla was acceptable for many consumers, particularly those in the United States who bought the most vanilla and had the highest percentage of industrial uses for it (Ecott 2004:223). Indonesia's share of the export market gradually grew from around 15 percent in 1985 to almost 45 percent by 1990. Meanwhile, from 1985 to 1995, Madagascar's share of the market for natural vanilla dropped from close to 80 percent to around 40 percent (Cadot et al. 2006).

Besides Indonesia, the main competition faced by Madagascar in the 1980s and 1990s was increasing reliance on synthetic vanilla. Although the taste is noticeably different—and to connoisseurs, inferior—many industrial users rely at least in part on a synthetic product to supply their vanilla flavor needs. The cost of extract made from synthetic vanilla is considerably cheaper than extract made from natural vanilla. For producers of vanilla-based products aimed at the mass market, the higher cost of natural vanilla could reduce their competitiveness (Ecott 2004:212).

By the mid 1990s, failed pricing and marketing policies had thus led Madagascar's vanilla sector nearly to collapse (Cadot et al. 2006). The response emphasized price and quality above all else. Under pressure from international lending agencies, Madagascar began loosening state control of its vanilla sector in 1995. By 1997, fixed prices had been eliminated, and instead were allowed to go up and down in accordance with market forces (Cadot et al. 2006). The government continues to intervene in the market, but primarily by working with groups of exporters to establish rules that are intended to secure the quality of Madagascar's vanilla. These rules include: requiring each farmer, collector, and exporter to have a government-issued license; setting the dates on which harvesting of vanilla can begin; setting market dates for each commune; and setting dates on which the sale of prepared vanilla can begin. It is important to note that these regulations are all intended to ensure the quality of the vanilla product by reducing theft and discouraging early harvesting by farmers. None of the regulations explicitly address farmer concerns other than theft and fraud.

The result of the liberalization of the vanilla market has been unprecedented price volatility. At the same time, Madagascar has reclaimed its command of the market, accounting in 2007 for around 60 percent of annual sales of natural vanilla. Under state control, farmers were paid about \$2.50/kg of green vanilla in 1995. In 1996, as state control was loosened, farmers were paid about \$1.00/kg, with some getting more, some getting less depending on their bargaining abilities and location. In 2000, farmers were earning over \$15/kg, and in 2004, when the price of vanilla was at its peak, some earned over \$100 per kilo. In 2004, however, some farmers who had previously been able to count on some income from vanilla made absolutely nothing. Exporters, importers, and government officials have expended a great deal of effort and many resources to improving the quality of Madagascar's vanilla in order to reclaim the country's market share. However, the focus on quality of the product has at times blinded them to one of the factors that make Malagasy farmers so desperate for income and willing to sell inferior vanilla on the informal market—the frequency and unpredictability of cyclones. Indeed, the price differences experienced by farmers in 2004 were largely caused by Cyclone Hudah, which completely destroyed some farmers' crops, making the remaining vanilla crop more lucrative.

Vanilla from Start to Finish

Vanilla in Madagascar is grown primarily by smallholders (Brown 1999; Laney 2004). There have been some recent attempts to develop large plantations as part of a process of "vertical integration" of the market, but these have yet to make a major impact (author's field notes July 2001; Ecott 2004). There are about 80,000 planters of vanilla in Madagascar, and the vast majority of them live in the SAVA region. Their plantations cover nearly 30,000 hectares of land, and farm sizes vary greatly. Almost all farmers are diversified to the extent that they also grow some rice for auto-consumption and for sale, but almost none undertake any non-agricultural activities for income (Brown 1999; Cadot et al. 2006; Laney 2004). Growing vanilla is very labor-intensive, and most work is done by family members.

As vanilla pods reach mature size, each farmer goes to his or her field to mark the beans with a unique stamp issued by the Malagasy government. This stamp identifies the owner of the beans, and is intended to reduce the risk of theft of beans or attempts by owners to sell beans illegally. After harvest, green vanilla must be sold relatively quickly. Sale is typically done in villages at communal markets on the dates set by the government. Farmers bring their beans to the market where collectors/middlemen gather to purchase. Most of these collectors are working under contract with one or more of the major exporters, although there are a few who are self-funded and a few who are employed directly by one of the exporting companies (author's field notes July 2001). There are more than 30 registered exporters, only 5 of which export more than 100 tons each.

Besides external market demand, collectors base their purchase price on the length and condition of the beans, relying on standards established by the exporter associations Groupement National des Exportateurs de Vanille (GNEV) and Groupement des Entreprises de la SAVA (GES). Purchases are made only on a cash basis. While collectors are generally given money in advance of going "to the bush" to buy vanilla, farmers are rarely provided any advance on their future harvest. Thus, farmers receive large sums of money in a short time period, but they have few ways to invest that money for long-term goals. That inability to invest hurts all farmers, but, as shown in the following section, not all of them suffer equally. Those who are more easily accessible have benefited from improved infrastructure and increased attention from exporters, buyers, and officials of national and international development organizations.

Location, Location, Location

I conducted my original fieldwork in Madagascar in a community about 40 kilometers south of Antalaha in 1995 and 1996, during the early period of vanilla market reforms. Approximately 90 of the 150 households in my study area farmed vanilla, along with other cash crops (coffee and cloves) and rice. In 1995, the average amount of vanilla sold was between twenty and twenty-five kilograms per seller in each village although variation in size and age of vanilla plantations and cultivation style had some sellers selling as little as one kilogram, with others selling as much as one hundred kilograms; one seller reported a sale of two hundred kilograms. Some cultivators did not sell any vanilla at all in 1995 because their vines did not flower, due either to immaturity or to lack of adequate care.

In 1995, the official price for green vanilla was set at 10,000 Malagasy francs/kg.[4] One year later, the price was dropping, and farmers were both anxious and suspicious, not quite certain why they were not being offered a higher price. In addition to the decline in value per kilogram of vanilla, farmers also experienced progressively stricter standards regarding the quality of their vanilla (author's field notes October 1996). This was connected to the effort at the national level to restore Madagascar's reputation as the producer of the best vanilla in the world (author's field notes July 2001), but it nevertheless hurt those farmers who were not accustomed to devoting a lot of effort into caring for their plants besides pollination, harvesting, and occasional weeding. Developing longer pods would require more input of time and resources, and perhaps more technical knowledge than many of these farmers possessed, and farmers in these remote villages were not receiving technical or financial assistance to develop those skills.

Perhaps the diverse experiences of farmers helps to explain why it has been difficult for them to organize themselves into collectives and to work within Madagascar's decision-making framework. Exporters and collectors have associations to help them coordinate their activities. Representatives of the exporter associations GNEV and GES participate in formal decision-making meetings with government officials. For their part, collectors compete against one another, but their association serves as a source of information regarding pricing and production. Farmers have indicated that they would like to develop a growers' association, and a couple of small regional attempts have been started. However, the localized nature of these associations means that farmers in different regions of Madagascar, each of which will have its own experience of a natural disaster such as a cyclone, do not band together to share the overall risk faced by all vanilla farmers in the cyclone-prone north and east. Take as an example the farmers in my original research community, which is in a relatively remote part of the vanilla growing region. While Befotaka and Andranobe were still recovering from Hudah and had little vanilla to sell, the more accessible farmers to the north who had not been hit as hard by Hudah—and who had enjoyed greater access to relief and development efforts—were reaping a strong harvest and expanding their vanilla plantations.

The general and farmer-centered activities of STABEX tend to focus on the more accessible areas, particularly on the 80-kilometer route between Sambava and Antalaha. This allows them to reach more farmers as direct participants (e.g., in training) and also to display their projects to more passersby. As regards cyclone recovery, farmers in less accessible areas generally must expend more effort to feed themselves and get their food crops restored than do those in more accessible villages, USAID gunnysacks notwithstanding. Even more remote villages than these have a tougher go of it, as noted in the many newspaper reports where government and NGO officials lament their inability to get aid to villages with no access by road (Africa News 2007, Kriner 2000).

An additional issue related to location is the technology of information flow. Although cell phones have become very common in Madagascar, they did not work at a very great range from major towns. At a distance of 20 kilometers north or south from Antalaha, my cell phone ceased working. Thus, farmers at even a modest distance from town could not benefit from the increased information flow made possible by cell phones. Further, because few villages have a generator, maintaining cell phone power would be nearly impossible. For all rural residents, this impeded information access not just about vanilla prices, but also about approaching cyclones and the status of rescue and aid efforts.

Finally, few farmers of vanilla, regardless of their location, have access to savings and credit institutions. Even those who are close to town find that banks will not lend to them to enable the purchase of inputs that might improve their bargaining position in the vanilla market. Few farmers have their own cars, for example, to take their crop to town to sell for potentially better prices, or to travel to pursue non-vanilla-related economic opportunities. Few farmers can afford the materials that would be necessary to reinforce their homes and build secure storage facilities so they could choose to prepare their own vanilla before selling. And when farmers do sell their crops, many do not have nearby savings institutions where they can build up their own money to make these types of larger investments. So, instead, what one sees are signs of money everywhere, but few signs of wealth. Many farmers who were able to take advantage of the rising prices between 1999 and 2003 (see below) used their income to buy fancy stereo systems, bicycles, and tennis shoes—things that can be bought quickly, with one year's income, but that have little positive impact on a farmer's future agricultural returns. Because of the absence of farmer or village cooperatives, there are few institutional means of pooling portions of each farmer's income in a community to put toward schools or other collective goods that could have long-term benefits. The inability and/or unwillingness to make such long-term investments is already beginning to hurt Malagasy farmers as they are faced with the reality that vanilla may not always be like money in their pocket. The lack of these types of security and alternative sources of income increase cyclone vulnerability of Madagascar's vanilla farmers.

External Competition for Vanilla

In this section, I place Madagascar's experience within the context of the global vanilla market from the late 1990s to 2007, a period marked by frequent cyclones. It has also been marked by extreme price fluctuations on the international vanilla market. In 2003, prices for cured vanilla reached $500/kg, compared with $50/kg in 1999. The world demand for vanilla in 1999 was 3,000 tons. However, as prices for natural vanilla were gradually increasing between 1999 and 2003, world demand was steadily declining, to around 1,500 tons in 2004. At the same time, the number of farmers producing vanilla both within Madagascar and in other parts of the world multiplied. Because of this, heading into the 2007 market, prices were anticipated to be almost 90 percent lower than they were in 2003 (Financial Times Information Limited 2006), and supplies were expected to be more than double world demand.[5]

A number of countries have taken advantage of the declining quality of Madagascar's vanilla in the 1980s and early 1990s, rising prices for vanilla post-1999, and the shortage of vanilla caused by cyclones and a brief spell of political instability in Madagascar in 2001.[6] India, Uganda, and Papua New Guinea are among the nations that increased their efforts in the vanilla sector post-2000. Uganda is of particular interest for the purposes of this study because that country's investment in vanilla was an explicit attempt to fight rural poverty through the production of vanilla for international export. Uganda produces high-quality bourbon-style vanilla similar to that produced in Madagascar. A major project of vanilla promotion in Uganda has been funded by USAID since 1995. Between 1995 and 2003, Uganda's exports of vanilla increased from fewer than 5 tons to at least 100 tons (Japan Economic Newswire 2004; USAID 2007).[7] Similar to the European Union's efforts through STABEX in Madagascar, USAID's work in Uganda has centered on increasing the quantity and quality of vanilla production. Their program helped pave the way for farmers to jump into the lucrative vanilla sector when prices began to surge upward in 2000. The number of farmers grew from just a few thousand in the 1990s to tens of thousands by 2004. The success stories of farmers who progressed from barely getting by through subsistence or coffee farming to having great wealth through vanilla production were widely reported (Japan Economic Newswire 2004; Nathu 2003; USAID 2007).

Like their counterparts in Madagascar, however, vanilla farmers in Uganda have been subjected to wild price swings in the past few years, price swings that are partially a result of their own success. Ugandan farmers entered the vanilla market in large numbers in the wake of Cyclone Hudah and the resulting rise in vanilla prices, but as Malagasy farmers themselves entered vanilla production in larger numbers spread over an expanding geographical range, the fate of Ugandan farmers continued to depend upon the fate of Malagasy farmers. As early as 2004, news stories began to surface of farmers in Uganda pulling up their crops amid reports of a record crop in Madagascar (Milmo 2004). Since then, the fortunes of Uganda's vanilla farmers have continued to decline along with the market price of vanilla. In January 2007, one story quoted a farmer saying, "If the prices don't pick up, many of us will quit. Most of the farmers are indebted. Some have run away from their homes because they can no longer sustain their families" (New Vision 2007).

Ugandan farmers perceive their fate to be linked to that of Madagascar's farmers, but in an inverse relationship. Ugandan farmers can trace their days of vanilla success back to Cyclone Hudah, which destroyed much of Madagascar's crop for 2001 and created an incentive for other countries to jump into the market. The peak of Uganda's success was in 2003, the first season post-Hudah when the vines planted in the wake of the cyclone bore fruit. However, the incentives confronted by Ugandan farmers were also faced by farmers in other parts of the world. In 2004, worldwide production of vanilla was up, due partly to the large number of new entrants to the market. This was true even in Madagascar, where vanilla production had begun to expand, post-2000, to areas beyond the traditional northeastern vanilla belt (author's notes October 2004; Ecott 2004). Furthermore, the record prices of the 2003 season prompted importers of vanilla to aggressively seek substitutes for vanilla, and technological advances in the production of synthetic vanilla have made it more palatable for some of them to turn away from natural vanilla (Althucher 2005; Ecott 2004). In 2007, the situation is that, for Uganda's farmers to receive high returns on their vanilla, they almost need a catastrophe in Madagascar. Otherwise, with current annual global consumption of natural vanilla at around 1,500 tons, and Madagascar's projected annual production (under expectations of good weather) exceeding 1,600 tons, Uganda's farmers will receive virtually nothing for their product unless they can convince potential buyers of Ugandan vanilla's superior and unique qualities. In this regard, they face a daunting challenge, as Madagascar itself continues its own campaign to maintain its reputation as producing the world's best vanilla.

Creating Value

Sourcing has come to play an important role in the marketing of many products, and vanilla is no exception (Freidberg 2004; Houghton 2007; Murdoch et al. 2000). Typically, vanilla beans' origins are mentioned in retail catalogues or on labels of retail products only if they come from Madagascar, Mexico, or Tahiti because each has unique flavor qualities. Otherwise, if a manufacturer wants to indicate high quality but cannot specify a single source, they might use the more generic term "bourbon" vanilla. This term connotes high quality, but it can apply to vanilla produced in Uganda and India, as well as in Madagascar and the other Indian Ocean islands. Increasingly, exporters of vanilla from Madagascar have come to believe that their future lies in being able to convince the market that their product is of exceptional quality. This involves not only an investment in actual improvements in vanilla production practices, but also improvements in marketing techniques. Hence, Madagascar's exporter associations and the Malagasy government have increased their efforts to promote Malagasy vanilla in the popular press, resulting in glowing coverage in international newspapers and magazines (Andreas 2006).

The focus on quality as a marketing tool makes economic sense. It allows exporting countries to draw on the language of food standards and the perceived value of single origins that has become pervasive in certain segments of the food industry, particularly those that appeal to wealthier, better-educated consumers.[8] However, overemphasis on perception of quality can have unintended negative consequences. At an international vanilla conference I attended in 2004 oriented toward exporters, large-scale producers, buyers, and scientists—but not attended by any small-scale farmers—discussion turned to how to manage the volatility of the vanilla market. Almost all of the talk focused on getting the prices right while maintaining a free market. Most discussion about farmers involved how to improve their techniques. No one discussed cyclones except as abnormal events that shocked the market, despite the fact that Madagascar had experienced so many such events in the preceding four years that it was becoming more difficult to consider them abnormal. The focus on prices and quality typically overshadows other concerns.

In another context, when the head of one of the world's largest vanilla-purchasing companies was asked by a writer about the situation for Malagasy farmers in the up-and-down vanilla market, he focused solely on price. If the price were to stabilize around $60/kg for cured vanilla, he said, then farmers would have a good deal (Ecott 2004:251). What he left out of this assessment, however, is how to ensure farmers a good deal in those years when they had no crop at all. He also failed, as have many Malagasy government officials and members of national and international nongovernmental organizations, to consider the lives of vanilla farmers in broader context. Besides vanilla, what do they have? What do these farmers have to fall back on if they lose both their food and cash crops in a single season, as usually happens when a cyclone strikes? What resources will be available to support them while they rebuild homes and regain their health after a natural disaster? These questions are all relevant to the ongoing viability of vanilla production in Madagascar if it is going to have the economic development effects the Malagasy government, exporters, and international donors desire (author's field notes June 2001). Yet, the belief seems to persist that, given enough time, improved techniques of vanilla production will themselves provide the foundation for the economic betterment of vanilla farmers.

Conclusion

My observations in this study have led me to the following conclusions regarding Malagasy vanilla farmers' vulnerability to cyclones. First, cyclones do not affect the entire vanilla-growing region equally. The villagers of

Befotaka and Andranobe give a human face to the statistics cited by Andreas (2006) after Cyclone Indlala, when he announced that 80 percent of the vanilla around Antalaha was destroyed but only 5–10 percent was destroyed elsewhere. Villages south of Antalaha had been hit hard by Cyclone Hudah, just as they were more recently hit hard by Cyclone Indlala. Because most farmers who lose vanilla also lose the rice harvests upon which they feed themselves, they are likely to be suffering doubly—they cannot feed themselves, and they will not be able to earn the money through their vanilla to buy food. Thus, while country-level statistics may make the shortfall in vanilla production seem almost negligible, the absence of insurance or farmers' cooperatives to help spread the wealth and the risk of vanilla farming means there will be many farmers who are absolute losers, even if the cyclone ends up having the effect of causing vanilla prices to rise.

Second, the unpredictability of cyclones, combined with the absence of savings and insurance institutions, means no farmer can ever feel secure in his or her position in the market. Madagascar may go two years without a cyclone, or a particular region may be missed by three successive cyclones. But in this region of Madagascar, no farmer can ever be sure he or she will not be in the path of the next cyclone. Indeed, those same farmers who were missed by Cyclone Indlala in 2007 were hit head-on by Cyclone Jaya two weeks later. Related to this, farmers in other parts of the world such as Uganda and India are subject to markets which respond primarily to conditions in Madagascar.

In terms of vulnerability, farmers in more isolated regions of the vanilla-growing regions of Madagascar seem more vulnerable both to the immediate effects of a local cyclone and to downturns in Madagascar's economy that might result from a cyclone. Because of their remoteness, it is more difficult for these farmers to access short-term post-cyclone assistance[9]—although it does arrive, as evidenced by the USAID-provided gunnysacks and water jugs. Of equal importance, however, it is difficult for these farmers to remain visible to receive longer-term assistance that might help them develop some cushion against the effects of future cyclones. As evidenced by STABEX's work on the road to Sambava, the more visible farmers are more likely to receive assistance that has the goal of promoting longer-term stability for vanilla farmers. Although those farmers, because they plant more, might lose more in absolute terms in the event of a cyclone, they are likely to recover more quickly because of their visibility and their accessibility. Exporters will have an interest in helping these farmers rebound if for no other reason than reaching them is cheaper than reaching the more remote farmers. For all vanilla farmers, however, it is apparent that continued investment in the vanilla sector to the exclusion of alternatives and also in the absence of risk-sharing institutions makes them highly vulnerable to cyclones.

Finally, poverty is a problem that faces all regions of Madagascar. Although it is beyond the scope of this study, the vulnerability of vanilla farmers in cyclone areas cannot be fully appreciated without putting them in the context of an island in which 70 percent of the population is living in poverty (less than $1/day, as defined by the World Bank). Each year, some portion of the island faces a food shortage, and in many years, there are multiple crises at once. When Cyclone Jaya crossed Madagascar, not only was the country dealing with the aftermath of 5 previous cyclones in 2007, but it was also dealing with a severe drought in the south which had affected nearly 600,000 people (Africa News 2007). With Madagascar's poor infrastructure and relatively low levels of foreign relief assistance, most of its millions of poor people are made more vulnerable when a natural disaster affects any part of the island. Although this is obviously true for those immediately affected by a disaster, it is also true of the poor in more distant regions because of the redirection of meager resources toward people in the affected area. So cyclones in the northeast have ripple effects throughout the country.

Specifically referring to vanilla farmers, vulnerability rests on the following characteristics:

- their physical location, which makes them a target for cyclones
- poverty, both their own and that of the country overall, which means there are few safety nets available in the event of a crisis
- lack of reliable alternatives to vanilla as a source of cash income
- the absence of financial institutions where farmers can both borrow and save

To explain why mechanisms have not been developed to protect vanilla farmers against disaster, it is important to recognize that the framework within which international development agencies, exporter associations, and the Malagasy government are operating has emphasized improving the quality of vanilla, potentially as a means to improve the quality of farmers' lives. Instead, the long-term security of Madagascar's vanilla production may rest on improving the quality of farmers' lives in some ways that are independent of vanilla. Such a focus on reducing social vulnerability would require an approach to development much along the lines of Amartya Sen's capabilities approach, which requires attention to what individuals need in a given context to flourish as human beings (Sen 1999). From this perspective, it cannot be enough simply to provide farmers with the opportunity to produce more and better-quality vanilla. Rather, they simultaneously must be provided with opportunities to access existing social and economic safety nets or to create their own.

Notes

1. One Malagasy newspaper did report that the vanilla harvest was not affected by Jaya, and that one vanilla operator had said that only 15 percent of the plantations had been touched by the two cyclones. This source had informed his own clients that "the natural vanilla of Madagascar will be present on the world market" (Madagascar Tribune 2007). Since the earlier report indicated a loss of 25 percent before Jaya, this estimate seems unlikely. What is more likely is that the source wanted to try to prevent his international clients from looking elsewhere to fill their vanilla needs.

2. These are not the real names of the villages.

3. See Ecott 2004, chapter 12 for a discussion of the wide variety of uses of vanilla.

4. At the time, the exchange rate was 4000 Malagasy francs (fmg)/US$1.

5. These estimates were made prior to the 2007 cyclones, which reduced Madagascar's production by an uncertain amount.

6. An election in 2001 was indecisive, leading Madagascar to the brink of civil war. Besides general instability in the country, a considerable amount of violence occurred in the northeast, causing the destruction of vanilla harvests.

7. Sources differ in reporting the exact amounts exported.

8. Here, one might think of parallels with other "exotic" commodities such as coffee (Roseberry 1996) and shea butter (Chalfin 2004). Just as these goods have become differentiated on the market from luxury to dime store brands, so has vanilla.

9. One news report after Cyclone Indlala in March 2007 contained the following: "Nicolas Blondel, project manager of the Masoala and Nosy Mangabe site [national parks] in Madagascar, says that it is unclear whether 'humanitarian help will reach the bush areas near the park'" (Butler 2007).

References

Africa News

2007 Madagascar: Vanilla Coast Devastated. Retrieved from LexisNexis. March 19.

Althucher, James

2005 Supply, Demand, and Edible Orchids. Financial Times (London), September 20, 2005, USA 1st ed.. Retrieved from LexisNexis.

Andréas, Claude
2006 Deuxième Atelier National Vanille: Pour Une Meilleure Compétitivité de la Vanille de Madagascar [Second National Vanilla Workshop: For Better Competitiveness for the Vanilla of Madagascar]. Electronic document, www.cirad.mg/fr/anx/idx_actu.php, accessed August 30, 2007.

Blarel, Benôit, and Diane Dolinsky
1995 Market Imperfections and Government Failures: The Vanilla Sector in Madagascar. *In* Marketing Africa's High Value Foods. Steven Jaffee and John Morton, eds. Pp. 255–318. Dubuque, Iowa: Kendall/Hunt.

Brown, Margaret
1999 Authority Relations and Trust: Social Cohesion on the Eastern Masoala Peninsula, Madagascar. PhD dissertation, Department of Anthropology, Washington University.
2004 Reclaiming Lost Ancestors and Acknowledging Slave Descent. Comparative Studies in Society and History 46(3):616–645.

Butler, Rhett A.
2007 Deadly Cyclones Set Back Conservation Efforts in Madagascar. Electronic document, news.mongabay.com/2007/0403-madagascar.html, accessed April 2, 2007.

Cadot, Olivier, Laure Dutoit, and Jaime de Melo
2006 The Elimination of Madagascar's Vanilla Marketing Board, Ten Years On. World Bank Policy Research Working Paper, 3979. Washington, DC: World Bank.

Chalfin, Brenda
2004 Shea Butter Republic: State Power, Global Markets, and the Making of an Indigenous Commodity. New York: Routledge.

CIRAD
2007 CIRAD Madagascar, "Filière Vanille." (Vanilla Sector) www.cirad.mg/fr/anx/dr004.php, accessed April 2, 2007.

CNN
2000 Cyclone Hudah Kills 13 in Madagascar. Electronic document, europe.cnn.com/2000/WEATHER/04/04/mozambique.floods.02/index.html, accessed April 4, 2007.

Cole, Jennifer
2001 Forget Colonialism? Sacrifice and the Art of Memory in Madagscar. Berkeley: University of California Press.

Cutter, Susan L., and Christopher T. Emrich
2006 Moral Hazard, Social Catastrophe: The Changing Face of Vulnerability along the Hurricane Coasts. Annals of the American Academy of Political and Social Science 604:102–112.

Cutter, Susan L., J. T. Mitchell, B. J. Boruff, and W. L. Shirley
2003 Social Vulnerability to Environmental Hazards. Social Science Quarterly 84(1):242–261.

Dairyreporter.com
2004 World Vanilla Crops Hit By Cyclone. Electronic document, www. dairyreporter. com/news-by-product/news.asp?id=50673&idCat=41&k=, accessed March 17, 2007.

Ecott, Tim
2004 Vanilla: In Search of the Ice Cream Orchid. New York: Grove Press.

Feeley-Harnik, Gillian
1991 A Green Estate: Restoring Independence in Madagascar. Washington, DC: Smithsonian.

Financial Times Information Limited—Asia Intelligence Wire
2006 Vanilla Prices Losing Flavour, December 14, 2006. Business Line. Electronic document retrieved from LexisNexis.

Freidberg, Susanne
2004 French Beans and Food Scares: Culture and Commerce in an Anxious Age. New York: Oxford University Press.

Houghton, Natalie
2007 Good Tastes: Single Origin Ice Creams. Daily News of Los Angeles, March 6, 2007, Valley Edition. Electronic document retrieved from LexisNexis.

Japan Economic Newswire
2004 SCOPE: Uganda's Vanilla Industry Blossoms after Cyclone Hudah, March 10. Electronic document retrieved from LexisNexis.

Kriner, Stephanie
2000 Cyclone Demolishes Madagascar Town, Heads for Mozambique. Electronic document, www.ReliefWeb.int, accessed August 30, 2007

Laney, Rheyna M.
2004 A Process-Led Approach to Modeling Land Change in Agricultural Landscapes: A Case Study from Madagascar. Agricultural Ecosystems and Environment 101:135–153.

Madagascar Tribune
2007 Passage du Cyclone Jaya: Deux Morts à Sambava [Rassage of Cyclone Jaya: Two Dead at Sambava], April 4. Electronic document, www .madagascar-tribune.com, accessed August 30, 2007.

MAEP (Ministère de l'agriculture de l'élèvage et de la peche)
2005 Atelier National de la Filière Vanille [National Vanilla Workshop], August 22, 2005. Electronic document, www.maep.gov.mg/fr/actuvanille1.htm, accessed April 2, 2007.

Metzel, Jeffrey, Emilienne Raparson, and Eric Thosun Mandrara

1999 The Political Economy of Trade Liberalization: The Case of Vanilla in Madagascar. Washington, DC: USAID Bureau for Africa.

Milmo, Cahal

2004 The Bitter Taste of Vanilla. The Independent, UK, August 19, 2004. Electronic document retrieved from LexisNexis.

Mongabay.com

2007 Madagascar Cyclones May Be Boon to Vanilla Market, March 27. Electronic document, news.mongabay.com/2007/0327-madagascar.html, accessed August 30, 2007.

Murdoch, Jonathan, Terry Marsden, and Jo Banks

2000 Quality, Nature, and Embeddedness: Some Theoretical Considerations in the Context of the Food Sector. Economic Geography 76(2):107–125.

Nathu, Samash A.

2003 Vanilla: the New Essence of Uganda. Uganda National Commission for UNESCO:145-147, copy provided by author. Also electronic document, www.ucil-uganda.com.

New Vision

2007 Vanilla Price Doom Continues. Africa News, January 11. Electronic document retrieved from LexisNexis.

Oliver-Smith, Anthony

1996 Anthropological Research on Hazards and Disasters. Annual Review of Anthropology 25:303–328.

ProVention Consortium

2007 ProVention Forum: Risk Assessment in Madagascar Powerpoint Presentation. Electronic document, www.proventionconsortium.org, accessed August 30, 2007.

Rakotoarivelo, Voahangy

2000 Eyewitness: Madagascar Left Reeling. BBC News online, March 10. Electronic document, news.bbc.co.uk/1/hi/world/africa/672813.stm, accessed August 30, 2007.

Reuters

2007 Another Cyclone Kills Three in Madagascar, April 5. Electronic document, www.alertnet.org/thenews/newsdesk/L0555398.htm, accessed August 30, 2007.

Roseberry, William

1996 The Rise of Yuppie Coffees and the Reimagination of Class in the United States. American Anthropologist 98(4):762–775.

Sen, Amartya
1999 Development as Freedom. New York: Knopf.

Sodikoff, Genese
2004 Land and Languor: Ethical Imaginations of Work and Forest in Northeast Madagascar. History and Anthropology 15(4):367–398.

USAID
2007 USAID. Telling Our Story: Vanilla is Uganda's Green Gold. Electronic document, www.usaid.gov/stories/uganda/ss_uganda_vanilla.html, accessed April 2, 2007.

POLITICAL ECONOMIC MITIGATION OF DISASTERS

V

12 Learning from Disaster? Mad Cows, Squatter Fires, and Temporality in Repeated Crises

ALAN SMART AND JOSEPHINE SMART

Research suggests that while communities generally recover from a single disaster, repeated crises within a short period can deplete resources used for coping and, at extremes, lead to economic, social, and cultural collapse (Dyer 2002; Oliver-Smith and Hoffman 2002). We consider how a series of disasters might allow insights into the processes of government response that differ from what would be concluded from restricting analysis to individual extreme events. A rapid succession of similar disasters, such as the fires in Hong Kong squatter settlements that left about 200,000 people homeless during the 1950s, may both create a sense that "something has to be done" and facilitate a learning process whereby policy-makers learn from their mistakes. In other cases, vulnerable populations may experience disaster again and again without effective responses. In this chapter, we use our empirical research on two sets of crises, squatter fires and outbreaks of mad cow disease, to consider the impact of temporality on the political economy of institutional responses to extreme events.

The majority of disaster research either concentrates on a particular extreme event or addresses a category of disaster more generally. The former research projects tend to emphasize the effects of a disaster in a specific time and place and the impact on a particular population of people. The latter kind of research is more likely to aspire to generalizations, and is likely to emphasize the appropriate policies that should be implemented to avoid negative impacts from that kind of extreme event (e.g., building-code revisions for earthquake-prone areas, or prohibiting rebuilding in floodplains). Both types of research are of great importance, but we suggest a less common focus for disaster research. We argue that we can learn a great deal by looking at series of related disasters in order to explore how institutional responses are influenced by the sequence. Garcia-Acosta (2002:49) has suggested that the mainstream of disaster studies tends to be "ahistorical and even antihistorical." We believe that this needs to be changed, that experiences and perceptions of past related events may have a major impact on later responses. In some cases, governments may learn from the inadequacies of their previous responses and reflexively modify their programs. This may lead to an institutional "learning curve" where future reactions are more effective, efficient, or have better outcomes in some other way. In other cases, a response "flatline" is more apparent than any real learning from past mistakes, so that little improvement in policy appears to occur over time. Our comparative effort here is intended as an exploration into the factors that influence such outcomes, for example, whether the interests of economic elites may prompt faster and more effective institutional learning when those interests are threatened.

Many factors influence the effectiveness of government responses to disaster, including whether there are real electoral contests, the economic costs and opportunities, state capacity to intervene, scientific knowledge about the problem and its possible solutions, interventions by NGOs, and the institutional value placed on the populations that are vulnerable to extreme events (Mitchell 1999:40). In this study, we turn attention to another factor: the temporality of disasters. We hypothesize that a series of disasters occurring within the same locale and within sufficiently frequent intervals may be more likely to generate good policy than those that are very infrequent, all other things being equal. In particular, when comparable events occur within a year or two, and government responses are seen as failing to resolve the problem or even exacerbating them, there may be a much greater demand that "something must be done" than when public memory has faded about previous events. In situations where disasters are expected to routinely, if unpredictably, recur, it may be that the intertwining of disasters and governance create "not only states of emergency but emergency-oriented states" (Clancey 2006:4).

When are Lessons Learned?

Given certain conditions, disasters can initiate changes in direction for a society. The Great Mississippi River Flood of 1927 prompted many direct responses but more broadly initiated sharp shifts in perceptions of the roles and responsibilities of the U.S. federal government, well before the more noted New Deal response to the Depression (Barry 1998:422). As another example, Meiji Japan has been famous for its wholehearted learning from the West, but the pattern was sharply changed for architecture after an earthquake in 1891 because the "modern" and "strong" Western-inspired buildings collapsed while older wooden buildings survived. This event became a "watershed in Japanese nationalism" (Clancey 2006:2).

Some disasters, such as monsoon flooding in Bangladesh, seem to repeat endlessly, with minimal or ineffective responses by governmental agencies (Doughty 1999; Zaman 1999). In such cases, either disaster is accepted as inevitable, or little effective learning has been generated by earlier failed responses. Local people may have adapted their social, cultural, and built environment arrangements to facilitate coping with repeated crises (e.g., McCabe 2002; Oliver-Smith 1999), but what are the circumstances under which governments learn from disasters and from the failures of their prior efforts, and in doing so produce policies and institutions that reduce vulnerability, mitigate effects, or more efficiently coordinate relief efforts? To what extent is that learning a result of the nature of the political economy?

Birkland (2006:5) argues that the larger the disaster in lives lost, property damage, and extent, "the larger the potential influence on the political and policy world, all other things being equal." While this claim may be valid, since other things are never equal, it has its limits in accounting for actual patterns of policy change. At the same time, policy domains prone to disaster may find learning more difficult because of their infrequent occurrence. We suggest that the timing of a disaster can be as influential as its scope or scale.

To say that the timing of an event influences its impact does not go far enough. What is it about timing that makes a difference? Sometimes it can be the straw that breaks the camel's back. Occurring shortly after a prior extreme event, it might raise questions about why policies had not changed, or if they had, why they were not effective. It might depend on the state of play of electoral politics: its effects on a vulnerable minority government could be quite different than on one that is more secure. Large extreme events are neither always necessary nor sufficient to focus political attention. Although catastrophic events almost inevitably attract attention, they do not always result in effective policy change. In an analysis of five major earthquakes in Turkey in the 1990s that cost 19,126 lives, Corbacioglu and Kapucu (2006:229) found that no significant organizational learning within Turkish disaster management as a result of the Erzincan (1992), Dinar (1995) and Ceyhand (1998) earthquakes. Responses tended to be linear: more resources applied to the same strategies. The 1999 Marmara earthquake, however, changed this pattern by making the continuation of past failures blatantly (and internationally) apparent. It resulted in new and more effective approaches to communication and cooperation with the nongovernmental sector, which showed great improvements in response to the next quake that followed by a few months.

Dissatisfaction with past disaster response records has led to considerable effort at developing new approaches which will facilitate learning, situation-specific appropriate responses, and optimal utilization of resources. Among the critical factors are the kind of extreme event, who is affected, and what circumstances might combine to encourage strenuous and thoughtful responses. One of the factors, the relevant prior sequence of disasters, has received relatively little attention in the disaster literature.

Squatter Fires in Hong Kong

This section focuses on the dynamics of a series of squatter fires in Hong Kong in the 1950s, and how they contributed to the emergence of one of the world's largest public housing systems, which houses nearly half of the territory's seven million people and has profoundly influenced the nature of its political economy. The analysis draws on research reported in Alan Smart's (2006) book *The Shek Kip Mei Myth*, but uses that material in a new way to consider the nature of the policy-making process in the context of disasters.

Although the project was stimulated by ethnographic research in Hong Kong in the 1980s and between 1999 and 2003, the dominant research method utilized was archival research in the Public Records Office in Hong Kong. All of the cases within the class of squatter fires in Hong Kong in the decade of the 1950s were investigated, along with any other remotely relevant file, rather than simply reinterpreting the one iconic event that has been seen as the beginning of squatter resettlement—the 1953 Christmas Eve fire in Shek Kip Mei. No quantitative treatment of this sample was attempted, because it is biased in terms of what officials considered significant. "Uneventful" fires, even ones that affected several thousand people, were only lightly documented, for example by the Fire Department's incident report. If they did not raise issues that needed to be dealt with at senior levels, they were not discussed, and any information collected remained primarily undocumented. Rather than simply a "bias," however, this process allowed concentration on a distinct subset of "squatter fires, 1950–1960," that is, "squatter fires that received attention in policy discussions in the 1950s." These were the pivotal events that moved policy discussions or policy decisions forward. Files were circulated for comment among relevant officials, and the appended memos provide remarkable insights into how decision-making proceeded, into what was considered feasible and practicable, in ways that the final documents produced rarely do.

Bonham Richardson (2004:xii) has commented that "as scholars of fire know, their subject is everywhere and nowhere." Fires have throughout history prompted urban restructuring and often served as the catalyst for the earliest planning regulations. They have prompted the banning of certain building materials and the creation of safety codes, and opened up large spaces for new forms of development.

Massive fires in Chicago (1871), Boston (1872), and Baltimore (1904), were welcomed by many as an opportunity to improve sanitation and urban planning (Rosen 1986:11). The same can be said about squatter fires in Hong Kong. Early responses used government resources without ending the risk of fire or resolving the squatter problem or improving the lives of squatters. The learning process initiated by repeated crises prompted the adoption of an initially very inexpensive, high-density multistory Resettlement Program. This Program eventually became a wider public housing system that houses half of the Hong Kong population and has had impacts on every dimension of the territory's economy, society, culture, and landscape. Manuel Castells et al. (1990) and others have attributed Hong Kong's economic miracle in large part to this public subsidization of labor power. Certainly, the economic, social, and cultural impacts of this transformation of housing practices have been immense. Just one year's (2005/2006) current expenditures of the Housing Authority total US$2.07 billion. But consequence does not entail causation, and this and other prominent explanations of the beginnings of public housing in a staunchly laissez-faire colony fail to stand the test of close examination based on archival documents (Smart 2006). The economic situation in the 1950s, an emerging economy based on manufactured exports, influxes of refugees from China, low incomes with almost no vestiges of social welfare, and the economic domination of the commercial elite (both expatriate and Chinese), conditioned but did not explain the rapid shift toward expenditure on the resettlement of squatters. Constraints on resolving the squatter problem combined with the interests of the Hong Kong government were more important as the determinants of the change of direction.

Squatter settlements in Hong Kong in the early 1950s created situations where fires were more or less inevitable. At least 200,000 people lost their homes during the 1950s through a continual series of many small fires punctuated by a number of much larger conflagrations. Because the colonial government wished to eradicate existing illegal building and to discourage it in the future, the provision of elementary fire safety precautions such as fire hydrants was rejected. Discouragement failed, and crises provoked by the larger fires eventually resulted in the government embarking on the construction of squatter resettlement estates.

The first big fire, on January 11, 1950, in Kowloon City left about 17,000 people homeless and ignited a high-level debate about squatter fire prevention. The Deputy Colonial Secretary responded to a proposal from Mr. Gorman, Chief Officer, Fire Brigade, shortly before the Kowloon Walled City fire. He stressed that Gorman's memorandum had to be considered:

> . . . against a background of practical politics. I do not see how we could put this area [Kowloon City] on a proper basis, from the point of view of fire risk, without tearing down a considerable number of these shacks . . . in order to make the necessary fire lanes. The effect of this would almost certainly be to upset the whole of the political apple-cart at a time when we are most anxious to avoid any incidents . . . if only for political reasons, we shall have to compromise on Fire Brigade requirements. Perhaps that compromise should take the form of making the necessary preparations to prevent any serious loss of life, and taking no further steps than this for the time being. It would not be unreasonable to say that that was our main concern in the whole matter. These shacks are entirely illegal structures and it is not up to Government to take measures for their protection. If a large number of them were burnt down we should probably have to take steps to assist the occupants in re-provisioning themselves, but we can hardly overlook the fact that the effect would be no bad thing politically.[1]

Gorman's proposal was to lessen the serious fire risks. He suggested that:

> If a large Fire does occur in this Squatter area as stated it will be nothing short of a tragedy and end up in a holocaust with a terrific loss of life . . . To my mind the onus for the Tragedy will have to be borne by some one, as an explanation will be called for to explain what precautions were taken.[2]

Gorman's subsequent memo argued for the expenditure of HK$245,990 to extend water mains, provide fire hydrants and water storage, extend fire breaks through squatter areas, and control dangerous industrial premises and goods. The Colonial Secretary's response was that fire precautions were becoming excessive, given the fact that the housing was ". . . illegal, a menace to public health and security, and which our major policy aims at eliminating altogether (e.g., immigration control, expulsion of undesirables, cheap building schemes, etc.)." He goes on to assert that the use of public funds can be justified to protect lives and adjacent property, but not the shacks themselves.[3] The Financial Secretary minuted that he agreed that it was "quite crazy" to be asked to spend a quarter million dollars to protect the illegal structures.[4] The Social Welfare Officer supported the construction of fire lanes, but insisted that

> no hydrants, extinguishers or other fire-fighting apparatus should be installed, and no official encouragement should be given to the formation of local voluntary fire-fighting units . . . To do otherwise would be to encourage the squatters to count on and to trade on official recognition of their "rights."[5]

Concern to avoid the appearance of legality recurred regularly. The Senior Health Officer suggested that in clearing fire lanes it was important not to try to displace too many squatters at the same time as "five to eight hundred huts at a time is about the safe limit if disturbance is to be avoided." Gorman's replacement commented that:

> so long as there are squatter huts there will be squatter fires, and so long as industrial activity is permitted to remain in such areas are there are on the Kowloon Peninsula so frequently, these fires will remain frequent and must inevitably continue to cause loss of life through the abnormal spread of fire due to the presence of volatile substances.[6]

This debate illustrates that the vulnerability of squatters to fires was in part a product of government unwillingness to adopt expert advice that might have ameliorated the hazardous situation. The construction of fire lanes (demolition of sections of homes to create an open space the fire was less likely to cross) was more attractive to government officials, but risked disturbances from squatters. The combination of this possibility of militant reactions and the vulnerability of Hong Kong's geopolitical situation made resolving the overall squatter problem impossible in the 1950s (Smart 2006).

The Chief Officer of the Fire Brigade pointed out that "statistically on the law of average, for every 800 to 900 people in a built-up area, one amongst them will cause a fire once per annum." What was distinctive about squatter fires was how quickly the flames traveled through the highly flammable huts.[7] In addition to the structural characteristics of many squatter settlements, there were two factors that helped cause greater damage after a fire started. First, there tended to be very substantial delays in receiving reports of a fire due to the absence of telephones. Second, the fire areas were virtually always 1,200–4,000 feet from water supplies. Under squatter conditions it took one pumping appliance and a team of six men 15 minutes to provide one jet of water over one thousand feet. With an average eight to ten jets required to extinguish a squatter fire affecting 200 huts, "it is understandable why such fires can, and do, spread rapidly without adequate deterrent for periods up to one hour." [8]

The topographic features of most squatter areas further increased the difficulty of fighting fires. Hills, valleys, and bottlenecks added to the problem of access. For example, in the 41 acres razed by the 1953 Shek Kip Mei fire, there had been only one road:

> which was wide enough for a jeep. No motor traffic used the road regularly because it was too crowded with people . . . All the other paths and alleyways were very narrow, ill-made and frequently bordered by deep drains.[9]

Local resistance often derailed the construction of fire lanes, and even when completed, re-squatting often erased the clearings. The Commissioner of Police reported on proposed fire lanes that would displace between 2,500 and 3,000 persons: "It is possible that they could be pushed further up the hillside . . . but it would be a costly business and unless Government either paid compensation or gave them houses on the same basis as is being done at Shek Kip Mei, it would be an intensely unpopular move and could create an explosive *situation*."[10]

Clearly very few useful lessons were learned from the Kowloon City fire, although fire official recommendations were sensible. The next big fire, adjacent in Tung Tau on November 21, 1951, left as many as 25,000 homeless. Its aftermath had repercussions that set the stage for the more innovative response to the Shek Kip Mei fire. Dissatisfaction with arrangements made for the fire victims lead to civil unrest when a mission sent from Communist China to aid the victims was stopped at the border. The disturbances were significant enough that Governor Grantham pointed to them as evidence for the precariousness of Hong Kong's internal security position. Simple demolition and control of squatting were discovered not to be viable, and new approaches were attempted. In particular, squatter resettlement was attempted, in the form of areas in which squatters could erect their own dwellings according to certain rules, "tolerated areas," and other areas where small stone cottages were erected by private contractors, "approved areas." These programs failed to resolve the problems, most fundamentally because in a city with among the world's highest population densities, resettlement areas had lower densities than the squatter areas that they replaced. They produced expenses for the government, were not satisfactory for the resettled squatters, and could not prevent the continual reemergence of new squatting, with its attendant fire risks.

Most narratives of the beginnings of public housing in Hong Kong accept to some degree what is described as the Shek Kip Mei myth (Smart 2006). The official version is that "we built simple, low-cost shelters to a minimum standard to meet emergency needs resulting from a tragic Christmas night fire in 1953 in Shek Kip Mei" (Suen 2003:xvii). Academic versions differ on the motivations underlying resettlement and its social and economic consequences, but do not question the basic features of the narrative: squatter resettlement began in response to this massive fire, and the genealogy of the contemporary public housing programs can be traced back to this founding moment. Given the low level of development of public participation and social welfare in colonial Hong Kong, public housing came to be both a key strategy for building a sense of citizenship and commitment by Hong Kong residents, and a symbol for the positive dimensions of the colonial legacy.

The immediate response to the Shek Kip Mei conflagration, despite its massive scale, did not involve an instant "momentous decision" that set Hong Kong on the path to providing homes for half its population. While the Governor or other key officials might have had in mind something like the multistory Resettlement Estates that were eventually produced, the Executive Council decision that enabled it envisioned something more congruent with what had been constructed in approved areas in the past. There was also no commitment to an ongoing program of permanent squatter resettlement at that time, and the decision was still an open and contested one in the early months of 1954.

The initial, provisional plans for the Shek Kip Mei fire site seem to have been a response to the scale of the disaster combined with the consequences, perhaps unintended, of the commitments that the Executive Council and Finance Committee did undertake in the days after the fire. The difficulty of resettling fire victims off the site created a context in which increased density was the natural way of dealing with the various constraints. This in turn encouraged an architectural solution that helped resolve the problem of resettling many low-income squatters while avoiding heavy financial commitments or politically inconvenient resistance. By adopting a scheme with minimal standards and maximal density, the costs of producing resettlement housing could be accomplished with both low rents and substantial space being opened up for more profitable forms of development. As the Commissioner for Resettlement said: "this is the only possible way of removing the fire risk and the constant risk to public health and public order which are presented by the remaining squatter areas, and also the only practical means of recovering for proper and permanent development the extensive areas of Crown land still sterilised by squatter colonies."[11] In January 1954, however, no such continuing program was yet in existence.

Government documents indicate that at least the initial response to the Shek Kip Mei disaster was not a sharp break from earlier policies. Previous fires had also prompted significant shifts in government legislation, policies, and practices. It was also only subsequent fires that turned a provisional experiment into a permanent program.

The adoption of multistory Resettlement was the eventual result of a learning process punctuated by a continuing series of crises. The 1950 Kowloon City fire showed the potential catastrophes that lurked behind the rapid growth of illegal settlements, but the lessons taken by the Chief Officer of the Fire Brigade on the need for greatly enhanced precautions against future fires were not adopted due to considerations of cost and the risk of giving the appearance of legality to squatter structures, resulting in casual neglect of the possible fates of thousands of people. The Tung Tau fire in 1951 made clear the political and diplomatic costs of inadequate arrangements for resettlement for fire victims. Important legal innovations for more effective squatter clearance followed, but the types of resettlement provided could not resolve the fundamental problems due to their inefficient use of scarce land. The Shek Kip Mei fire did not immediately result in an ongoing multistory Resettlement Program. Instead, commitments undertaken by the Executive Council to rehouse all the fire victims and to set the rent at ten dollars a month produced a context in which new architectural models were needed if these commitments were not to demand the provision of housing off the fire site to tens of thousands.

The consequences of the sequence of squatter fires for Hong Kong's subsequent development have been momentous. With real estate being central to Hong Kong's political economy, both private and public, the provision of housing for half of the population has tremendous impact on the economy (Smart and Lee 2003). Beyond the economic, however, there is no dimension of Hong Kong's politics, society, culture, or landscape that can be understood without an examination of the impact of public housing on it. Manuel Castells et al. (1990) and others have attributed Hong Kong's economic miracle, based on manufactured exports, in large part to this public subsidization of labor power. While this argument fails as an *explanation* of public housing in Hong Kong, it does not mean that the promotion of export-oriented development might not have been an *effect* of public housing (Smart 1989). The low rents, combined with policies that did not punish those who improved their condition by evicting them, seem to have contributed to the significant amounts of upward mobility experienced by Resettlement Estate residents in the decades after 1953.

Bovine Spongiform Encephalopathy

The story of bovine spongiform encephalopathy (BSE, also commonly known as mad cow disease) began with the first confirmed case in Britain in November 1986. Subsequently BSE was found in many countries in Europe and in Japan. Canada joined the ranks with its first confirmed indigenous case in May 2003. The following section is a discussion of the Canadian government's handling of the BSE crisis and how the international reality of BSE crisis management informed the Canadian policy responses.

The International Reality of Transmissible Spongiform Encephalopathy

BSE is a neurological disease in cattle that causes progressive loss of motor coordination and appetite, and eventually death. The condition has no known treatment and as yet there is no effective diagnostics on live animals. The disease is believed to be caused by the misfolding of protein units known as prions, and the infective agents in the brain, spinal cord, and other tissues are known to be highly resistant to decontamination. There remain significant gaps in our knowledge about the biology of the various kind of transmissible spongiform encephalopathy (including BSE), the links between transmissible spongiform encephalopathy and human health, and the socioeconomic implications of transmissible spongiform encephalopathy (Deslys and Picot 2001).

The current understanding points to the presence of BSE-infected meat and bone meal in feed as the primary cause of BSE. A ban on ruminant-to-ruminant feed was recommended by World Health Organization in 1996. Subsequent emergence of atypical BSE and new cases after the feed ban raise serious questions about the possibility of other, unknown modes of transmission and about disease causation.

Table 12.1. Variant Creutzfeldt-Jakob Disease (vCJD) Up to June 2007

Country	*Total # (alive)*	*6+ Months Residence in UK in 1980–1996*
United Kingdom	165(4)	165
France	22(2)	1
Ireland	4(1)	2
Italy	1(0)	0
USA	3(0)	2*
Canada	1(0)	1
Saudi Arabia	1(1)	0
Netherlands	2(0)	0
Portugal	2(1)	0
Spain	1(0)	0
Japan	1(0)	0**

*3rd patient most likely infected as a child in Saudi Arabia.
**Resided in the UK for 24 days in the 1980–1996 period.
Source: EUROCJD/NEUROCJD 2007Table 12.3. Timeline—Bovine Spongiform Encephalopathy (BSE) and Government Response in Canada, 2003–2007, by Event

A possible link between BSE and variant Creutzfeldt-Jakob Disease in people was first acknowledged in March 1996 based on research findings in the United Kingdom (Packer 2006:75–98). Ongoing research is under way to identify the full implications of transmissible spongiform encephalopathy in blood products and pharmaceuticals for human use. Other known human forms of spongiform encephalopathy include Creutzfeldt-Jakob Disease, which occurs worldwide and in a random fashion, and kuru, which occurred among the Fore people in the highland Papua New Guinea up to the 1960s (Beasley 2006; Lindenbaum 2001).

Spongiform encephalopathy in sheep and goat has been known since 1732. It is known as "scrapie" and is common worldwide. Infected animals will scratch against a fence or tree until the wool and sometimes the skin is worn away, giving the animal a half-naked appearance. New Zealand is the only country today that is scrapie-free.

Some animals in the deer family *Cervidae* are also susceptible to spongiform encephalopathy, a condition known as chronic wasting disease. Chronic wasting disease in North America was first found in Colorado in 1967 and is known to affect many areas in the United States (Commonwealth of Virginia 2007). The northward expansion of chronic wasting disease reached Canada in 1996. Farmed and wild mule deer and white-tailed deer in Saskatchewan and Alberta are known to be infected.[12] As of April 2008, a total of 49 cases of chronic wasting disease were confirmed in Alberta, of which 34 were found in wild deer (Alberta Sustainable Development 2008). In Saskatchewan, a total of 145 mule deer and 49 white-tailed deer were

Table 12.2. Bovine Spongiform Encephalopathy (BSE) Cases up to July 2007 in selected Countries

Country	*Year of 1st Case*	*Total*
Belgium	1997	133
Canada	1993	12
France	1991	984
Ireland	1989	1,602
Japan	2001	33
Portugal	1990	1,029
Spain	2000	681
United Kingdom	1987	184,508
United States of America	2005	2

Source: OIE 2007

tested positive for chronic wasting disease from 1997 to 2008 (Government of Saskatchewan 2008). The BSE situation in the United Kingdom is the best studied and possibly most discussed case since the disease was first identified in November 1986 (table 12.2). The devastation—measured in financial, social, and political terms—cut a wide path across all spectrums of British society. In the end, the British cattle industry was decimated, beef consumption fell by more than 50 percent, consumer confidence fell to rock bottom in 1996 after the announcement of a link between beef consumption and the human disease variant Creutzfeldt-Jakob Disease, which is a form of spongiform encephalopathy, and the British government suffered a near moral bankruptcy under the public's perception that the government had acted irresponsibly in dismissing the growing public concern and anxiety about BSE-linked food safety after 1986 (Leiss and Powell 2004). The United Kingdom's experience becomes a prominent source of information for other countries in their design of BSE policies.

The First BSE Case in Canada—December 8, 1993

The media reports on the December 1993 BSE case highlight the tremendous financial stake in BSE in the context of international trade and the political will to take extreme measures to safeguard public and international perception of food safety in Canada. They also point to the political expediency of downplaying the Canadian relevance of this first BSE case by attributing it to the British origin of the animal, thereby diverting a potential "crisis" into an "inconvenience." In reaction to the federal government's decision to destroy all cattle imported from the United Kingdom between

1982 and 1990 in the aftermath of the December 8, 1993 confirmation of BSE in one animal on a farm in Alberta, one Agriculture Canada official comments that "[y]ou might say this is science influenced by trade" (Campbell 1993a:C2). The lack of any significant trade loss arising from this 1993 case explained why none of the media reports made front-page news as shown below:

- Killer Mad Cow Disease Strikes in Alberta (Campbell 1993b:D1)
- Alberta Herd Quarantined: Mad Cow Disease Found (Romahn Dec-9-1993:B6)
- Mad Cow Disease Creates Problems for Area Exporters (Romahn Dec-14-1993:B7)
- Fate of 270 Cattle Uncertain in Wake of "Mad Cow" Alert (Campbell 1993c:C6)
- Entire Herd to be Destroyed to Prevent Mad Cow Spread (Campbell 1993a:C2)
- Imported Cattle to be Killed (Canadian Press Dec-31-1993:C3)

Two pieces of information emerging from this series of media coverage are worthy of note. First, Canada imported 175 cattle from the United Kingdom before 1990, one of which came down with BSE in 1993 at the age of six. The United States imported 459 cattle from the United Kingdom before 1990 and since then bought some of the 175 cattle imported by Canada (Romahn Dec-14-1993:B7). Second, Mexico was the first and only country to impose a ban on live cattle, meat, and semen from Canada after December 9, 1993. However, under the pressure from the Canadian government, and the fact that the United States did not impose a beef and cattle trade ban on Canada, Mexico rescinded the ban shortly afterward. Three other countries imposed a ban on Canadian bovine semen after December 9, 1993—Japan, Australia, and the Czechoslovakian Republic (Romahn Dec-14-1993:B7).

Slow-But-Steady Learning (1993–2003)

The cattle industry is a multibillion-dollar business in Canada—it is a highly diversified sector that includes livestock, meat, semen, embryo, and other by-products. It is also highly dependent on export. In 2002, 23 percent of the Canadian total feed and cull cattle marketings were for export, of which 70 percent went to the United States. On average, 4.2 million heads of cattle are slaughtered and/or exported every year, which translates into

81,000 animals per week. The Prairie Provinces (Alberta, Saskatchewan, Manitoba) account for over 80 percent of the country's cattle population. Alberta alone supplied 71 percent of the total cattle slaughter capacity in 2002 (Grier 2005).

While BSE is a medical problem that affects animals and people in lethal manners, its more immediate and politically sensitive threat to the well-being of the cattle industry and various affiliated sectors is rooted in the uncertainty of the financial fallouts arising from trade bans. The Canadian federal government has a vested interest in doing the right thing to ensure the safety of Canadian beef by keeping BSE out of the Canadian cattle population. This trade-driven mandate is borne out by a series of BSE-relevant policy introductions prior to the 1993 case (Canadian Food Inspection Agency 2006):

- 1990—BSE became a reportable disease under the *Health of Animals Act*.
- 1990—Importation of cattle from the United Kingdom was further restricted.
- 1992—The Canadian Food and Inspection Agency (CFIA) implemented the National Bovine Spongiform Encephalopathy surveillance program.

Even though the 1993 case was successfully neutralized as a non-Canadian case on grounds of the United Kingdom origin of the infected animal, the government had not softened its efforts to keep BSE out of Canada. While Canada was criticized by some as moving too slowly and not having done quite enough in optimizing its coping capacity before the BSE crisis hit in May 2003 (Leiss and Powell 2004:232–235), an institutional learning perspective would suggest that the government of Canada was learning slowly but steadily from the international BSE experience. It was gentle in its policy reform approach, which likely was informed by a political conservatism rooted in the merit of maintaining the status quo as far as possible. While the government of Canada underestimated the risk of a full-blown BSE scenario, the government's strategic capacity-building efforts leading up to the 2003 BSE crisis are indicated by the following actions (Canadian Food Inspection Agency 2006):

- 1997—The CFIA amended the National Feed Inspection Program to include mammalian-to-ruminant feed ban regulations in response to World Health Organization recommendations.

- 2000—The CFIA suspended importation of rendered animal protein products of any species from any country not recognized by Canada as free of BSE.
- 2001—The Canadian Cattle Identification Program was instituted for cattle and bison, enhancing Canada's ability to trace individual animals from herd of origin to slaughter.
- 2001—Health Canada produced a 37-page report titled *Transmissible Spongiform Encephalopathy Action Plan* (Bob Hills, Director, Transmissible Spongiform Encephalopathy Secretariat, personal communication, April 3, 2007).
- 2001—The Transmissible Spongiform Encephalopathy Secretariat created and housed within Health Canada, later moved to the CFIA (Bob Hills, Director, Transmissible Spongiform Encephalopathy Secretariat, personal communication, April 3, 2007).
- 2002—Transmissible Spongiform Encephalopathy Risk Assessment Template in place (Bob Hills, Director, Transmissible Spongiform Encephalopathy Secretariat, personal communication, April 3, 2007).

May 20, 2003—The First Indigenous Case of BSE in Canada

When the first indigenous Canadian BSE case was confirmed by the World Reference Laboratory in Weybridge on May 20, 2003, the reaction around the world was instantaneous and dramatic. First, there were blanket trade bans on Canadian beef, live cattle, embryos, and semen around the globe. Predictions of anticipated economic devastation in the billions of dollars range were widespread in local and international media coverage. The delay in the re-opening of the American market became a symbol of the unequal political relationships between Canada and the United States.

Even after the government of the United States announced a partial lift of the ban in early August 2003 to allow the import of boxed meat of boneless cuts from animals less than 30 months, and boneless veal from animals less than nine months, the perception remained that Canada was treated unfairly. Mexico followed three days later to announce a similar partial lift of the trade ban on August 11, 2003. Over time, the trade ban was partially or completely lifted in many countries around the world. Japan resumed beef trade with Canada in December 2005, much to the relief of the beef industry given the importance of the Japanese market for high-value Canadian beef. Canadian live cattle (less than 30 months of age) shipments to the United States resumed in July 2005. Taiwan reopened its border to selected Canadian beef imports in late June 2007. The People's Republic of China and South Korea remain closed to Canadian beef import as of May 2008.

It was estimated that for each day the trade ban was imposed on Canada due to BSE, the Canadian economy suffered a loss of $11 million and many jobs in a wide range of sectors linked to the cattle industry (CBC 2006; Government of Alberta 2006; Le Roy and Klein 2005). The federal government and some provincial governments responded promptly to the BSE crisis (see table 12.3). The Canadian government responses to BSE fall into several major categories: risk management (e.g., feed ban, animal tracking protocol, testing for BSE), post-BSE assistance programs (e.g., CAIS, FIP, TISP), and research (e.g., APRI, PrioNet; see table 12.4).

The total financial outputs by federal and some provincial governments are well over 4 billion dollars, most of which were channeled into various recovery programs intended to help producers and the cattle industry cope with the financial hardship arising from the trade bans. The responses to these government interventions were mixed, but the regular financial injections throughout 2003, 2004, and 2005 were instrumental in blunting some of the extreme consequences of the loss of export markets for Canadian beef due to BSE. It is evident that Canada's BSE crisis management was largely reactive in nature, partly guided by a competence built on lessons learned and knowledge accumulated from BSE situations in other parts of the world, and partly driven by a real concern about the financial implications of an international trade ban on Canadian products.

Conclusions

This chapter is only a preliminary attempt to learn from two series of disasters. The two cases demonstrate, we believe, a better understanding of the process of why which governments and responsible agencies learn or fail to learn from disasters can be achieved by examining sequences of related extreme events. Going beyond the single case, or ahistorical generalizations about particular types of disasters, allows for greater attention to the dynamics that influence governmental responses and planning for future disasters.

In this conclusion, we will briefly spell out some of the significant similarities and differences between the cases, and the questions that these patterns raise. First, in both cases learning actually occurred, and happened over a relatively short period of time. This similarity suggests that

Table 12.3. Timeline—Bovine Spongiform Encephalopathy (BSE) and Government Response in Canada, 2003–2007, by Event

2003—2nd and 3rd Quarters	*2003—4th Quarter*	*2004*	*2005*	*2006*	*2007*
May 20—1st BSE case	Oct 9—Alberta Slaughter Market Adjustment Program for Other Ruminants[1,2]	Jan 9—$92.1 million federal funds for BSE risk management[4]	Jan 2—2nd BSE case	Jan 23—4th BSE case	Feb 7—9th BSE case
Jun 12—Federal program for BSE-affected workers[1,2]	Oct 9—Alberta Steer and Heifer Market Transition Program[1,2]	Mar 22—Federal $995 million Transitional Industry Support Program[4]	Jan 11—3rd BSE case	Feb 7—Alberta animal disease lab opens[2]	Mar 9—$1 billion to Canadian farmers[7]
Jun 18—$460 million National BSE Recovery Program[1,2]	Oct 24—Alberta Beef Product and Market Development Program[1,2]	Jul 9—Feed restrictions require SRM removal[4]	Feb 28—$38 million Alberta Prion Research Institute[1,2]	Apr 16—5th BSE case	Mar 12—Feed ban transition, $80 mil.[8]
Jul 4—Specified Risk Materials (SRMs) banned from human food[2,3]	Nov 4—British Columbia offers $7 million for BSE assistance[5]	Sep 10—$718 million for second federal-provincial BSE recovery strategy and assistance program[1,2,5]	Mar 7—$37 million for Alberta BSE recovery initiatives[4]	Jun 26—SRMs banned from animal/pet food, fertilizer[4]	May 2—10th BSE case

Jul 18—Must remove SRMs from older cattle carcasses[4]	Nov 21—$200 million federal-provincial Cull Animal Program[4]	Dec 29—$8 million Alberta Industrial Transition Program for Other Ruminants[1,2]	Mar 29—$1 billion federal Farm Income Payment Program[4]	Jul 4—6th BSE case	Jul 12—Fed. feed ban begins[9]
Jul 25—Alberta provides $79 million in interim support to producers[4]	Dec 11—Canadian Agricultural Income Stabilization (CAIS) program[4]		Apr 7—$2 million marketing Alberta's other ruminants[4]	Jul 13—7th BSE case	Dec 18—11th BSE case
Aug 6—Manitoba $100 million loan program for producers[1]	Dec 23—BSE case: Canadian cow found in USA		May 18—Cattle tracking & tracing program[2]	Aug 23—8th BSE case	
Aug 12—$36 million extension to federal recovery program[4] Aug 12—Advanced disaster assistance funds announced[4] Sep 23—Alberta BSE recovery program for other ruminants[1,2]			Jul 13—$16 million added to Alberta BSE Recovery Plan[2] Nov—$20 million PrioNet federal research program[6]	Sep 6— Integrated Traceability Program[4]	

[1]Le Roy and Klein (2005), [2]Government of Alberta (2006), [3]CFIA (2006), [4]LeBlanc (2006), [5]CBC (2006), [6]PrioNet Canada (2006), [7]Office of the Prime Minister (2007), [8]Howard (2007), [9]CFIA (2007).

Table 12.4. Timeline and Number of Yearly Government Responses to Bovine Spongiform Encephalopathy Cases in Canada, 2003–2007.

	Risk Management	*Recovery Programs (number and cost)*	*Research Foci*
2003	2 (SRM removal, livestock disease surveillance)	15 ($1 billion) Federal $603.3 million Alberta $179 million Other: $289 million	
2004	2 (import rules, SRM removal)	5 ($1.85 billion) Federal $1.63 billion Alberta $230 million	3 (Pricing, Assistance audit [Alberta], future lessons)
2005	3 (track & trace, fines, import rules)	6 ($1.1 billion) Federal $1.09 billion Alberta $16 million	4 (slaughter capacity, trade, $20 million [PrioNet], $38 million [APRI])
2006	5 (lab, SRM ban, trace, import rule, feed investigation)		
2007	3 (enhanced feed ban, feed investigation, feed ban awareness)	2 ($1.08 billion) Federal 1 billion Federal/provincial: $80 million	

Adapted from: CBC (2006), CFIA (2006), Dunn (2004), Government of Alberta (2006), Howard (2007), LeBlanc (2006), Le Roy and Klein (2005), Office of the Prime Minister (2007).

an expanded comparative study should include disasters that have lower frequency. In Hong Kong, a series of responses had been attempted, and it was these failures and the pressing need to deal with the situation that eventually prompted the adoption of a political economic initiative that was initially scorned, but in a number of ways served to resolve the situation, although serious fires continued afterward. In the BSE case, the series of outbreaks in different countries created a situation where policy learning occurred quietly in Canada and facilitated a quick set of responses after the first domestic case was discovered.

Second, the two cases are very different in many respects. The most significant difference is that the squatter fires occurred within a political territory, while the other case involved an international series of outbreaks. The nature of the actors, particularly the involvement of supranational agencies such as the *Office International des Epizooties*, the consequences, and the policy processes that were possible in a colony versus a number of rich democratic countries, all play into the distinctive dynamics described above.

Third, despite these many differences, the two cases possess some similarity because in both situations powerful external forces could impose sanctions on the governments that in turn influenced their policy directions. For Hong Kong, it was the diplomatic and geopolitical vulnerability that turned the housing need and problems of squatters who occupied much-needed urban space into a major governmental program and continuing set of expenditures. For BSE, the problem was not so much the consequences of the disease itself, but the risk that other countries would close their borders to beef exports that forced rapid response to BSE outbreaks. In both cases, the ability of powerful outsiders to impose serious risks and costs forced learning to occur in a way that long-standing domestic hazards such as monsoon flooding might not. This suggests that it is not simply the economic costs of a series of disasters that motivates institutional learning, but the concentration of these costs in forms that demands the attention of economic and political elites.

Our analyses above focused exclusively on institutional responses to crisis. A broader perspective on disaster sequences and temporality would need to acknowledge that there are other responses to the same crises and thus also different social constructions of the relevant temporal sequence. The temporality of crisis and disasters should not be understood as a singular reality rooted in a series of chronological events; rather there are multiple streams of temporality rooted in the particular experiences among key stakeholders. Whether the pivotal events or crises are defined as part of the history of confrontational social movements or of social progress, organized social agents are the prime movers. Our suggestion in this chapter is that events that are outside of the control of such social actors can drive forward institutional change, and that this is more likely when disasters occur in a series that are close enough in time to each other to make clear the failings of past responses, or the failure of institutions to respond at all.

Acknowledgments

The authors wish to thank the Alberta Prion Research Institute and the Social Sciences and Humanities Council of Canada for funding that support the research and data collection included in this study. Special thanks to the Prion Disease and Social Research Lab at the University of Calgary for the use of Lab resources and databases in the preparation of this manuscript. The reviewers' comments on an earlier draft of this chapter are greatly appreciated and we thank them for their thoughtful and constructive contributions. Vince Terstappen and Elly Carlson were instrumental in the final editorial processes in this creative project.

Notes

1. DCS to CS (12/14/1949). Record ID HKRS163-1-1231. "Fire Services–Squatter Areas–Correspondence re fire precautions to be adopted in. . . ." Archived material, Hong Kong Public Records Office, Hong Kong, China.

2. Chief Officer, Fire Brigade to Colonial Secretary (12/09/1949). Record ID HKRS163-1-1231. "Fire Services–Squatter Areas–Correspondence re fire precautions to be adopted in. . . ." Archived material, Hong Kong Public Records Office, Hong Kong, China.

3. Deputy Colonial Secretary (05/19/1950). Record ID HKRS163-1-1231. "Fire Services–Squatter Areas–Correspondence re fire precautions to be adopted in. . . ." Archived material, Hong Kong Public Records Office, Hong Kong, China.

4. 05/20/1950. Record ID HKRS163-1-1231. "Fire Services–Squatter Areas–Correspondence re fire precautions to be adopted in. . . ." Archived material, Hong Kong Public Records Office, Hong Kong, China.

5. 11/08/1950. Record ID HKRS163-1-1231. "Fire Services–Squatter Areas–Correspondence re fire precautions to be adopted in. . . ." Archived material, Hong Kong Public Records Office, Hong Kong, China.

6. COFB Cox to Colonial Secretary (03/12/1959). Record ID HKRS41-1-8858. "Squatter Fires." Archived material, Hong Kong Public Records Office, Hong Kong, China.

7. COFB Gorman to Colonial Secretary (10/20/1954). Record ID HKRS163-1-1747. "Squatter fire at Li Cheng Uk on 1st October 1954–Reports re. . . ." Archived material, Hong Kong Public Records Office, Hong Kong, China.

8. Record ID HKRS41-1-8858. "Squatter Fires." Archived material, Hong Kong Public Records Office, Hong Kong, China.

9. Governor Grantham to Secretary of State for the Colonies (01/26/1954). Record ID HKRS163-1-1578. "Shek Kip Mei fire–1. early policy decisions (26.12.53–31.12.53) 2. early reports re . . . 3. letters of thanks. 4. miscellaneous general matters." Archived material, Hong Kong Public Records Office, Hong Kong, China.

10. Record ID HKRS163-3-64. "Squatter clearance and resettlement.–1. general questions on . . . 2. programme of." Archived material, Hong Kong Public Records Office, Hong Kong, China.

11. Commissioner for Resettlement to Colonial Secretary (11/26/1954). Record ID HKRS156-1-4429. "Squatter fire at Tai Po road on 20th, 24th and 26th November, 1954–resettlement of persons made homeless by–establishment of Tai Wo Ping resettlement area." Archived material, Hong Kong Public Records Office, Hong Kong, China.

12. Chronic wasting disease was first identified in a herd of farmed elk in Saskatchewan in 1996. Subsequently 40 game farms in Saskatchewan and three in Alberta were found to be infected with chronic wasting disease. Chronic wasting

disease in wild mule deer was first identified in 2000 in Saskatchewan (Bollinger et al. 2004).

References

Alberta Sustainable Development

2008 Chronic Wasting Disease—electronic document, www.srd.gov.ab.ca/fishwildlife/livingwith/diseases/chronicwastingdisease.aspx, accessed May 5, 2008.

Barry, John M.

1998 Rising Tide: The Great Mississippi Flood of 1927 and How It Changed America. New York: Simon & Schuster.

Beasley, A. N.

2006 The Promised Medicine: Fore Reflections on the Scientific Investigation of Kuru. Oceania 76(2):186.

Birkland, Thomas A.

2006 Lessons of Disaster: Policy Change after Catastrophic Events. Washington, DC: Georgetown University Press.

Bollinger, T., P. Caley, E. Merrill, F. Messier, M. W. Miller, M. D. Samuel, and E. Vanopdenbosch

2004 Chronic Wasting Disease in Canadian Wildlife: An Expert Opinion on the Epidemiology and Risks to Wild Deer. Final Report. Electronic document, wildlife1.usask.ca/Publications/CWD%20Expert%20Report%20Final%20-%2020040804.pdf, accessed September 8, 2006.

Campbell, Donald

1993a Entire Herd to Be Destroyed to Prevent Mad Cow Spread. Calgary Herald, December 30:C2.

1993b Killer Mad Cow Disease Strikes in Alberta. Calgary Herald, December 9:D1.

1993c Fate of 270 Cattle Uncertain in Wake Of Mad Cow Alert. Calgary Herald, December 17:C6.

Canadian Food Inspection Agency

2006 Chronology of BSE Events. Electronic document, www.inspection.gc.ca/english/anima/heasan/disemala/bseesb/chronoe.shtml, accessed May 30, 2006.

2007 Canada's new animal feed regulations come into force. Electronic document, www.inspection.gc.ca/english/corpaffr/newcom/2007/20070712e.shtml, accessed July 16, 2007.

Canadian Press

1993 Imported Cattle to Be Killed. Calgary Herald, December 31:C3.

Castells, Manuel, Lee Goh, and R. Yin-Wang Kwok
1990 The Shekkipmei Syndrome: Economic Development and Public Housing in Hong Kong and Singapore. London: Pion.

CBC
2006 CBC News Online Indepth: Mad Cow. Electronic document, www.cbc.ca/news/background/madcow/timeline.html, accessed April 9, 2007.

Clancey, Gregory
2006 Earthquake Nation: The Cultural Politics of Japanese Seismicity, 1868–1930. Berkeley, CA: University of California Press.

Commonwealth of Virginia
2007 What is Chronic Wasting Disease? Electronic document, www.dgif.state.va.us/hunting/cwd.asp#What_is_Chronic_Wasting_Disease_(CWD), accessed September 8, 2006. Also see Canadian Food Inspection Agency website above.

Corbacioglu, Sitki, and Naim Kapucu
2006 Organisational Learning and Self-Adaptation in Dynamic Disaster Environments. Disasters 30(2):212–233.

Deslys, Jean-Philippe, and André Picot
2001 Mad Cow Disease: The Risks to Humans. France: Dominos-Flammarion.

Doughty, Paul L.
1999 Plan and Pattern in Reaction to Earthquake: Peru, 1970–1998. *In* The Angry Earth: Disaster in Anthropological Perspective. Anthony Oliver-Smith and Susanna M. Hoffman, eds. Pp. 234–256. London: Routledge.

Dyer, Christopher L.
2002 Punctuated Entropy as Culture-Induced Change: The Case of the Exxon Valdez Oil Spill. *In* Catastrophe & Culture: The Anthropology of Disaster. Susanna M. Hoffman and Anthony Oliver-Smith, eds. Pp. 159-85. Santa Fe, NM: School of American Research Press.

EUROCJD/NEUROCJD
2007 Electronic document, www.eurocjd.ed.ac.uk/vcjdworldeuro.htm, accessed July 20, 2007.

Garcia-Acosta, Virginia.
2002 Historical Disaster Research. *In* Catastrophe & Culture: The Anthropology of Disaster. Susanna M. Hoffman and Anthony Oliver-Smith, eds. Pp. 49–66. Santa Fe, NM: School of American Research Press.

Government of Alberta
2006 BSE in Alberta—A Timeline. Electronic document, www.gov.ab.ca/home/index.cfm?Page751, accessed May 4, 2006.

Government of Saskatchewan

2008 Chronic Wasting Disease. Electronic document, www.environment.gov.sk.ca/adx/aspx/adxGetMedia.aspx?DocID=1766,300,254,94,88,Documents&MediaID=978&Filename=Summary+of+the+Provincial+CWD+Sampling+Effort+1997-2008.pdf, accessed May 5, 2008.

Grier, Kevin

2005 Analysis of the Cattle and Beef Markets Pre and Post BSE: Final Report to the Competition Bureau. Guelph, ON: George Morris Centre.

Howard, J.

2007 $39.8 million to implement enhanced feed ban in Alberta. Electronic document, www.agr.gc.ca/cb/index_e.php?s1=n&s2=2007&page=n70312a, accessed July 9, 2007.

Le Roy, Danny G., and K. K. Klein

2005 Mad Cow Chaos in Canada: Was It Just Bad Luck or Did Government Policies Play a Role? Canadian Public Policy 31(4):381–399.

LeBlanc, Marc

2006 Chronology of BSE-Related Events and Government Initiatives. Ottawa: Parliamentary Information and Research Service, Library of Parliament.

Leiss, William, and Douglas Powell

2004 Mad Cows and Mother's Milk: The Perils of Poor Risk Communication. Montreal: McGill-Queen's University Press.

Lindenbaum, S.

2001 Kuru, Prions, and Human Affairs: Thinking About Epidemics. Annual Review of Anthropology 30:363–85.

McCabe, J. Terrence

2002. Impact of and Response to Drought among Turkana Pastoralists: Implications for Anthropological Theory and Hazards Research. *In* Catastrophe & Culture: The Anthropology of Disaster. Susanna M. Hoffman and Anthony Oliver-Smith, eds. Pp. 213–236. Santa Fe, NM: School of American Research Press.

Mitchell, James K.

1999 Natural Disasters in the Context of Mega-Cities. *In* Crucibles of Hazard: Mega-Cities and Disasters in Transition. James K. Mitchell, ed. Pp. 15–55. New York: United Nations University Press.

Office of the Prime Minister.

2007 Prime Minister announces $1 billion for Canadian farmers. Electronic document, www.pm.gc.ca/eng/media.asp?category=1&id=1566, accessed July 9, 2007.

OIE
2007 Electronic document, www.oie.int/eng/info/en_esbmonde.htm, accessed July 20, 2007.

Oliver-Smith, Anthony
1999 What Is a disaster? Anthropological perspectives on a persistent question. *In* The Angry Earth: Disaster in Anthropological Perspective. Anthony Oliver-Smith, Susannah M. Hoffman, eds. Pp. 18–34. New York: Routledge.

Oliver-Smith, Anthony, and Susanna M. Hoffman
2002 Introduction: Why Anthropologists Should Study Disasters. *In* Catastrophe & Culture: The Anthropology of Disaster. Susanna M. Hoffman and Anthony Oliver-Smith, eds. Pp. 3–22. Santa Fe, NM: School of American Research Press.

Packer, Richard
2006 The Politics of BSE. New York: Palgrave Macmillan.

PrioNet Canada
2005/2006 Annual Report 2005/2006. Electronic document, www.prionet canada.ca/landing.aspx?landing=About&menu=2&app=100, accessed August 25, 2007.

Richardson, Bonham C.
2004 Igniting the Caribbean's Past: Fire in British West Indian History. Chapel Hill, NC: University of North Carolina Press.

Romahn, Jim
1993 Alberta Herd Quarantined: Mad Cow Disease Found. Kitchener-Waterloo Record, December 9:B6.
1993 Mad Cow Disease Creates Problems for Area Exporters. Kitchener-Waterloo Record, December 14:B7.

Rosen, Christine Meisner
1986 The Limits of Power. Cambridge: Cambridge University Press.

Smart, Alan
1989 Forgotten Obstacles, Neglected Forces: Explaining the Origins of Hong Kong Public Housing. Environment and Planning D. Society and Space 7:179–96.
2006 The Shek Kip Mei Myth: Squatters, Fires and Colonial Rule in Hong Kong, 1950–63. Hong Kong: Hong Kong University Press.

Smart, Alan, and James Lee
2003 Financialization and the Role of Real Estate in Hong Kong's Regime of Accumulation. Economic Geography 79(2):153–171.

Suen, Michael Ming-Yueng

2003 Foreword. *In* Fifty Years of Public Housing in Hong Kong: A Golden Jubilee Review and Appraisal. Y. M. Yeung and Timothy K. Y. Wong, eds. Pp. xvii–xviii. Hong Kong: Chinese University Press.

Zaman, Mohammad Q.

1999 Vulnerability, Disaster, and Survival in Bangladesh: Three Case Studies. *In* The Angry Earth: Disaster in Anthropological Perspective. Anthony Oliver-Smith and Susanna M. Hoffman, eds. Pp. 192–212. London: Routledge.

"Hurricanes Did Not Just Start Happening": Expectations of Intervention in the Mississippi Gulf Coast Casino Industry

13

JENNIFER TRIVEDI

On August 29, 2005, Hurricane Katrina made its third landfall striking the American Gulf Coast, near the Louisiana-Mississippi border (Graumann et al. 2005:2). Biloxi, Mississippi, including the thriving casino industry that largely supported the local economy, was hit hard by the storm. Thousands were affected as lives, homes, and livelihoods were destroyed. In the aftermath of the storm, the *Sun Herald* (the area's local newspaper) distributed 80,000 free papers daily over six weeks (Boyer 2006). *Sun Herald* articles framed economic discussions directly related to the impact of local casinos (both proven and potential) not primarily as an issue of sinfulness, but rather one of "social responsibility."

Casinos are important because of their primacy within the local economy—compared to shrimping and Keesler Air Force Base—and because casinos have a prominent place in debates about morality of business at the local and national level. This chapter is a study of how the *Sun Herald* portrayed the plight of the casino industry vis-à-vis Hurricane Katrina and other hurricanes.

Framework and Study Design

This study is based on three data collection methodologies: text analysis, semiformal interviews and formal interviews via mail and telephone, and participant observation. I analyzed 139 newspaper articles from the Biloxi *Sun Herald* dealing with both hurricanes and casinos appearing between

May 1, 2003 and September 30, 2006. I also conducted more cursory work with other articles, editorials, and Letters to the Editor. I chose to analyze the *Sun Herald* because of its role as a key local news source. During the week following Katrina, the newspaper had a sister newspaper produce a reduced run for it, and coverage was available online (Moozakis 2006). While further analysis of how audiences read these articles would complement this study, generally, the opinions of community members seemed to parallel frames presented by the articles.

Semiformal and formal interviews were conducted as mail-response surveys with a variety of officials and businesspersons during the summer of 2006, and followed up with phone discussions with some individuals between August 2006 and February 2007. I conducted fieldwork for six weeks in June and July 2006 in Biloxi, Mississippi. While there, I conducted participant observation within the community, paying special attention to public meetings related to Katrina and the city's future. I spent time in establishments frequented by locals including grocery stores, popular restaurants, and casinos.

My analysis of articles appearing in the *Sun Herald* between 1996 and 2006 revealed an increase in discussion of casinos and hurricanes in May 2003 following comments made by the then-chairman of the Mississippi Gaming Commission regarding the need to protect casinos from hurricanes. September 30, 2006 was selected as a concluding date in order to allow for articles related to the one-year anniversary of Katrina to appear. Each of the 139 articles that I analyzed discuss both casinos *and* hurricanes (see table 13.1). These fall into a broader range of articles regarding *either* casinos *or* hurricanes. During this same time, approximately 823 articles dealt solely with the issue of casinos and 2,185 solely with hurricanes (excluding the 139 here).

My analysis is based on Druckman's (2001:1042) framing effect where the speaker emphasizes a subset of an issue's characterizations, causing the listener (or reader) to focus on those particulars in creating their own opinions. In my research, the speaker is an article's author via the *Sun Herald*, emphasizing particular frames to the audience. A study of framing helps us understand the influence of stories on public attitudes and on governmental and employer policies (Druckman 2001:1043–1044; see also Goffman 1981:257). For my purpose, words like "vulnerable" are signs of a social responsibility frame.

Framing not only depends on the wording of the article but also on sources employed by journalists. The vast majority of sources used in these articles had ties to casinos (e.g., employees, etc.) or the state or local government (e.g., the Mississippi Gaming Commission, state senators, etc.). People less directly related to these organizations (e.g., gambling analysts) were also cited. Average citizens were less frequently cited and usually in the context of their own gaming. Religious organizations and representatives were also less frequently cited or quoted. Many articles were written by a small number of reporters.

Trends in media sources revealed by framing analysis allow for a consideration of disaster and vulnerability as socially influenced *processes*, not isolated events. A fundamental component to framing risk and recovery is that the environment is separate from humans, resulting in disasters being perceived as isolated events, not part of ongoing processes of vulnerability (cf. Oliver-Smith 2002:32). These articles reveal vulnerability as long-term economic processes citizens expect the government to take steps to prevent. I have followed Oliver-Smith's (2002:28) use of Blaikie et al.'s definition of vulnerability as:

> the characteristics of a person or group in terms of their capacity to anticipate, cope with, resist, and recover from the impact of a natural hazard. It involves a combination of factors that determine the degree to which someone's life and livelihood is put at risk by a discrete and identifiable event in nature or in society.

Throughout my stay in Biloxi, the role of the *Sun Herald* in community discussions about casino vulnerability became clear. Local residents discussing casinos often predicated their claims with statements that what they were about to say was influenced by (or even pulled directly from) *Sun Herald* articles. While standing near a *Sun Herald* sales case, I overheard one middle-aged woman asking others in her group to "wait up a minute" as she wanted "to buy the paper to figure out what was going on with the whole casino thing."

Comments of individuals revealed the interaction present between them and the media. This was not a simple process of citizens simply absorbing whatever the media told them. Rather, there seems to be an active interaction between people and the media as individuals selected what in the media they listened to and then interpreted through personal circumstances. Individuals seemed to consider the casino issue not simply a case of religious morality versus casinos' economic profits, but rather as related issues on a spectrum of ideas about how casinos should care about the community.

The articles reviewed in this study generally fall into categories of casinos as framed by: 1) social responsibility, 2) sinfulness, or 3) other (see table 13.1).

Table 13.1. Articles Featuring Frames of Social Responsibility, Sinfulness, or Neither

Frame	*Number of Articles*	*Percent of Total Articles*
Social responsibility	86	61.9%
Sinfulness	5	3.6%
Other	48	34.5%

I define social responsibility as citizens' understandings of governments' and employers' responsibility to protect citizens, communities, and underlying economic structures, such as taxation systems, opportunities for ongoing or stable employment, and the general economic stability of the area.[1] In this case, benefits for the casino industry ultimately help residents rebuild through economic benefits like continued employment and taxes. During the summer of 2006, a number of local residents used arguments framed in terms of social responsibility to support casinos, arguing that the industry played an important role economically (through jobs and taxes) and were thus essential to rebuilding. These claims support the idea that casinos are important due to their primacy in the local economy, demonstrating with social responsibility a discussion regarding the morality of business at the local and national levels. Social responsibility may be subdivided into 3 subcategories of public expectations about government and employers: 1) responsibility to maintain local economic viability (including reducing economic vulnerability and rebuilding costs); 2) planning and foresight to avoid economic disruptions (particularly as related to taxation and employment); and 3) expectation of consideration of local history and culture related to economic decisions.

Articles that apply the frame of sinfulness present casinos as immoral or sinful. Those articles that simply mention "God" are not considered. This frame focuses the discussion on a portrayal of casinos as sinful which, for many, equates to going against God's will. While I believe that in some situations this may extend to other religions, here the primary objections seem centered in Christianity and frequently as a sentiment of evangelicalism. In this context, behavior that goes against biblical and church teachings, then, is "sinful" and depictions of casinos and gambling as such constitute the frame I refer to here as "sinfulness."

The "other" category here includes a variety of articles which fall into two primary subcategories. One simply mentions terms like "casinos" or "Katrina" in relation to where or when a meeting occurred or where a person worked and when they died. The other subcategory deals with economics, but in a way different from those articles classified as social responsibility. For example, some of these deal with the amount of funds garnered by casinos. Rather than dealing with these funds in terms of taxability or employment base vis-à-vis the community, these other articles often describe casinos in terms of economic output or profits.

Background

The focus on social responsibility in this chapter is better understood when contextualized by knowledge of the long local history of illegal gambling prior to casinos. Biloxi is located in Mississippi's Harrison County on the Gulf of Mexico. The city has just over 50,000 residents, and population estimates from January 2005 and 2007 indicate similar population sizes for the area before and after Katrina (U.S. Census Bureau 2000; Arbitron, Inc. 2006:8). Casinos are a crucial part of one of Biloxi's three economic mainstays: tourism. The other major economic sectors are Keesler Air Force Base and the seafood industry (Community Development Department 1999:28). Throughout the 1990s and early 2000s, casinos became the economic engine for Biloxi and Harrison County (Brinkley 2006:82).

The final report of the Mississippi Governor's Commission on Recovery, Rebuilding, and Renewal (hereinafter GCRRR) points out the centrality of tourism, and casinos in particular, to the economy, society, and culture of the Mississippi Gulf Coast. Thirty-seven percent of the 30.7 million visitors (over 11 million) to Mississippi in 2004 went to the Gulf Coast, spending $1.9 billion (GCRRR 2005:78, 81). Prior to Katrina, 22 percent of jobs available in the three coastal counties were classified as part of the leisure/hospitality sector—including jobs related to casinos (GCRRR 2005:83). This percentage was about twice the state average and more than twice the national average (GCRRR 2005:83). In Harrison County, jobs tied to tourism made up 28.7 percent of all jobs in the county (GCRRR 2005:83).

The degree to which community control can be exerted over local economic mainstays—casinos, Keesler, and the seafood industry—varies. While clearly a part of the community, Keesler is ultimately controlled by the federal government. In the seafood industry, community members control the work regimen, but may feel little control over sales or prices of catches. In comparison, casinos are seen as local employers, such that labor relations and working conditions become the focus of community interest (although that does not necessarily translate into local control). Conceptualizing these issues as part of a broader decision-making process by the community framed by the media reveals the importance of the relationship between different economic systems and issue of cultural values they may present.

Gambling became widespread in Biloxi as early as the 1920s when, despite illegality, slot machines were used openly in hotels and clubs (Nuwer 2005). By the 1950s the legalization debate began, along with local media coverage. The Ministers Association lobbied to keep gambling illegal and crack down on gaming; however, there was support for legalization, demonstrated in the *Daily Herald* as resident Paul Skrmetti countered:

> I think if a person wants to play [a] machine or any other game[,] it should be their privilege and no one should have a right to prevent others from enjoying themselves (Nuwer and O'Brien 2006:20).

Discussions among local residents in the summer of 2006 reveal this history of illegal gambling was by no means hidden. A number of residents stated they knew there was a history of gambling in the area long before legalization, and several people remembered illegal slot machines in many businesses—"everywhere but clothing stores," as one local resident described. The history of gambling in the area is a part of local knowledge that informs ideas about current gambling.

When Mississippi legalized gambling in the 1990s, it was seen as a means to stimulate the tourism-based economy suffering since 1969's Hurricane Camille (Nuwer 2005; see also Nelson and Lyman Mason 2006:27). Casinos were described by residents as the key component of *finally* rebuilding local economic viability after Camille, and were seen as key to rebuilding post-Katrina. The legality of casinos was made conditional upon the fact that they be on water, and that surrounding communities approve. In 1992, the Isle of Capri opened, named for the Isle of Caprice (a site of illegal gambling in the 1920s) in an attempt to tie the company to community history (Mapp 2003). The emergence of the legal casino industry has been referred to as "a renaissance," sparking not only some level of economic prosperity but also a renewed sense of pride (Holloway 1999:2). Casinos were presented as a means to improve the local economy, and reducing their vulnerability to hurricanes appears as a way to ensure continued economic viability. The *Sun Herald* discussed vulnerability in these terms, marking casinos as legitimate local business:

> moving Coast casinos onto land, out of the direct path of hurricanes, is no different than the city of Biloxi working to ensure that tall buildings don't encroach on Keesler Air Force Base's flight path (Boone 2003a:A1).

The comparison was also made by the then-state gaming chairman:

> they're sensitive about the economic value of Keesler because there are 15,000 jobs over there [. . .]. We're sensitive about the thousands of jobs created by the gaming industry (Boone 2003a: A1).

Katrina's widespread destruction brought into question the future of casinos. Other hurricanes have forced casinos to close for days or weeks. Some suffered repeated damage from hurricanes, such as Treasure Bay. However, most casinos survived these storms relatively unscathed, leading to claims that they had proved their ability to withstand hurricanes.[2] The *Sun Herald* described how:

> casino executives said they anxiously watched Georges' approach since they had never had the opportunity to see whether the casinos were hurricane-proof as required. All of them passed (Bridges 1998).

Discussions of casino vulnerability increased substantially in spring 2003 when then-chairman of the Mississippi Gaming Commission, Leonard "Len" Blackwell III, commented on the situation[3]:

> it's time to become amphibious [. . .]. There are definite safety drawbacks to the requirement that casino vessels float (Boone 2003a:A1).

Since legalization, Mississippi had become the third largest American gambling market, behind only Las Vegas and Atlantic City (Federal Reserve Bank of Atlanta 1996). Vincent Creel, then-spokesman for Biloxi, explained that 35 percent of the city's budget comes from gaming taxes. Gaming Commission Chairman Blackwell said moving casinos onto land was in the state's best economic interests and failure to do so would result in the public blaming the Gaming Commission (as a part of the government) for not doing their part to reduce casino vulnerability (Tortorano 2003:C9). Such sentiments were echoed by residents during the summer of 2006 and many commented that the government's steps to move casinos onto land were important to protect taxes and jobs (Mississippi Gaming Commission 2007). These comments further support the centrality of the casino industry to the local economy, insofar as individuals saw casinos as key to employment.

In January 2004, discussion began of a hypothetical major hurricane that could shut down the entire casino industry, crippling the government for lack of funds. There was local and widespread awareness that storms that had hit since casinos opened largely lacked the power of storms like Camille:

> the destruction along the Mississippi Gulf Coast from Ansley to Biloxi was incredible, as antebellum homes, restaurants, motels, apartments, schools, churches, and other structures were swept off their foundations and deposited in mountains of rubble together with trees and automobiles. The atomic-bomb effect of Camille's 200-mile-per-hour wind gusts and 25-foot storm surge destroyed 100 years of growth and progress along the Mississippi Gulf Coast in just three hours (Hearn 2004:124).

With the potential for repetitive destruction, an ongoing debate regarding protecting this key economic component seems less surprising. Discussion in 2004 focused on not moving casinos on land, but protecting them in place. In early 2005, the legislature allowed casinos to anchor to concrete pilings, similarly to oil rigs.

Initially, casinos were not allowed on land as they were seen as a greater moral threat if not buffered by water. Putting casinos on water was "a concession to the political perils of trying to legalize gambling deep in the Bible Belt" (Parker 2005:3A). After Katrina, Mississippi state law changed to allow casinos to rebuild on land, citing concerns that otherwise the casinos would refuse to rebuild in the area or would be destroyed again by the next hurricane. These fears illustrate the centrality of casinos to the local economy.

Prior to Katrina there were nine operating casinos in Biloxi and another about to open; all were affected by the storm, ranging from moderate interior damage to total destruction. In Katrina's aftermath, as featured in *Sun Herald* articles, some felt there were enough potential profits to guarantee casinos' quick returns. At the beginning of 2006, Biloxi Mayor A. J. Holloway claimed the gambling industry would be "the tide that lifts all the boats in the city of Biloxi," thus clearly revealing the portrayed role of the casinos in the area (Wilemon 2006a:A1).

The state legislature debated in September and October 2005 about conditions under which casinos would be willing to rebuild. After extended discussion and a close vote, casinos were allowed to rebuild on land, providing they remained within 800 feet of the same waterways to which they were previously restricted (Pender and Wilemon 2005:A4). Several coastal senators equated allowing casinos to rebuild on land with the region's continued economic viability. Katrina's impact seems to have forced the Mississippi legislature to come to a decision in previously drawn-out debates about moving casinos onto land. While this move can be framed as decreasing vulnerability, this is not to say such a move objectively *eliminates* vulnerability.

Framing the Debate in the *Sun Herald*

Throughout the time considered here, articles framing the discussion for casinos and hurricanes in terms of social responsibility far outweigh those framing it in terms of sinfulness (see table 13.2). Of 37 articles on casinos and hurricanes in the *Sun Herald* between May 1, 2003 and August 28, 2005 when Katrina hit, 83.3 percent framed the discussion in terms of social responsibility, while 2.7 percent framed it in terms of sinfulness. This leaves 13.5 percent that feature neither frame (e.g., "God" as an exclamation without reference to sin, how much money casinos make without reference to community benefit, or discussion of measures taken to protect casinos but not as action necessary by government or corporation).

Between Katrina and the law change allowing casinos to rebuild on land (i.e., August 29, 2005 and October 16, 2005), 19 articles appeared: 73.7 percent framed as social responsibility, 10.5 percent framed as sinfulness, and 15.8 percent of neither frame. For almost a year after casinos were allowed to rebuild on land, from October 18, 2005 to September 30, 2006, articles featuring social responsibility reduced to 49.4 percent, while those featuring neither frame increased to a whopping 48.2 percent. Those featuring sinfulness returned to a low 2.4 percent.

The following passage places social responsibility (represented by the need for state aid and incentives to keep casinos in the area) in sharp contrast with the frame of sinfulness:

> propositions of state aid and incentives to the industry and regulations changes such as allowing them to build on land, even after Katrina's devastation, would probably still be a hard sell. The Bible Belt state's uneasy relationship with gambling and the Coast's still relatively weak power at the Capitol could prevent such measures. But such formerly taboo political issues will at least be broached, lawmakers said. (Pender and Wilemon 2005:A4)

Table 13.2. Articles Featuring Frames of Social Responsibility, Sinfulness, or Neither in Relation to When They Were Published

Frame	*Before Katrina*	*Between Katrina and the Law Change*	*After the Law Change*
Social Responsibility	31	14	41
Sinfulness	1	2	2
Other	5	3	40
Total	37	19	83

This chapter demonstrates what is perhaps more obscure in other studies—the contrast between the two frames, as well as the idea that one is used to counter the other. While not articulated in so many words, this theme continued among the members of the general public with whom I spoke in 2006. However, it seems difficult to discern if the media reflects public sentiment or if public sentiment reflects media coverage. There is, clearly, an interaction wherein individuals discussed these issues in dynamic ways. Some individuals went on to ask direct questions of the contrast between economics and religion.

Before Katrina (August 29, 2005), articles featuring social responsibility were more common than after the law changed allowing casinos on land (as of October 17, 2005; see table 13.2). After the law changed, the decrease in articles framing the discussion around social responsibility may indicate that people felt there was no longer a need for debate since land-based casinos were now a fact of life, that people agreed with the portrayal of casinos as no longer vulnerable, or acceptance by the community that the government has fulfilled its obligations under social responsibility to protect the casinos and community economic viability. Thus, the issue of the vulnerability of casinos and local economic viability seems almost forgotten. Several community members commented they thought moving the casinos onto land would help rebuild casinos and provide increased jobs and tax funds security to help the community. A number of people qualified this as a good move by the state government.

Sinfulness and Social Responsibility: A More Detailed Examination

As previously mentioned, there are three subcategories of public expectations of government and employers in the frame of social responsibility: (1) maintenance of economic viability; (2) planning ahead; and (3) consideration of cultural impacts. Examples of each are in the *Sun Herald* articles considered here. Close examination will tell us something of how people look to structure and restructure their local economy given the threat or impact of disasters.

A *Sun Herald* article, published in 2004, considers Len Blackwell's push to move casinos onto land; "Blackwell worries that one strong hurricane could knock out almost half of the state's casino revenues" (Wilemon 2004a:A1). Such an article characterizes the first subcategory of social responsibility insofar as the government is expected to maintain economic viability and reduce vulnerability. This argument was presented by several local citizens in the Biloxi area during my time there in the summer of 2006 as a counterargument to the idea that casinos are sinful. Furthermore, the comment demonstrates the importance of casinos to the local economy, showing the depth of concerns related to the threat of lost casinos.

In 2006, the *Sun Herald* ran an article featuring the second subcategory of planning ahead. Leigh Anne Biggs, director of the Committee for Long Beach (and as such, a representative of local government), was quoted:

> the city was having financial troubles before the storm. . . . When things got devastated and things got destroyed, you have to look around and say what are you going to do. This is our opportunity now to say yes to gaming (quoted in Wilemon 2006b:A1).

Finally, while the frame of sinfulness appears in relatively few articles, they are nevertheless important. One September 2005 article began:

> Rep. Jim Simpson of Pass Christian couldn't believe the e-mails he received from some self-proclaimed Christians this week attempting to lobby against casinos rebuilding after Hurricane Katrina. "I've gotten more than half a dozen from people who said, 'I prayed for the destruction of Katrina. This is God's wrath,'" said Simpson. "I got so mad I wanted to scream at them. But I didn't respond. Not yet." (Pender 2005a:A4)

The same article later offers a more positive example wherein:

> the Rev. Mickey Dalrymple of Columbus cornered Rep. Steve Holland, D-Plantersville, near the Capitol elevator Thursday. "Sir, we want you to stand strong against the casinos," the preacher told Holland. "Don't let the camel's head under the tent." Holland replied: "Sorry, Parson, I just don't agree with you. I'm for lettin' them up on shore. I just don't see the harm. But I respect your position. Pray for me." Dalrymple said: "I understand. We love you, either way" (Pender 2005a:A4).

Religious opposition to casinos is underscored by individuals associated with Christian religious organizations. For example, in response to the question in a survey with the author "have the casinos changed the discussion or morality in Biloxi," one religious opponent replied that "casinos are associated with excessive drinking, drugs, prostitution, organized crime, addicted gamblers, bankruptcies, pawn shops, et cetera. YES—there is a moral downside to casinos."

Similarly, in June 2006, the *Sun Herald* featured an article discussing a vote to allow casinos to build on land in nearby Long Beach, Mississippi. The article points out that

> most of the people who voted against it said they did so out of moral obligations. 'It's against my Christian values,' said Angela Taylor, adding that even if she was not morally opposed, she still would not vote for gaming in Long Beach. 'It's a quick fix.' (quoted in Norman 2006:A1).

Some articles present religious groups as leading the call to block casinos from land; for example, "the Mississippi Baptist Convention, with a membership of 2,100 churches and more than 718,000 people, is lobbying lawmakers to 'oppose all measures that would benefit the gambling industry'" (Pender 2005b:A6). By framing religious opposition as a part of ongoing governmental debates, these articles underscore the role of government policy in determining casino vulnerability and viability.

In an October 2005 article, Sen. Trent Lott (R-MS) was quoted on his agreement to provide federal tax cuts "to area casinos as it would any other business, pointing out that 'I'm saying that as a Southern Baptist,' he said, 'but also as someone who understands that [the casino] industry employed 17,000 people in the Gulf Coast'" (Nanos 2005:A4). By presenting Lott's statements thusly, the article ultimately placed social responsibility and sinfulness in a dichotomous position.

The Future

Hurricanes are environmental phenomena that affect economic viability. The assumption now is that moving casinos on land will make them less vulnerable to hurricanes. However, illegal gambling establishments wiped out by Hurricane Camille were in many of the same areas as Katrina-damaged casino hotels and parking lots, where casinos are now rebuilding on land.[4] In an area routinely affected by hurricanes, smaller episodes between major hurricanes like Camille and Katrina offer opportunities for complacency.

Casino vulnerability is a concern because it increases *community* vulnerability. The casinos' and government's economic obligations to the community seem essential to the public. By discussing casinos as economic investments in the community (jobs or taxes) rather than as company profits, the debate focuses on casinos as part of the community, rather than as corporations. Community discussions emphasize casino protection as part of the economy alongside important issues like housing, local history, and the cultural importance of the seafood industry. Such discussions seem to act as conversations between local media and the public, demonstrating how the public is not simply repeating what the media tells them, nor is the local media simply repeating national news coverage. Additionally, subdivisions of the social-responsibility frame offers insight into how the Biloxi community negotiates the impact of disasters on the local economy.

Oliver-Smith and Hoffman (2002:6) point out that disasters reveal how people frame the disaster—specifically through their perceptions and explanations—and how they look to the future. The sheer variety of ideas held by local people about community can allow a range of people to claim to speak for the community, although some will be seen as more legitimate than others (Olson 2005:253). Thus, the frame of social responsibility appears implicated by the *Sun Herald* as more legitimate than the other frames.

Articles presenting losses that casinos and governments could experience due to closure (temporary or permanent) existed before and after the storm. Prior to Katrina, these discussions center primarily on temporary closures for repair or of a hypothetical major hurricane. The state of Mississippi and local governments would lose at least $385,747 each day that Coast casinos:

> are closed because of a hurricane, according to a report presented Wednesday to the Mississippi Ad Hoc Gaming Committee. A sizable portion of the 16,643 people who work in the casino industry also could be without paychecks (Wilemon 2004b:B7).

After Katrina, discussions focused on previous and potential future economic contributions by casinos. References were made to contributions prior to Katrina:

> "The loss is going to clip us a direct half-million dollars in taxes to the state a day, and a tremendous loss in casual sales tax along with that," said House Gaming Chairman Bobby Moak, D-Bogue Chitto. "But then you're also looking at the 16,000 jobs" (Pender and Wilemon 2005:A4).

Spitulnik (1993:293–294) states that mass media "are at once artifacts, experiences, practices, and processes" that "represent and shape cultural values." Thus, analyzing mass media from an anthropological perspective allows for an analysis of, as Spitulnik describes, "forces that provide audiences with ways of seeing and interpreting the world, ways that ultimately shape their very existence and participation within a given society"

(1993:293–294). I believe there is ongoing interaction between the community and the media, each shaping the other. Given the influence of the mass media and the effect of disasters on culture, it seems that a consideration of both from an anthropological perspective could contribute not only to anthropology as a whole, but specifically to the subfield of disaster studies within anthropology.

My focus here on perceptions of casinos as an economic force in the community follows a long record of anthropologists considering the role of the economy. However, by contrasting this view with one of the role of religion in the same decision-making process and by conceptualizing the entire debate as framed in the media, my work here offers a means of considering vulnerability as part of cultural processes. While anthropologists study the everyday lives of many cultures and communities, it is important to include studies on everyday perspectives of both what are often seen as the extraordinary (the impact of hazards such as hurricanes) and of the vulnerabilities that contribute to disasters (ongoing discussions of the local Biloxi economy) as part of these studies. Social responsibility is innately tied to *local* understandings of vulnerability and local expectations of the degree to which government and local employers are responsible for the protection of citizens from the impact of disasters. Thus, it is essential to understand and consider local social responsibility when investigating disasters.

Acknowledgements

Thank you to my Master's Committee Virginia Dominguez, Michael Chibnik, and Meena Khandelwal. Thanks also to Amish Trivedi, Juli Williams, Cerisa Reynolds, Christina Ortiz, and the Skrmetti Family (especially John Patrick, Lehwmon, John, Joshua, Jacob, Peter, Jean, Margaret, Thomas, Ann Renee, and Thomas Sr.).

Notes

1. There is relatively little mention of nongovernmental organizations in the articles considered here, but it is conceivable that they may be understood similarly.
2. This may be due to a variety of reasons including the casinos' location, as well as the speed and direction of the impact of the storm, etc.
3. This commentary may have been sparked by a seeming need to expand coastal casinos, due to the relative profit margins of those and river casinos, given that "Coast casino winnings increased a modest 2 percent during the first three months of the year, and revenues for the rest of the state dropped by 2 percent as

the sluggish national economy continued to take a toll on the gambling industry" (Boone 2003b:C8). Critiques of the state's decision to allow casinos emerged from people such as University of Mississippi professor Denise von Herrmann, who argued that "state legislatures, including Mississippi's, weren't asking the important questions when they were discussing legalizing casino gambling in the early 1990s" (Boone 2003c:C8).

4. Most of the area contained within the 800-foot rule seems to be within hurricane Evacuation Zone A, the area most prone to hurricane damage and flooding.

References

Arbitron, Inc.

2006 Gulf Coast White Paper: Reestablishing Arbitron Surveys in Biloxi-Gulfport-Pascagoula and New Orleans. Electronic document, www.arbitron.com/downloads/2006GulfCoastWhitePaper.pdf, accessed February 5, 2007.

Blaikie, Piers, Terry Cannon, Ian Davis, and Ben Wisner

1994 At Risk: Natural Hazards, People's Vulnerability, and Disasters. New York: Routledge.

Boone, Timothy

2003a Mississippi Gaming Chairman Says State Ready for Amphibious Casino Industry. Sun Herald, May 8:A1.

2003b Coast Casino Profits Up a Bit. Sun Herald, April 25:C8.

2003c Professor: State Rushed Into Casino Market. Sun Herald, April 11:C8.

Boyer, Mark

2006 The Newspaper that the National Media Forgot? Electronic document, www.cjrdaily.org/behind_the_news/the_newspaper_that_the_nationa.php, accessed November 11, 2006.

Bridges, Tyler.

1998 Biloxi, Miss., Casinos Await Approval to Reopen After Georges. Sun Herald, September 30. LexisNexis, web.lexis-nexis.com/universe, accessed August 30, 2007.

Brinkley, Douglas

2006 The Great Deluge. New York: William Morrow (Harper Collins Publishers).

Community Development Department

1999 General Market Analysis (Biloxi, Mississippi). Electronic document, www.biloxi.ms.us/development/marketinganalysis/biloxibook.pdf, accessed November 10, 2006.

Druckman, James N.
2001 On the Limits of Framing Effects: Who Can Frame? The Journal of Politics 63(4):1041–1066.

Federal Reserve Bank of Atlanta
1996 Mississippi Set to Emerge from Its Post-Boom Hangover; Regional Update 9(4). Electronic document, wwwfrbatlanta.org/invoke.cfm? objectid=572D8943-8FE1-11D5898000609459DBE6&method=display, accessed on July 17, 2006.

Goffman, Erving
1981 Forms of Talk. Philadelphia: University of Pennsylvania Press.

Governor's Commission on Recovery, Rebuilding, and Renewal
2005 After Katrina: Building Back Better Than Ever. December 31, 2005. Electronic document, www.governorscommission.com/final/Main.asp, accessed February 5, 2007.

Graumann, Axel, Tamara Houston, Jay Lawrimore, David Levinson, Neal Lott, Sam McCown, Scott Stephens, and David Wuertz
2005 Hurricane Katrina: A Climatological Perspective, Preliminary Report. Technical Report, 2005-01. Asheville, NC: U.S. Department of Commerce, NOAA/NESDIS.

Hearn, Philip D.
2004 Hurricane Camille: Monster Storm of the Gulf Coast. Jackson, Mississippi: University Press of Mississippi.

Holloway, A. J.
1999 Introductory Letter to the Reader. *In* 1999 General Market Analysis (Biloxi, Mississippi). Electronic document, www.biloxi.ms.us/development/marketinganalysis/biloxibook.pdf, accessed November 10, 2006.

Mapp, Christopher
2003 Customer Satisfaction Key to Success in Gaming Industry, Casino Executive Tells Southern Miss Students. Electronic document, www.usm.edu/pr/prnews/feb03/gallaway03.htm, accessed December 5, 2006.

Mississippi Gaming Commission
2007 History. Electronic document, http://www.mgc.state.ms.us/, accessed February 19, 2007.

Moozakis, Chuck
2006 The Untangled Web. Electronic document, www.newsandtech.com/issues/2006/05-06/nt/05-06_moozakis.htm, accessed November 11, 2006.

Nanos, Brian P.
2005 Treasury Chief Calls Business-Tax Relief 'Surest Way to Rebuild' Gulf Coast. Sun Herald, October 7: A4

Nelson, Michael and John Lyman Mason

2006 Mississippi: The Politics of Casino Gambling. *In* Resorting to Casinos: The Mississippi Gambling Industry. Denise von Herrmann, ed. Pp. 26–46. Jackson, MS: University Press of Mississippi.

Norman, Joshua

2006 Long Beach Approves Casinos: Outside Polls, Battle Goes On. Sun Herald, June 28: A1.

Nuwer, Deanne

2005 Gambling in Mississippi: Its Early History. Electronic document, mshistory.k12.ms.us/features/feature61/gambling.htm, accessed August 28, 2006.

Nuwer, Deanne Stephens, and Greg O'Brien

2006 Mississippi's Oldest Pastime. *In* Resorting to Casinos: The Mississippi Gambling Industry. Denise von Herrmann, ed. Pp. 11–25. Jackson, MS: University Press of Mississippi.

Oliver-Smith, Anthony

2002 Theorizing Disasters: Nature, Power, and Culture. *In* Catastrophe and Culture: The Anthropology of Disaster. Susanna M. Hoffman and Anthony Oliver-Smith, eds. Pp. 23–48. School of American Research Advanced Seminar Series. Santa Fe, NM: School of American Research Press.

Oliver-Smith, Anthony, and Susanna M. Hoffman

2002 Introduction: Why Anthropologists Should Study Disasters. *In* Catastrophe and Culture: The Anthropology of Disaster. Susanna M. Hoffman and Anthony Oliver-Smith, eds. Pp. 3–22. School of American Research Advanced Seminar Series. Santa Fe, NM: School of American Research Press.

Olsen, Julia

2005 Re-Placing the Space of Community: A Story of Politics, Policies, and Fisheries Management. Anthropological Quarterly 78(1): 247–268.

Parker, Laura

2005 Casinos Ponder Move to Dry Land. USA Today, September 11: 3A.

Pender, Geoff

2005a Christians Lobby Against Rebuilding Mississippi-Coast Casinos. Sun Herald, September 30:A4.

2005b Mississippi Casinos May Rebuild on Dry Land. Sun Herald, September 24:A6.

Pender, Geoff, and Tom Wilemon

2005 Biloxi Blues: Mississippi Economy Hit Hard by Casino Losses. Tallahassee Democrat, September 2: A4. Reprinted from Sun Herald, November 1, 2005. Electronic document, LexisNexis, web.lexis-nexis.com/universe, accessed August 30, 2007.

Spitulnik, Debra
1993 Anthropology and Mass Media. Annual Review of Anthropology 22:293–315.

Tortorano, David
2003 Land Casino Idea Makes Waves, Gaming Panel Chairman Defends Studying the Options. Sun Herald, May 16: C9

U.S. Census Bureau
2000 U.S. Census 2000. Electronic document, www.census.gov/, accessed February 5, 2007.

von Herrmann, Denise
2006 Afterword. *In* Resorting to Casinos: The Mississippi Gambling Industry. Denise von Herrmann, ed. Pp. 167–173. Jackson, MS: University Press of Mississippi.

Wilemon, Tom
2004a Biloxi, Miss., Officials to Study Gambling Barges in Upland Areas. Sun Herald, June 25:A1.
2004b Mississippi, Governments to Lose Money When Casinos Are Closed Due to Storms. Sun Herald, September 9:B7.
2006a From Small to Big: Island View Owners Go for Ambitious Overhaul. Sun Herald, August 19:B8.
2006b Boom or Doom? Residents Question What's Right in Long Beach Casino Dispute. Sun Herald, June 25:A1.

From the Phoenix Effect to Punctuated Entropy: The Culture of Response as a Unifying Paradigm of Disaster Mitigation and Recovery

14

CHRISTOPHER L. DYER

THERE ARE DEEP-SEATED, CULTURALLY BOUND CAPACITIES and associated responses to disaster in every society and community which can be challenged and evolve when disaster occurs (Hoffman 1999; Torry 1978). Our contemporary reality is such that hazards resulting in disasters are no longer isolated events but are becoming frequent life experiences for billions of people, and the awareness of and capacity to respond to these events "is always present at a level just below that of consciousness and that can be summoned swiftly to the surface at the slightest provocation" (Bankoff 2003:179). These capacities or *cultures of response* (Dyer and McGoodwin 1999:214) to disaster can also reach boundaries and limits beyond which adaptations fail, economies collapse, and communities die (Adams 1965; Diamond 2004). On the other hand, given appropriate adaptive responses, including internal as well as external economic aid, a community can not only recover from a disaster event but in some cases flourish (Dyer 1995, 1999).

I propose here a unifying model of human reaction to disaster as the *culture of response*—the traditionally embedded resilience to a disaster event of a population as influenced by: 1) prior social memory of disasters, 2) available social-economic response and recovery resources, and 3) the local political economy of disaster aid, that is, the application or withholding of disaster aid as purposeful strategies of external power brokers. At one extreme on a culture of response continuum is the Phoenix Effect—a sustainable improvement in the social and economic resilience of a community or organization arising from the strategic investment of capital resources after a disaster event (Dyer 1995:35). The other extreme, Punctuated Entropy, is a permanent decline in the adaptive flexibility of a human ecosystem and its associated economies brought on by lack of appropriate aid combined with the cumulative impact of periodic and repeated disaster events (Dyer 2002:164). As Smart and Smart (2006) note of fire disasters in Hong Kong in the 1950s: "A rapid succession of similar disasters . . . may . . . create a sense that 'something has to be done' and facilitate a 'learning curve' where policy makers can learn from mistakes."In other cases vulnerable populations may experience disaster again and again without effective responses although Bankoff (2003) notes that what is accepted as "effective" disaster response in modern developed polities carries with it an inherent cultural bias.

Using case studies of natural and technological disaster events, specifically the Exxon Valdez oil spill (Punctuated Entropy) and Hurricane Andrew (the Phoenix Effect), I show that different cultures of response can occur in different regions of the same nation, and even within different communities sharing a similar geographic area. A significant point is that cultures of response can be unique to a population, owing to their historical ecology and economies, their long-standing relationship to their environment, their utilization of its resources for economic means, and their subsequent adaptations to its perturbations. Technological disasters, which can disrupt the ability of communities to reconnect to their environment, can lead to prolonged non-recovery and secondary disasters if not properly responded to. However, large natural disasters, such as Hurricane Katrina and the tsunami of 2007 that devastated parts of Indonesia and coastal areas of Near East and Southeast Asia, can also result in a prolonged recovery.

The model proposed here assumes that multiple social constructions of disaster and resultant coping and resilience mechanisms have evolved; embraces diverse cultural interpretations, explanations, and responses to disaster; and tries to make no judgment as to what is most effective and what is not. Multiple strategies of disaster response within the same community or culture follow *paths of resilience* that represent tried and effective strategies that have been shown to have the best possible outcomes in the face of repeated disaster events. These culturally and economically bound paths of resilience need not conform to a view of disaster management as a top-down affair where outsiders come to rescue powerless victims of nature with a superior knowledge that posits them as masters of response and recovery (Bankoff 2003; Schneider 1992; Siegel 1985). Diversity of responses is possible.

The Culture of Response Model

Figure 14.1 describes the culture of response model for disaster. The baseline (sustainable state = š) point is the normative, pre-disaster state. The baseline state describes the point where the community or region is stable—there is no disruptive change occurring beyond normal seasonal or occasional small-scale disruptions (the wave cycles around š) with associated hazards to social, ecological, and economic processes and functions. For example, a severe rainstorm may cause some flooding and loss of infrastructure, but is not on the scale of a major flood that takes lives and can destroy whole communities. At š, then, individuals of a community consider conditions to be "normative" (not in disaster mode). I am not talking about homeostasis, but a perceived semblance of normalcy.

A disaster event creates disruption at T1, and the outcome is a negative impact to social, cultural, and economic conditions. The depth of the curve—how far down it drops below the stable state—is in part determined by: 1) the severity of the disruption/disaster event; and 2) pre-disaster resilience of the impacted community. A disaster response, whether external, internal, or a combination of the two, can either improve or hinder restoration or reconfiguration of relatively stable patterns of human-environment interaction. In some cases, particularly with state-assisted (external resource) recovery of a localized disaster, the final stable state may

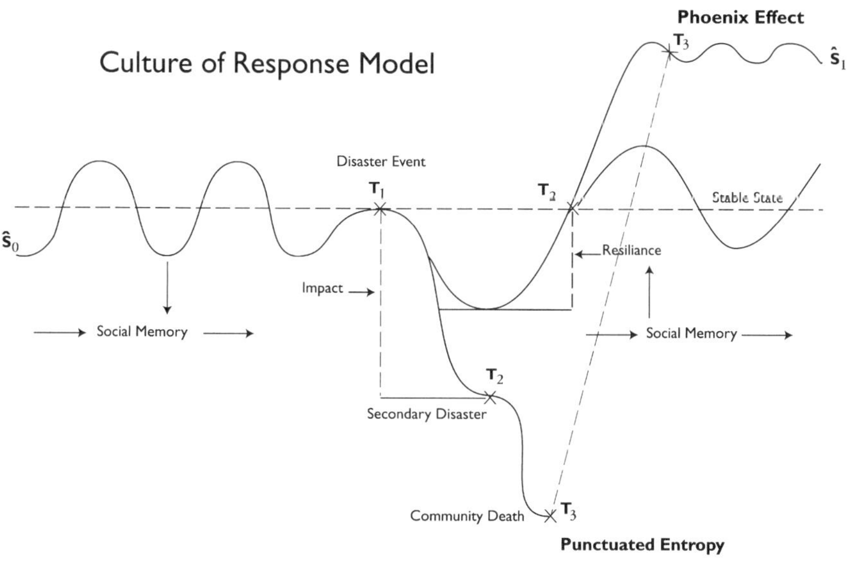

Figure 14.1. Culture of Response Model

exceed the pre-disaster level, resulting in an economy and quality of life that exceeds what existed in the pre-disaster state (the Phoenix Effect).

At all points on a culture of response continuum, communities can proactively demonstrate (McCabe 2002) or, with proper planning, develop (Paton and Johnston 2006; Ronan and Johnston 2005) varying degrees of economic resilience, which is the capacity to resist and absorb periodic disruptive changes in the environment. Resilience adaptations that become culturally embedded, or that take on the mantle of tradition, can be conceptualized as "equilibrations" on existing adaptations to changed environmental conditions (Brookfield and Brown 1967; Dirks 1980; Lee 1969; Spencer 1959; Torry 1978; Waddell 1975). These equilibrations—as they become tradition and thus modify any existing culture of response milieu—are sustained through social memory (McIntosh et al. 2000:25).

Capacity to recover is determined by the severity of the event, the internal capacity to recover (resilience), and the degree of internal/external assistance. Cumulative natural disasters can also induce punctuated entropy, and may include serial earthquakes followed by landslides or tsunamis, extended droughts followed by sandstorms, or repeated severe flooding events (Blaikie et al. 1994:27). Technological disasters may tend to have more cumulative effects known as secondary disasters (Erikson 1976), such as legal action, demolition, resettlement, and so forth (Dyer 1993; Freudenberg and Jones 1991; Hirsch 1997; Rodin et al. 1997).

The Community Continuum: From Natural Resource Community to Dominant Social Paradigm

Two basic and polar community types can be used to define a continuum of cultures of response to disaster, and that also thrive under diametrically opposed political economies—The Natural Resource Community (Dyer et al. 1992) and the Dominant Social Paradigm community (figure 14.2).

Communities can be placed on a continuum between these polar types. States are more highly entropic and tend toward the Dominant Social Paradigm pole, particularly in modern extraction community models, while the Natural Resource Community type is more natural resource dependent (agricultural or hunting-fishing societies). The Dominant Social Paradigm is tied to a value set of extraction and use that does not integrate sustainability or renewability into community function. Rather, it promotes free (unlimited) and open-ended use of the environment and resources—use

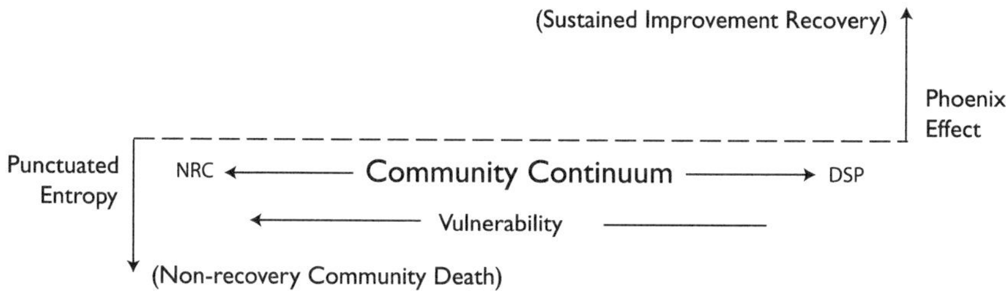

Figure 14.2. Community Vulnerability Continuum.

that can dominate other community patterns of a less developed, but more sustainable, nature when they are in competition. These community types can still coexist within a state, as do the agrarian peasant communities of China next to their burgeoning urban-centered elites, as long as urban development needs don't overlap with those of their less powerful agrarian community counterparts. The first world economy is best modeled on the Dominant Social Paradigm end of the continuum (Catton and Dunlap 1980). Douglas and Wildavsky (1982:21) describe the unmitigated risk of uncontrolled change promoted by the Dominant Social Paradigm strategy as irreversible risk: "Irreversible changes are explosive and unstable, each deviation growing larger until the environment is so altered it can never return to its original state."

The Natural Resource Community (Dyer 1993; Dyer et al. 1992) is defined as a population of individuals living in a bounded area whose primary existence depends upon the direct utilization or extraction of renewable natural resources (Dyer et al. 1992). Disasters which threaten and disrupt the nature-culture bond have the most severe and lasting social and economic impacts.

Given the economic networking and interdependence which is the signature of the contemporary global economy, pure Natural Resource Communities are practically nonexistent. Despite this, most of the world's population (83 percent) is still reliant on a direct nature-culture link as peasants and agriculturalists (Dyer et al. 1992). This community model is one of accepted limits, and disasters have the potential to stress and exceed these limits. Economic marginalization leads to increased vulnerability to disaster, and greater numbers of people living in more frequently marginal habitats results in decreased community resilience and increased vulnerability. In fact, ordering and incorporating resilience—the equilibration process—is tenuous to nonexistent in the increasing number of refugee and poverty-based communities in the developing world.

The human-nature relationship described by the Natural Resource Community model builds on the ecological-symbolic approach (Kroll-Smith and Couch 1991), which recognizes the existence of culturally based responses to environmental disruptions (Dyer 1993). In figure 14.1, these disruptions are represented by the wave cycles around š. The key to resilience for Natural Resource Communities is the application of local, community-based solutions to environmental challenges. Indeed, local resources are the first and primary source of adaptation for Natural Resource Communities and expectations of outside aid are not a well-defined part of the culture of response. As Dynes (1976:24) asserts: "Such societies possess such a delicate relationship with the environment that when it is disturbed the whole social and cultural structure is threatened." This is why, as in the case of the Exxon Valdez oil spill and its impact on salmon and herring fisheries economies, this kind of toxic disaster results in decline or even collapse of localized natural resource based economies. Toxic technological disasters are the most difficult to recover from for Natural Resource Communities, but less so for Dominant Social Paradigms which can reorder economies by modifying them through external assistance or have the community residents out-migrate to seek opportunities in unaffected clone-like community settings (Picou et al. 2004).

Placing our community models in context, Natural Resource Communities are more likely to suffer punctuated entropy outcomes in the face of repeated disaster events, while the resources available to the Dominant Social Paradigm community type means recovery is much more rapid, with greater dependence on state-level interventions and external (extra-communal) resource input. While the costs in terms of loss of life might be greater in the smaller Natural Resource Communities–type (developing world), the potential for infrastructure damage is greater in the Dominant Social Paradigm. For example, the loss of infrastructure and related economic and social capital resulting from Hurricane Katrina exceeded $120 billion. In comparison, the economic recovery costs attributed to the 2005 tsunami that devastated parts of Indonesia and coastal areas of 17 other countries was $13.6 billion. By comparison, the death toll for the South Asian tsunami was over 250,000 and for Katrina over 2,000 (1,836 confirmed dead and 705 missing as of May 2006). Hurricane Katrina was tracked for days before making landfall, while no warning at all was available for the tsunami victims.

The case study material that follows is from prior work on primary and secondary disasters spawned by the Exxon Valdez oil spill (1989) and Hurricane Andrew (1992). The hope is that this model provides a template for the integration of theory and practice, and—like the Indian fable of the elephant and the blind men who, when they each examined a different part of the elephant, accurately describe that part and assumed it represented the whole—pulls together parts examined through different theoretical lenses into a model that provides predictive explanation, but also lays the foundation for more effective social, economic, and cultural mitigation and recovery strategies.

Case 1: The Exxon Valdez Oil Spill as Punctuated Entropy

On March 24, 1989, the Exxon Valdez tanker collided with Blight Reef in the Valdez Narrows, resulting in the worst oil spill in U.S. history. This event was a technological disaster that broke the culture-nature link between the commercial fishing and subsistence communities in Prince William Sound and their supporting environment. The economy was driven by fishery extraction activities that were severely and permanently disrupted by the spill event.

Before it was contained, at least 11 million gallons of North Slope crude had poured from the hold of the Exxon Valdez into the waters of Prince William Sound. Cumulative impacts of this event persist as "secondary disasters" (Dyer 1993; Picou et al. 1997; Picou et al. 2004). Tracking the economic, cultural, and psychological impacts of the spill has continued over the last 19 years and the initial impacts continue for many of the original residents present during the 1989 disaster event. As of August 2008, the ongoing litigation battle with the Exxon Corporation had yet to bring economic relief to many of those originally affected, although the terms of the distribution of the $507.5 million Supreme Court judgment were finalized. Impacts have been described as "resource loss spirals," in which there is an ever downward trend away from recovery pushed by lingering economic woes and PTSD effects including "litigation stress" (Picou et al. 2004). Some 40 to 45 percent of residents still endure social and psychological effects of the spill, and have not been able to dissociate themselves from it (Picou 2006). As of 2007, original (1989) residents of Cordova, Alaska still suffer from the following secondary disaster events, making this a true case of punctuated entropy:

- Ongoing debt
- Litigation-related stress and hardship

- Economic resource loss
- Loss of culture capital involved with a diminished subsistence economy
- Loss of community cohesion and social capital
- Post-traumatic stress disorder directly keyed to the 1989 disaster event

The vulnerability of these communities is linked to their reliance on renewable natural resources (Dyer 1993; Picou and Gill 1996). The nature of technological and natural resource disasters has been shown to be particularly harmful to Natural Resource Communities because of the uncertainty it creates in the marketplace and among locals as to the safety of the environment and the natural resources extracted from it (Dyer et al. 2000). Uncertainty created by biophysical contamination events (Davidson and Baum 1991; Erikson 1991; Vyner 1988) impacts local residents in their subsistence and commercial harvesting activities, and can result in lingering stress—PTSD, that can last decades (Picou and Gill 1996). Economic impacts of the Exxon Valdez oil spill event in the period immediately following were estimated to reach $155 million, and economic recovery of the initial value of natural resources and permits to extract them has not occurred. As of 2000, these natural resources had shown some recovery, but most are slow to recover or remain depressed. In 2000, only 2 of the 24 commercially and recreationally valuable species impacted by the spill had recovered. Among the animals "not recovered" were harbor seals, Pacific herring, sea otters, harlequin ducks, killer whales, and three species of cormorants.

The 2006 report of the Exxon Valdez Oil Spill Trustee Council indicates that 17 years after the spill only 9 of those 24 species/resources are classified as "recovered" while the remainder are classified as either still "recovering" (8 species), status "unknown" (5 species), or "not recovering" (2 species). Commercial fishing, recreation, and tourism, and subsistence human services are still "recovering."

The Pacific herring population, which is a major cash fishery and source of herring roe for subsistence, completely collapsed in 1993 and shows no sign of recovery. Pink salmon runs failed in 1992 and 1993 and were depressed in 1994, while sockeye have not recovered to their pre-spill numbers as of 2006. Although the runs of hatchery salmon are recovered or recovering, wild stocks are still down from years prior to the spill. Evidence exists that potentially lethal doses of oil are still leaching into Prince William Sound streambeds where wild stocks of salmon spawn, and aquatic birds and marine mammals continue to show signs of exposure.

Table 14.1. Prince William Sounds Fisheries Year-End Permit Value (Alaska Commercial Fisheries Entry Commission 2008).

	Herring Purse Seine (G01E)*	*Herring Gillnet* (G34E)*	*Herring Spawn on Kelp/Pound* (L21E)*	*Salmon Purse Seine* (S01E)*	*Salmon Drift Gillnet* (S03E)*	*Salmon Set Gillnet* (S04E)*
1989	$245,000	$92,667	$47,884	$236,333	$141,115	$64,167
1998	$120,000	$69,400	$28,500	$36,600	$69,300	$51,000
Percent of Prior Value	49	75	60	15	49	79

*Prices are based on the last four permits of the particular fishery that have been sold.

Killer whales, some seabirds, and deer populations are still in a "recovering" mode from the spill. Harbor seals, an important subsistence resource, declined about 6 percent per year between 1989 and 1996 (Exxon Valdez Oil Spill Trustee Council 1996), but recovered by 2006 (Exxon Valdez Oil Spill Trustee Council 2006).

One measure of economic impact is reflected in the decline in fishing permit values (table 14.1). From the year of the Exxon Valdez oil spill in 1989 until 1998, the value of fishing permits for Prince William Sound declined dramatically. The value is predicated on the measure of return one gets from the fishery, and declining permit values parallel declining health and value of local fishery stocks.

The most dramatic value losses are in the herring purse seine permits (a 51 percent decline from $245K to $120K) and salmon purse seine permits (an 84 percent decline from $236,333 to $36,600; table 14.1). These dramatic declines reflect the non-recovery of both the ecology and the economy of the fishery—strong indicators of a Punctuated Entropy scenario.

This non-recovery/decline has not allowed community residents to reorder their ecological relationships with their environment—to regain their original stable state—and it appears that for those inhabitants they will never recover what was lost—what one key respondent described as their "Nirvana on Earth." As of 2008, the human impact of this technological disaster continues. It continues to directly impact two forms of the economy—a formal economic activity that is the commercial fishing of salmon and other marine species and an informal economy of subsistence activities based on local natural resources use by households. The human relationships that thrived through shared subsistence harvesting were broken; as a consequence, an entire generation has lost that connection with the natural resource base and its social and cultural values. A survey of Alaska natives in Cordova indicated that by 1991 over half of all respondents could no longer obtain previously consumed subsistence foods (Picou and Gill 1996). The documented ongoing compromise of the natural resource base of Prince William Sound by the Exxon Valdez oil spill disaster fulfills the condition of the punctuated entropy model.

The cultural and community impact of non-recovery from natural resource loss is further compounded by a series of ongoing secondary disasters (points T-2 and T-3 in figure 14.1), including the misdirection and withholding of external assistance, and lack of resolution of the legal battle. Culturally appropriate and economically effective external assistance is critical for a community to recover its adaptive flexibility after a disaster event (Davis 1984; Dyer 1993; Picou et al. 1997; Rodin et al. 1997). Withholding such external assistance, in the face of the overwhelming impact of this technological disaster to the primary natural resource base, has had long-term consequences to the quality of life and general community cohesiveness in Cordova and other impacted communities in Prince William Sound.

The exogenous economic capital that arrived in the form of oil spill cleanup funds and payoffs has done little, if anything, for long-term community recovery. While the capital infusion was directed at cleaning the environment, the social and cultural needs of communities received inadequate support, resulting in the depression of adaptive flexibility and the occurrence of "secondary disasters" (points T2 and T3 of figure 14.1). The money spill that followed from the cleanup created an artificial economy that corroded the social capital of affected communities, particularly the Native communities (see Bates 1982 for another example). Economic capital provided short-term improvements in material culture and comforts (Dyer 2002), but was socially destructive, fueling social instability by allowing for unmitigated access to drugs, alcohol, and firearms. Residents who traditionally generated social capital by sharing resources from commercial fishing and subsistence were engaged in cleanup—an activity which generated no social responsibilities, but only inordinate amounts of cash. As noted in Dyer (2002:183):

> Restoration, as agreed upon by Exxon and the federal government, in no way meant restoration of community, culture, social networks, or physical and mental health of disaster victims. Instead, it was restricted to settlement of "damages to publicly owned natural resources affected by the spill. Also excluded were various claims from private parties, the most prominent being Alaska Natives and Alaska Native corporations, owners of nearly all the private land in the spill area." (Piper 1993:261).

Vulnerability levels are much higher (figure 14.2) for Natural Resource Communities than those reliant on nonrenewable resources. Cordova and other nearby fishing communities are not at the extreme end of a Dominant Social Paradigm/Natural Resource Communities continuum (figure 14.1), but on the Natural Resource Community side of the midpoint. Other communities are even more vulnerable to a Punctuated Entropy scenario, including peasant communities of Asia, Africa, and the Near East.

Dyer (2002) has argued the Exxon Valdez oil spill is a technological disaster arising from a process of punctuated entropy for severely impacted commercial fishing and subsistence-based communities of Prince William Sound. The culture of response to this event—trying to clean up the spill, engaging in legal warfare with a mega-corporation, or trying to cope and recover the pre-disaster formal (commercial fishing) and informal (subsistence) economies—has not worked. The resilience evident in the years of struggle to regain what was lost could not overcome the devastating economic, biophysical, and cultural losses inflicted on the region.

Case 2: Hurricane Andrew, the Economic Development Administration, and the Phoenix Effect

Hurricane Andrew struck south Florida with 170 mph plus winds and rain on the morning of August 24, 1992, and was (until Hurricane Katrina in 2005) the costliest storm in United States history. The storm killed at least 41 people and caused over $24 billion in damage. Impacts included loss of 125,000 homes damaged or destroyed, and loss of 12 percent of Dade County's $38 billion personal income plus 14 percent of its 878,000 jobs and 59,000 businesses. By mid-1993 some 70 percent of those businesses were still closed; many were small firms/establishments employing fewer than five persons each, with 96 percent employing fewer than fifty people (Metropolitan Dade County 1993:9). The towns of Homestead and Florida City were the worst hit, followed by Perrine, Goulds, Princeton, Naranja, and Cutler Ridge. Out-migration for lack of jobs and housing reduced the population in South Florida 40 to 50 percent over the next three years, and by 1997 the population was still down 3,000 persons from pre-Andrew levels of 31,000. The culture of response I focus on here is that of the Economic Development Administration (EDA) whose mandate after a presidentially declared disaster is to provide assistance to effect long-term recovery.

The data used here comes from an assessment conducted three years after the event to determine the effectiveness of the EDA assistance strategy in supporting economic recovery (Dyer 1995). Unlike the Exxon Valdez oil spill, impacts in South Florida from this natural disaster did not create a toxic environment that threatened the natural resource base, dependent fisheries, and subsistence communities and economies. However, severe fisheries impacts were documented in Florida and Louisiana that reveal how differing state-level cultures of response resulted in the rapid economic recovery in one population of fisherfolk (Louisiana) and non-recovery in another (Florida; Dyer and McGoodwin 1999).

Interviews with regional EDA managers revealed a response strategy designed, in effect, to achieve the Phoenix Effect—"a sustainable improvement in the social and economic condition of a community or organization because of the strategic investment of capital resources after a disaster event" (Dyer 1995:35). This is not a mandated goal, but is rather a best scenario outcome of a strategic investment process:

> The phoenix effect is not an explicit goal or mandate of Economic Development Administration Economic Development Administration funding, but occurs spontaneously as a socioeconomic process when project planning efficiently identifies key areas of strategic investment. Economic Development Administration disaster experts select projects that in their judgment have long-term impact potential and that are good candidates for generating development funds from other public—and private-sector sources. (Dyer 1999:238)

Initial EDA investments may act as *economic triggers* to achieve results that might otherwise not have been realized in the post-disaster environment. The strategic aspects of investment come from prior knowledge by EDA managers of the economic landscape in a region. This is possible because areas that are eligible for aid are identified as "development regions" and a separate (normative) economic development investment program exists, the result being that managers can have a good idea of economic trends, capacities, and players if and when disaster strikes. Such prior knowledge is brought into play in the enrollment and selection of post-disaster projects deemed to have the highest potential benefit. This pre-disaster knowledge includes on-the-ground knowledge of stakeholders that helps in identifying major elements/organizations of the local economy in the post-disaster environment.

When Andrew struck South Florida, the EDA set up shop in Miami and met with various stakeholders and encouraged them to submit proposals for funding. The officials become well known to communities, organizations, and local leaders by staying on-site and developing close social bonds with their clients. These ties facilitated communication of economic needs in the post-disaster environment, as with this example where the City Manager of Hialeah referred to the Economic Development Administration–funded Industrial Teaching Factory Project:

> "The people we dealt with from Economic Development Administration—were exemplary. They worked with us no matter what we asked of them. They know we had something special with this project in particular—that it had the potential to be a model for the rest of the nation. Economic Development Administration was the trigger mechanism for the whole project. They didn't just have an idea—what they had was a vision."

Some seventy-six proposals were ultimately generated, and thirty-four of these received funding, with a total investment of $50,875,100. The original 34 projects were selected in 1992 after submission and review by the EDA staff in Atlanta. There are four broad project categories funded by the EDA: 1) infrastructure/public works grants; 2) revolving loan funds; 3) technical assistance/business incubator grants; and 4) economic planning grants (Dyer 1999 table 14.2).

The hope was that these diverse projects would contribute to the EDA's stated goals in the region, which were to: 1) reestablish viable, balanced, residential communities in south Dade; 2) protect, restore, and build upon economic strengths and strategic assets; 3) expand and improve delivery of governmental relief and assistance; 4) rebuild support services; and 5) improve upon the built environment that was destroyed (Dyer 1999:285). Seventeen out of the original thirty-four funded projects (50 percent) were selected for inclusion in the 1995 assessment (table 14.3)

Interviews with key managers, project personnel, and beneficiaries resulted in the conceptualization of a "development quotient" to measure project success (Dyer 1995). This measure of project success included the following components: 1) effective leadership; 2) effective progress; 3) job creation; 4) generating secondary resources; 5) creating economic partnerships; and 6) stakeholder empowerment. All of the case studies were ranked using these six variables, with assigned values based on individual case studies that include both quantitative and qualitative data from across and within the four project categories. Assessment of case studies occurred within the EDA's anticipated five-year development window. If funds were not engaged beyond this five-year window, they could be withdrawn and the project cancelled. Table 14.3 ranks the assessed projects using the six measurement criteria. Assigned values range from 1 (poor)

Table 14.2. List of Assessed Economic Development Administration Projects

Project Categories/Titles	*Project Descriptions*
Planning Grants	
Miami Convention and Visitors Bureau	$1 million to develop promotional materials/tourist marketing
South Florida Regional Planning Council	$200,000 for a strategic economic plan to promote international trade
Public Works/Infrastructure Grants	
Florida City	$5 million to replace and upgrade the city water treatment and distribution system
City of Homestead	$7 million for construction of water and sewer line and road improvements
City of Hialeah	$1.5 million for the rehabilitation of the old Hialeah train station into a farmers' market
Miccosukee Corporation	$1 million for public works infrastructure for commercial/residential development
Seminole Tribe	$1 million for public works infrastructure for commercial/residential development
Metro Dade County Parks and Rec Dept	$4.5 million to improve infrastructure of existing facilities and to provide additional facilities for expansion of the Metro Zoo
Metropolitan Dade County	$5 million to install water and sewer mains, sewage pumping stations, and fire hydrants along the Dixie Highway
Technical Assistance/Business Incubator Grants	
Tools for Change	$550,000 to provide technical assistance to minority businesses
Center for Health Technologies	$1 million to renovate a building for use as an incubator for health technology companies
City of Hialeah	$1 million for conversion of a manufacturing facility into a training center for computer technology transfer
Florida International University	$250,000 to fund training/technical assistance by the Small Business Development Center
Goodwill Industries	$1.5 million for building renovations and purchase of equipment to train and employ low-income/handicapped persons
Revolving Loan Funds	
Beacon Council	$2 million for a revolving load fund to help businesses in Dade County recover
Beacon Council	$1.5 million recapitalization of existing revolving loan fund
City of Homestead	$2 million to establish a revolving load fund to assist businesses in Homestead

Table 14.3. Comparative Development Quotients for Assessed EDA Projects*

Project	*Leadership*	*Progress*	*Jobs*	*Leveraging*	*Partnerships*	*Empowerment*	*Total*	*Quotient*
Development Quotients Planning Projects								
SFRPC**	5	5	NA	3	5	5	23	0.92
GMCVB**	4	5	NA	NA	5	4	18	0.90
Public Works/Infrastructure Projects								
Florida City Water	4	5	5	5	4	4	27	0.90
Hialeah Farmers	4	3	3	3	3	3	19	0.63
MetroZoo	5	4	4	5	4	4	26	0.87
Homestead Water	4	5	4	5	5	5	28	0.93
Dixie Highway	3	3	2	4	2	2	16	0.53
Miccosukee	2	2	2	2	2	1	11	0.36
Seminole	2	1	1	1	1	1	7	.23
Technical Assistance/Business Incubator Projects								
Goodwill	5	5	5	5	5	5	30	1.0
Tools for Change	5	5	5	5	5	5	30	1.0
HITPIC**	4	3	3	5	4	3	22	0.73
CHT**	3	1	1	3	5	1	14	0.47
FIU**	4	5	NA	3	4	4	20	0.80
Revolving Loan Funds								
Homestead RLF**	2	3	2	NA	3	3	13	0.52
Beacon Council	3	3	3	NA	4	2	18	0.72

*Scores can be interpreted as follows:
5 = outstanding—well above expectations at this point
4 = excellent—above expectations at this point
3 = good—about as expected at this point
2 = fair—somewhat below expected at this point
1 = poor—below expected at this point
**SFRPC (South Florida Regional Planning Council)
GMCVB (Greater Miami Convention and Visitors Bureau)
HITPIC (Hialeah Technology Production and Information Center)
CHT (Center for Health Technologies)
FIU (Florida International University)
Homestead RLF—Homestead Revolving Loan Fund

to 5 (excellent). The final development quotient value is calculated as the sum of the assigned values across the six categories divided by the total possible (30). Thus, a value of 1.0 (30/30, all 5s, "outstanding") indicates a perfect score, while a value of 0.20 (6/30, all 1s, "poor") is the lowest possible score. An average development quotient would be 0.60 (all criteria ranked as 3, "good" or "about expected at this point").

Overall, 11 of the 17 assessed projects (69 percent) exceed expectations three years (1995) after initial investments. Five projects had development quotients below 0.60, indicating they were not meeting expectations. However, only two of these projects (both public works projects—with the Miccosukee [0.36] and Seminole [0.23], respectively) were in the poor-fair range. Both of these projects had experienced changes in tribal leadership and indecision resulting in delayed or no progress. The other lower scores were anticipated to improve by project leaders, with the score reflecting a slower project start up and resulting project benefit. Eleven projects were scored around the expected value of 0.60, and ranged from 0.63 to 1.0 (a perfect score) with an average of 0.84, indicating measurable progress and impact (table 14.3). To exemplify the Phoenix Effect, I'll now turn to the Goodwill Industries project in Hialeah, one of the 34 projects assisted by the EDA after Hurricane Andrew.

A Model Project: Goodwill Industries and the Phoenix Effect

Goodwill Industries is a nonprofit organization dedicated to training, employing, and placing people with disabilities in community jobs. After Hurricane Andrew, the Goodwill chief executive office faced the immediate loss of around 300 service jobs when the Homestead Air Force Base commissary was destroyed, and when other Goodwill outposts in Florida City and Homestead were damaged or destroyed by the hurricane. Many of those they served were in areas without power, road signs, or any other public services. Many were experiencing anxiety, stress, and even post-traumatic stress disorder from their loss of jobs and homes. The chief executive officer worked quickly to account for his employees, get them resettled, and organized them the best he could. The rest of the year (1992) was one of reorganization and assessment, and Goodwill did not really get back on track until 1993 after receiving a $1 million infrastructure grant from the Economic Development Administration.

Goodwill Industries was fortunate in that their chief executive officer had a served a long career in the military and had 13 years experience in a leadership and leadership training capacity. He maximized the one million dollars in grant funds from the EDA by using refurbished building materials to get his operations back up and running. He leveraged his connections with the military into a contract with the U.S. Department of Defense for the installation of equipment in the main Goodwill facility in Hialeah to manufacture military garments, including fatigue pants, naval caps, and memorial flags.

Because of the difficulty some clients had in mastering the sewing and production operations, the process was broken down into numerous small steps, which, once learned, could be done very efficiently. The outcome was a level of quality control that was so outstanding the results were praised by their Department of Defense contractors, and demand for their high-quality products soon exceeded their production capacity.

This leader's effective planning and organization also resulted in a 13-year contract with the *Miami Herald* for doing Sunday inserts. This included over 150 new jobs and $7 million in equipment investments in the basement floor of the Hialeah facility. The initial jobs generated by the Department of Defense contract numbered over 300 jobs. As Goodwill president and chief executive officer Dennis Pastrana (2007, personal communication) has remarked:

> The recovery of the Goodwill Industries of south Florida from Hurricane Andrew has been noted by the Economic Development Administration as being remarkable. I have received repeated calls over the years from them (EDA) praising us as one of their truly great success stories.

With 1993 began the first full calendar year back in operation after the EDA's $1 million investment, there were steady and significant increases in all economic and job creation parameters on which Goodwill keeps records. For example, persons served increased 218 percent from 1,414 in 1993 to 4,493 in 2006; successful rehabilitations 156 percent from 1,096 to 2,806, total revenues 577 percent from $9.9 million to $67 million, and total earnings 391 percent from $7.0 million to $34.4 million. These economic indicators point to a steady and progressive recovery that goes well beyond the original stable state.

The Goodwill Industries example is one of the several projects with outstanding results from the initial post-disaster capital investment (table 14.3). Unlike the Exxon Valdez oil spill example that focused on a specific socio-geographic community (Cordova) the Goodwill Industries example describes a *regional* culture of response for a dispersed community of disabled individuals in South Florida. Goodwill Industries is, in fact, an organization embedded within a wider urban complex of multiple adjacent communities. To understand the EDA's response as a state-level (federal aid) response, we include all of the Administration's funded projects and their combined social and economic impacts. Certain Economic Development Administration projects did not fare as well as others in this analysis. Disaster recovery, even in best-case scenarios, can be a patchwork process where resilience and return to a pre-disaster quality of life (either stable or improved states) can be highly variable.

The potential also exists for any impacted region and its embedded populations and communities to experience *both ends* of the culture of response continuum, with punctuated entropy events co-occurring alongside pockets of remarkable (phoenix effect) recoveries. For example, after Hurricane Andrew, commercial fishing communities, support businesses, and enclaves in South Florida could have received recovery aid under the National Fisheries Act, but did not. They were ignored by local fisheries managers, and pleaded with us to convey the message: "Tell them we are hurting" (Dyer and McGoodwin 1999:211).

Fishing communities hold marginal status in South Florida in that they lack decision-making power over management of fisheries and the economic resources that would flow from such power. To receive targeted aid under the National Fisheries Act after Andrew, a request for such federal aid would have to come from the state government. No such request came from Florida, but in Louisiana such a request for federal aid was made. In Florida, fishing communities are geographically dispersed and ethnically and economically marginalized, whereas in Louisiana they represent a strong component of traditional coastal culture, and are more important politically and economically to the state's communities and quality of life.

Discussion and Conclusions

The case of the Exxon Valdez oil spill disaster and response and the Economic Development Administration's response to Hurricane Andrew exemplify different end points on the culture of response continuum, illuminating how polities, groups, and populations respond economically to hazards and calamity. The EDA case, as highlighted by the Goodwill Industries example, demonstrates how strategic investment of economic capital can result in effective post-disaster recovery, as measured by the development quotient and its variable set of 1) effective leadership, 2) effective progress, 3) stakeholder empowerment, 4) job creation, 5) leveraging of secondary funding, and 6) creation of economic networks. The Exxon Valdez oil spill disaster portrays the opposite result, where no Economic Development Administration type agency gives impulse to the recovery process. Instead, punctuated entropy displays non-recovery and secondary disasters through 1) compromise of the natural resource base; 2) misdirection or withholding of external economic aid; and 3) post-disaster legal maneuvering and culturally inappropriate intervention that hinder restoration of traditional patterns of human-environment interaction.

Disasters, as culturally interpreted phenomena, are adapted to by the extent they become part of a transgenerational cycle of memorable environmental events (e.g., Crumley 2000). As we cope with more frequent and widespread disasters, and economies and communities adjust to their outcomes, shifts in cultures of response are inevitable. At some point the semiotics of "normative disaster" may be institutionalized as part of an understood cultural order and parallel potential for disorder and recovery. However, if the frequency and scope of economic impacts of disaster increases, it can strain the sustainability of the most economically and culturally sophisticated disaster responses. Extreme (non-normative), infrequent, and slow-onset events can be beyond the capacity of a group's culture of response, resulting in economic collapse and non-recovery.

To understand any group's culture of response to a disaster, one must understand the phases of disaster: 1) what puts a population or group at risk of exposure, 2) what makes them vulnerable to impact from the hazards associated with disaster, and 3) how they—and those external aid agents responsible to them—invoke and engage recovery opportunities. In cases where local adaptations fail and significant recovery cannot proceed without external aid, recovery itself becomes a question of economic resources—those who have them and those who do not—and the extent that those resources are applied to the mitigation and recovery from the downturns associated with disaster events. What are "normative" responses to disaster can become ineffective in the face of overwhelming change and, in communities prone to rely on external assistance, disasters come to reveal the capacities of state and community-level response and planning (Brinkley 2006; Dyson 2006).

Acknowledgments

Many individuals' lives were disrupted by the disasters described here, and their courage and sacrifice are to be acknowledged in their demonstrated cultures of response. Special thanks go to Mr. Dennis Pastrana for his aid in documenting Goodwill Industries' recovery success after Hurricane Andrew, and Mrs. Cynthia Dyer, for her expert illustrations and enduring patience.

References

Adams, W.

1965 The Dead Community: Perspectives from the Past. *In* The Dying Community. A. Gallaher, Jr. and H. Padfield, eds. Pp. 1–23. Santa Fe, NM: School of American Research.

Alaska Commercial Fisheries Entry Commission

1994 Fishery Statistics. Electronic document, http://www.efec.state.ak.us/menus/mnu_permit_values.htm, accessed May 6, 2008.

Bankoff, Greg

2003 Cultures of Disaster: Society and Natural Hazards in the Philippines. London and New York: Routledge Curzon.

Bates, F.

1982 Recovery, Change and Development: A Longitudinal Study of the 1976 Guatemalan Earthquake. New York: Bantam.

Blaikie, P., T. Cannon, I. Davis, and B. Wisner

1994 At Risk: Natural Hazards, People's Vulnerability, and Disasters. London: Routledge.

Brinkley, Douglas

2006 The Great Deluge; Hurricane Katrina, New Orleans, and the Mississippi Gulf Coast. New York: William Morrow.

Brookfield, H. C., and Paula Brown

1967 Struggle for Land. New York: Oxford University Press.

Catton, W. R. Jr., and R. E. Dunlap

1980 A New Ecological Paradigm for Post-Exuberant Sociology. American Behavioral Scientist (24):15–47.

Crumley, Carol L.

2000 From Garden to Globe: Linking Time and Space with Meaning and Memory. *In* The Way the Wind Blows; Climate, History, and Human Action. R. J. McIntosh, J. A. Tainter, and S. K. McIntosh, eds. Pp. 193–208. New York: Columbia University Press.

Davidson, L. M., and A. Baum

1991 Victimization and Self-Blame following a Technological Disaster. *In* Communities at Risk. S. R. Couch and J. S. Kroll-Smith, eds. Pp. 33–52. New York: Peter Lang.

Davis, Nancy Yaw
1984 Contemporary Pacific Eskimo. *In* Handbook of North American Indians, vol. 5. D. Dumas, ed. Pp. 198–204. Washington, DC: Smithsonian Institution.

Diamond, Jared
2004 Collapse: How Societies Choose to Fail or Succeed. New York: Penguin Books.

Dirks, Robert
1980 Social Responses During Severe Food Shortages and Famine. Current Anthropology 21(1): 21–44.

Douglas, Mary, and Aaron Wildavsky
1982 Risk and Culture: An Essay on the Selection of Technological and Environmental Dangers. Berkeley: University of California Press.

Dyer, Christopher L.
1993 Tradition Loss as Secondary Disaster: The Long-Term Cultural Impacts of the Exxon Valdez Oil Spill. Sociology Spectrum 13:65–88.
1995 Assessment of the Economic Development Administration's Post-Disaster Recovery Program after Hurricane Andrew. Bethesda, MD: Aguirre International.
1999 The Phoenix Effect in Post-Disaster Recovery: An Analysis of the Economic Development Administration's Culture of Response after Hurricane Andrew. *In* The Angry Earth: Disaster in Anthropological Perspective. Pp. 278–300. New York: Routledge.
2002 Punctuated Entropy as Culture-Induced Change: The Case of the Exxon Valdez Oil Spill. *In* Catastrophe and Culture: The Anthropology of Disaster. Susanna Hoffman and Anthony Oliver-Smith. eds. Pp. 159–186. Santa Fe, NM: School of American Research Press.

Dyer, Christopher L., Duane A. Gill, and John S. Picou
1992 Social Disruption and the Valdez Oil Spill: Alaskan Natives in a Natural Resource Community. Sociology Spectrum 12:105–126.

Dyer, Christopher L., John J. Poggie, and David Allee
2002 Assessment of the Impact of the 1999–2000 Lobster Die-Off on the Fishing Families and Communities of Western Long Island Sound. Economic Development Administration Report. Ithaca, NY. Cornell University Agricultural Extension Office. 104 pp.

Dyer, Christopher L., and J. R. McGoodwin
1999 The Culture of Response: the Political Ecology of Disaster Assistance and its Impact on the Fishing Communities of Florida and Louisiana after Hurricane Andrew. *In* The Angry Earth: Disaster In Anthropological

Perspective. Anthony Oliver-Smith and Susanne Hoffman, eds. Pp. 211–231. New York: Routledge.

Dynes, Russell
1976 The Comparative Study of Disaster: An Organizational Approach. Mass Emergencies 1(1):21–32.

Dyson, Michael E.
2006 Come Hell or High Water: Hurricane Katrina and the Color of Disaster. New York: Basic Books.

Erikson, Kai
1976 Everything in Its Path: Destruction of Community in the Buffalo Creek Flood. New York: Simon and Schuster.
1991 A New Species of Trouble. *In* Communities at Risk. S. R. Couch and J. S. Kroll-Smith, eds. Pp. 11–29. New York: Peter Lang.

Exxon Valdez Oil Spill Trustee Council
1996 Exxon Valdez Oil Spill Restoration Plan: Update on Injured Resources and Services, September 1996. Anchorage, AK: Exxon Valdez Oil Spill Trustee Council. Electronic document, http://www.evostc.state.ak.us/recovery/status.cfm, accessed May 6, 2008.
2006 Exxon Valdez Oil Spill Restoration Plan: Update on Injured Resources and Services, November 2006. Anchorage, AK: Exxon Valdez Oil Spill Trustee Council. Electronic document, http://www.evostc.state.ak/recovery/status.cfm accessed May 6, 2008.

Freudenberg, William R., and Timothy R. Jones
1991 Attitudes and stress in the Presence of Technological Risk: A Test of the Supreme Court Hypothesis. Social Forces 4(69): 1143–1168.

Hirsch, William B.
1997 Justice Delayed: Seven Years Later and no End in Sight. *In* The Exxon Valdez Disaster: Readings on a Modern Social Problem. J. S. Picou, P. A. Gill, and M. J. Cohen, eds. Pp. 271–307. Kendall Hunt Publishing Company.

Hoffman, Susanna M.
1999 After Atlas Shrugs: Cultural Change or Persistence After Disaster. *In* The Angry Earth: Disaster In Anthropological Perspective. Anthony, Oliver-Smith and Susanne Hoffman, eds. Pp. 302–325. New York: Routledge.

Kroll-Smith, J. S., and S. R. Couch
1991 What is a Disaster? An Ecological Symbolic Approach to Resolving the Debate. International Journal of Mass Emergencies and Disaster 9:355–366.

Lee, Richard B.
1969 Kung! Bushman Subsistence: An Input-Output Analysis. *In* Environment and Culture Behavior. A. P. Vayda, ed. Pp. 47–79. New York: Natural History Press.

McCabe, Terrance J.
2002 Impact of and Response to Drought among Turkana Pastoralists: Implications for Anthropological Theory and Hazards Research. *In* Catastrophe and Culture: The Anthropology of Disasters. Susanna M. Hoffman and Anthony Oliver-Smith, eds. Pp. 213–236. Santa Fe, NM: School of American Research Press.

McIntosh, Roderick J., Joseph A. Tainter, and Susan Keech McIntosh
2000 Climate, History and Human Action. *In* The Way the Wind Blows. Roderick J. McIntosh, Joseph A. Tainter, and Susan Keech. Pp. 1–42. New York: Columbia University Press.

Metropolitan Dade County
1993 Economic Recovery Strategies Phase One: Short Term Strategies. Miami: Research Division of the Metro Dade County Planning Department.

Pastrana, Dennis
2007 Personal communication. President, Goodwill Industries of South Florida.

Patton, David, and David Johnston, eds.
2006 Disaster Resilience: An Integrated Approach. Springfield, IL: Charles C. Thomas, Publisher Ltd.

Picou, J. Steven.
2006 Personal Communication. Chair, Department of Sociology and Anthropology, University of South Alabama, Mobile.

Picou, J. Steven, and Duane A. Gill
1996 The Exxon Valdez Oil Spill and Chronic Psychological Stress. American Fisheries Society Symposium 18:879–893.

Picou, J. Steven, Duane A. Gill, and Maury J. Cohen, eds.
1997 The Exxon Valdez Disaster: Readings on a Modern Social Problem. Dubuque Iowa: Kendall/Hunt Publishing Co.

Picou, J. Steven, Brent K. Marshall, and Duane A. Gill
2004 Disaster, Litigation, and the Corrosive Community. Social Forces 82(4):1493–1522.

Piper, Ernest
1993 The Exxon Valdez Oil Spill: Final Report, State of Alaska Response. Anchorage: Alaska Department of Environmental Conservation.

Rodin, Mari, Michael Downs, John Petterson, and John Russell
1997 Community Impacts of the Exxon Valdez Oil Spill. *In* Picou, J. Steven, Duane A. Gill, and Maury J. Cohen, eds. 1997. The Exxon Valdez Disaster: Readings on a Modern Social Problem. Pp. 193–210. Dubuque Iowa: Kendall/Hunt Publishing Co.

Ronan, Kevin R., and David M. Johnston
2005 Promoting Community Resilience in Disasters: The Role for Schools, Youth, and Their Families. New York: Springer.

Schneider, S.
1992 Governmental Response to Disasters: The Conflict Between Bureaucratic Procedures and Emergent Norms.

Siegel, G.
1985 Human Resource Development for Emergency Management. Public Administrative Review 45:107–117.

Smart, Alan, and Josephine Smart
2007 Learning from Disaster? Mad Cows, Squatter Fires and Temporality in Repeated Crises. Department of Anthropology, University of Calgary. Abstract. The Society for Economic Anthropology Annual Meeting. Greensborough, NC. April, 2007.

Spencer, Robert F.
1959 The North Alaskan Eskimo: A Study in Ecology and Society. Washington: U.S. Government Printing Office.

Torry, William I.
1978 Bureaucracy, Community, and Natural Disasters. Human Organization 37:302–308.

Vyner, H. M.
1988 Invisible Trauma: The Psychosocial Effects of Invisible Environmental Contaminants. Lexington, MA: Heath.

Waddell, Eric
1975 How the Enga Cope with Frost: Responses to Climactic Perturbations in the Central Highlands of New Guinea. Human Ecology 3(4):249–273.

Index

About the Authors

Margaret L. Brown (PhD, Washinton University) is visiting assistant professor and administrator for the Program in Politics, Philosophy, and Economics at Duke University. Prior to that, she was Academic Coordinator for the Program in American Culture Studies and ran the graduate seminar for the interdisciplinary Center for New Institutional Social Sciences at Washington University in St. Louis. Her research interests are in the areas of informal and formal social and cultural institutions, including family and kinship, property rights, economic production and distribution, and education. She wishes to acknowledge the Russell Sage Foundation and Washington University in St. Louis for their generous support for the research published in this volume.

Shelly L. Brown-Jeffy (PhD, University of Michigan) is assistant professor in the Sociology Department at the University of North Carolina at Greensboro. Her teaching and research interests focus on the sociology of inequality. Dr. Brown-Jeffy's major research interests in the area of race and education focus on the influence of race, poverty, and school composition effects on the achievement of students in pre-kindergarten through twelfth grades. Her current research projects examine the effects of poverty and race on education (including the early years) to examine when, why, and how the educational gap in achievement begins and widens as students continue through the educational process.

Christopher L. Dyer (PhD, Arizona State University) serves as professor of anthropology and the Dean of Arts and Sciences at Mount Olive College in North Carolina. He is a seventeen year veteran of disaster research, natural resource management, and evaluation and assessment research. He has directed field research teams in the United States, as well as seventeen countries in Latin America and the Caribbean, as well as Kenya and the Philippines. His 1992 paper on the Natural Resource Community, won an award from the journal *Sociology Spectrum* as "An outstanding contribution to social science." He is a Fellow of the Society for Applied Anthropology, a member of the National Association of Practicing Anthropologists, the American Anthropological Association, American Association of University Professors, and the Council for Advancement and Support of Education.

Timothy J. Finan (PhD, University of Arizona) is director of the Bureau of Applied Research in Anthropology at the University of Arizona. He has worked in Africa, Europe, Latin America, Saudi Arabia, Bangladesh, and Albania. Most of his research has been in Portugal, the Cape Verde Islands, Chad, and Brazil. Finan's research interests focus around various dimensions of world poverty, including vulnerability, livelihood security, climate and society interactions, natural resource management, and public policy. He has worked to develop new, more effective methodological tools in participatory research and development. Currently his main research activities are a five-year study of climate vulnerability in Ceará, and a study of freshwater prawn farming in Southwest Bangladesh.

Vishwas D. Gogte is professor in archaeological chemistry at Deccan College Post Graduate and Research Institute in Pune, India. His main research interest is provenance studies by mineralogical and chemical analysis of pottery and related material from sites across the Indian Ocean. He has been acclaimed for establishing interrelationships between sites in Oman and India of the period of the Indus Civilization. Often cited is his work on the Rouletted Ware found on the coastal sites of India and Southeast Asia. He has discovered and excavated a number of coastal sites on the west coast of India in view of understanding the dynamics of ancient maritime trade.

Olaf Jaime-Riverón is a Mexican archaeologist. His undergraduate degree from the National School of Anthropology focused on archaeology. He obtained his MA at UNAM. Jaime-Riverón has conducted excavations and surveys at the National Palace of Mexico, Teotihuacán, Tula, Montagua River in Guatemala, and the Mayan area in Campeche. Since 1993, he has been conducting research in the Olmec area in the sites San Lorenzo, La Venta, El Manatí, and Tres Zapotes. He has participated in survey archaeology in the Alps. Currently he is a PhD candidate at the University of Kentucky.

Eric C. Jones (PhD, University of Georgia) is a research scientist in anthropology at the University of North Carolina at Greensboro. Jones' general research interest is in cross-cultural vulnerability and resilience under conditions of major environmental or social change. He has received a Fulbright fellowship and a fellowship from the University of Florida's Tropical Conservation and Development program in support of his research. His work focuses on understanding the development and structuring of social relations following extreme events (e.g., natural disasters, pioneer colonization, and immigration), exemplified by the 2004 *World Development* piece "Wealth-Based Trust and the Development of Collective Action." Jones serves as a Principal Investigator and project coordinator for two NSF studies on social networks and disaster mitigation in Mexico and Ecuador.

Steve Kroll-Smith (PhD, University of Pennsylvania) is professor of sociology at the University of North Carolina at Greensboro and editor of *Sociological Inquiry*. His latest book with Valerie Gunter is *Volatile Places, A Sociology of Communities and Environmental Controversies* (2007). Steve has written and edited four additional books on communities, people, and environmental troubles. His work on human-made hazards was recognized by the American Sociological Association's Distinguished Contribution Award for research on environments and technologies. He is currently working on two Katrina-related projects: (1) a comparative study of the 1906 San Francisco earthquake and fire and the 2005 flooding of New Orleans and (2) a comparative neighborhood study of post-flood recovery in New Orleans.

Chapurukha M. Kusimba (PhD, Bryn Mawr College) received his undergraduate education at Kenya University, Nairobi, majoring in African history and linguistics in 1986. He is currently curator and professor of anthropology at the University of Illinois, Chicago, and previously served for seven years as research scientist at the National Museums of Kenya. His books include *The Rise and Fall of Swahili States* (1999), *East African Archaeology: Foragers, Potters, Smiths, and Traders* (co-edited with Sibel B. Kusimba [2003]), and *Unwrapping a Little Known Textile Tradition: The*

Field Museum's Madagascar Textile Collection (2004). Chapurukha directs or co-directs archaeological research in East Africa, India, Madagascar, and Illinois. His research has been supported by the National Science Foundation, Fulbright Fellowship, the National Geographic Society, and the Field Museum. He is currently the vice chairman of the anthropology department at the Field Museum.

Ty Matejowsky (PhD, Texas A&M University) is assistant professor of anthropology at the University of Central Florida. His fieldwork in the Philippines has examined a number of issues related to globalization, including fast food, urban development, disasters, and international migration.

Arthur D. Murphy (PhD, Temple University) is professor and head of anthropology at the University of North Carolina at Greensboro. His research has focused on urban economic systems, and on cultural dynamics of disaster recovery. His disaster research has covered floods, volcanic eruptions, and hurricanes in the United States, Mexico, and Ecuador. He has served as president of the Society for Economic Anthropology. Murphy is the co-author of *Social Inequality in Oaxaca, The Mexican Urban Household, Estudio Primario de la Cuenca Hidrografica del Rio de los Perros*, and editor of *Latino Workers in the Contemporary South*. He is author or co-author of numerous articles dealing with Mexican households, migration, and the impact of disasters on individuals and households in Mexico.

Donald R. Nelson (PhD, University of Arizona) is assistant professor of anthropology at the University of Georgia, and Visiting Fellow at the Tyndall Centre for Climate Change Research. His research interests include vulnerability, adaptation and resilience studies, participatory GIS, political ecology, food security, remote sensing, and water resources management. His current research is in watershed management in Brazil defining the relationship between resilience and adaptation—specifically using both physical and social insights to understand limits to adaptation in vulnerable places.

Rahul Oka (PhD, University of Illinois–Chicago and the Field Museum) received his BA cum laude from Lawrence University in Appleton, Wisconsin in 2000, specializing in anthropology and biology. He is currently adjunct assistant professor of anthropology at the University of Notre Dame, and research associate at the Field Museum, Chicago. His ongoing research aims to understand the role of regional and intercontinental commerce in shaping the rise and fall of societies and settlements. His current research is focused on understanding the relationship between commercial trade and resource distribution in conflict zones in Western Kenya, specifically the Mt. Elgon region and the Pokot district. He is also co-directing archaeological research on the western coast of India and the Kenyan coast.

Anthony Oliver-Smith (PhD, Indiana University) is professor emeritus of anthropology at the University of Florida. He currently holds the Munich Re Foundation Chair on Social Vulnerability at the United Nations University Institute on Environment and Human Security in Bonn, Germany for 2008. Dr. Oliver-Smith has done anthropological research and consultation on issues relating to disasters and involuntary displacement and resettlement in Peru, Honduras, India, Brazil, Jamaica, Mexico, Japan, and the United States. He is the author, editor, or co-editor of 6 books and over 50 articles and book chapters on these topics.

Christopher Pool (PhD, Tulane University) is associate professor and chair of the Department of Anthropology at the University of Kentucky. He has conducted archaeological fieldwork in southern Veracruz since 1983 and has directed investigations of long-term change in political economy at Tres Zapotes, Veracruz, since 1995. His research interests focus on the evolution of complex societies, economic and political organization, and craft production. His publications include *Olmec Archaeology and Early Mesoamerica* (2007), *Pottery Economics in Mesoamerica* (2008, edited with George J. Bey III), and *Economies and the Transformation of Landscape* (2008, edited with Lisa Cliggett).

Alan Smart (PhD, University of Toronto) is professor of anthropology at the University of Calgary. His research has focused on urban issues, housing, foreign investment, and social change. He is author of *Making Room: Squatter Clearance in Hong Kong* (Hong Kong University Press, 1992), *Petty Capitalists and Globalization* (co-edited with Josephine Smart, SUNY Press, 2005), *The Shek Kip Mei Myth: Squatters, Fires and Colonial Rule in Hong Kong, 1950–1963* (Hong Kong University Press, 2006), and numerous articles in journals and edited volumes. He is also co-investigator on projects on "The Social and Economic Impact of BSE in Alberta."

Josephine Smart (PhD, University of Toronto) is professor of anthropology at the University of Calgary. Her research and teaching interests are economic anthropology, food and emergent infectious diseases, social and economic impact of prion diseases, social and economic development in post-1978 China, Chinese international migration, globalization, immigrant entrepreneurs, and the international mobility of capital and labor. She was project leader of the Alberta Prion Research Institute core project (2006–2008) "TSEs and Social-Economic Impact in Alberta—Module 1." She is co-editor of *Petty Capitalists and Globalization: Flexibility, Entrepreneurship and Economic Development* (SUNY Press, 2005) and *Plural Globalities in Multiple Localities* (University Press of America, 2001), and author of *The Political Economy of Street Hawkers in Hong Kong* (Centre of Asian Studies–HKU, 1989). She has authored numerous journal articles and book chapters.

Graham A. Tobin (PhD, University of Strathclyde) is professor of geography at the University of South Florida. His research specialties include natural hazards, water resources policy, and environmental contamination. He has published several books, chapters, and many refereed papers and other articles. He has been department chair at two universities and an associate vice chancellor, and he has served on committees at the national, university, college, and department levels; he has also held office in professional organizations, including co-executive director of the annual Applied Geography Conference. He has received a number of research awards, including the Research Honors Award from the Southeastern Division of the Association of American Geographers, the Askounes-Ashford Distinguished Scholar Award from the University of South Florida, and the Research Award from the College of Liberal Arts at the University of Minnesota Duluth.

Jennifer Trivedi is a graduate student at the University of Iowa currently working on her PhD and is interested in disaster studies, media studies, and issues of vulnerability. Currently she is researching bias in the media in relation to Hurricane Katrina and long-term recovery issues, focused primarily on Katrina's impact in the Biloxi, Mississippi, area and national perceptions of the event. Trivedi earned her Master's degree from the University of Iowa in anthropology in May 2007 studying the casino industry and hurricanes as discussed in the local media and her bachelor's degree in history (with a minor in anthropology) from the University of Georgia in May 2004 considering the use of historical and archaeological symbols by political leaders and nationalists in the former Yugoslavia.

Megan Underhill is an anthropology graduate student at Colorado State University. As part of her thesis research, Megan embarked upon a study of the disaster resettlement experience of Hurricane Katrina evacuees in Colorado from 2005 to 2006. In addition to her thesis investigation, she has also worked as a research assistant on several Katrina-related research projects that received support from the National Science Foundation.

Linda M. Whiteford (PhD, University of Wisconsion, Milwaukee; MPH, Public Health University of Texas), professor of anthropology at the University of South Florida, is an applied medical anthropologist and past president of the Society for Applied Anthropology. She has fieldwork experience in Ecuador, Bolivia, Costa Rica, Mexico, Guatemala, and Nicaragua focusing on waterborne and water-washed diseases, disaster planning, and mitigation strategies using participatory methods. She is author of *Primary Health Care in Cuba: The Other Revolution* and three other books, as well as *Vulnerability of Displaced Persons: Relocation Park Residents in the Wake of Hurricane Charley* and more than 50 other peer-reviewed articles.

Leslie Lea Williams is a doctoral student at the Ohio State University. She received her BA in anthropology from the University of California at Berkeley in 2002 and her MSc in human osteology and funerary archaeology from the University of Sheffield in 2004. She is a bioarchaeologist studying the health impacts of mass mortality and warfare on civilian populations, as well as the interaction of culture and infectious disease. Her current research focuses on the effects of climate change on medieval human health, and gender and status distinctions in infectious disease prevalence in prehistoric east Tennessee.